Optomechanical Design

Critical Reviews of Optical Science and Technology

The Critical Reviews provide a forum for presentation and publication of invited overview and review papers in key developing areas of optics and optoelectronics.

Vol.	Title	Editor(s)	Year
387	Technology of Stratified Media	R. F. Potter	1983
406	Optical Specifications: Components and Systems	W. J. Smith, R. E. Fischer	1983
442	Robotics and Robot Sensing Systems	D. Casasent, E. L. Hall	1983
443	Infrared Detectors	W. L. Wolfe	1984
456	Optical Computing	J. A. Neff	1984
475	Remote Sensing	P. N. Slater	1984
512	Fiber Optic Communication Technology	C. W. Kleekamp	1985
516	Diagnostic Imaging Applications	E. S. Beckenbach	1985
527	Applications of High Power Lasers	R. R. Jacobs	1985
528	Digital Image Processing	A. G. Tescher	1985
531	Geometrical Optics	R. E. Fischer, W. H. Price, W. J. Smith	1985
532	Holography	L. Huff	1985
541	Radiation Effects on Optical Materials	P. W. Levy	1985
543	Photovoltaics	S. K. Deb	1985
604	Holographic Nondestructive Testing: Status and Comparison with Conventional Methods	C. M. Vest	1986
607	Optical Component Specifications for Laser-Based Systems and Other Modern Optical Systems	R. E. Fischer, W. J. Smith	1986
609	Flashlamp-Pumped Laser Technology	F. Schuda	1986
611	Laser Processing of Semiconductors and Hybrids	E. J. Swenson	1986
614	Highly Parallel Signal Processing Architectures	K. Bromley	1986
738	Free Electron Lasers	B. E. Newnam	1988
741	Design of Optical Systems Incorporating Low Power Lasers	D. C. O'Shea	1987
755	Image Pattern Recognition: Algorithm Implementation, Techniques, and Technology	F. J. Corbett	1987
762	Electro-Optical Imaging Systems Integration	R. Wight	1987
795	Characterization of Very High Speed Semiconductor Devices and Integrated Circuits	R. Jain	1987
816	Interferometric Metrology	N. A. Massie	1988
928	Modeling of the Atmosphere	L. S. Rothman	1988
930	Infrared Detectors and Arrays	E. L. Dereniak	1988
935	Gradient-Index Optics and Miniature Optics	J. D. Rees, D. C. Leiner	1988
959	Optomechanical and Electro-Optical Design of Industrial Systems	R. J. Bieringer, K. G. Harding	1988
960	Real-Time Signal Processing for Industrial Applications	B. Javidi	1989
961	Industrial Optical Sensing	J. Carney, E. Stelzer	1988
1000	Laser Wavefront Control	J. F. Reintjes	1989
1082	Applications of Electronic Imaging	J. C. Urbach	1989
1150	Spatial Light Modulators and Applications III	U. Efron	1989
CR35	Digital Optical Computing	R. Athale	1990
CR36	Machine Vision Systems Integration	B. G. Batchelor, F. M. Waltz	1991
CR37	Standards for Electronic Imaging Systems	M. Nier, M. E. Courtot	1991
CR38	Infrared Optical Design and Fabrication	R. Hartmann, W. J. Smith	1991
CR39	Infrared Thin Films	R. P. Shimshock	1992
CR40	Optical Pattern Recognition	J. L. Horner, B. Javidi	1992
CR43	Optomechanical Design	P. R. Yoder, Jr.	1992

Critical Reviews
of Optical Science
and Technology

Volume CR43

Optomechanical Design

Paul R. Yoder, Jr.
Consultant in Optical Engineering
Editor

Proceedings
of a conference held
22-23 July 1992
San Diego, California

Sponsored by
SPIE—The International Society for Optical Engineering

SPIE OPTICAL ENGINEERING PRESS

A Publication of SPIE—The International Society for Optical Engineering
Bellingham, Washington USA

Library of Congress Cataloging-in-Publication Data

Optomechanical design : proceedings of a conference held 22–23 July
1992, San Diego, California / sponsored by SPIE—The International
Society for Optical Engineering : Paul R. Yoder, Jr., editor.
p. cm. — (Critical reviews of optical science and technology
; v. CR43)
"A publication of SPIE—The International Society for Optical
Engineering."
ISBN 0-8194-0952-9 (hardcover). — ISBN 0-8194-0953-7 (softbound)
1. Optical instruments—Design and construction—Congresses.
I. Yoder, Paul R. II. Society of Photo-optical Instrumentation
Engineers. III. Series.
TS510.0668 1992
681'.4—dc20 92-16833
CIP

Published by
SPIE—The International Society for Optical Engineering
P.O. Box 10, Bellingham, Washington 98227-0010 USA
Telephone 206/676-3290 (Pacific Time) • Fax 206/647-1445

Printed in the United States of America.

Contents

Conference Committee

Conference Chair

Paul R. Yoder, Jr., Consultant in Optical Engineering

Cochairs

James B. Breckinridge, Jet Propulsion Laboratory
R. Barry Johnson, Center for Applied Optics/University of Alabama in Huntsville
Roger A. Paquin Advanced Materials Consultant
Daniel Vukobratovich, Optical Sciences Center/University of Arizona

Session Chairs

Session 1—Optomechanical System Design
Paul R. Yoder, Jr., Consultant in Optical Engineering

Session 2—System Stability Considerations
James B. Breckinridge, Jet Propulsion Laboratory

Session 3—Optomechanical Component and Mechanism Design
Daniel Vukobratovich, Optical Sciences Center/Unversity of Arizona

Session 4—Design of the Optomechanical Interface
Roger A. Paquin, Advanced Materials Consultant

Session 5—Analytical Design Methods
R. Barry Johnson, Center for Applied Optics/University of Alabama in Huntsville

Preface

An SPIE "Critical Review" is always a special event. This one comprises 17 invited papers by well-known scientists and engineers who have made significant contributions to the field of optomechanical design. These papers provide in-depth analysis and review significant aspects of this key optical engineering technology. Prepared in advance for initial distribution at the San Diego '92 Annual Conference, this book includes papers addressing vital issues in the following general categories:

- Optomechanical System Design (4 papers)
- System Stability Considerations (5 papers)
- Optomechanical Component and Mechanism Design (4 papers)
- Design of the Optomechanical Interface (2 papers)
- Analytical Design Methods (2 papers).

I believe that in bringing these speakers together in a concerted effort to summarize technical progress in this broad field and in publishing their papers in a single volume, SPIE has provided a valuable service to the optical instrumentation design community. Those who hear the oral presentations and participate in the discussions that naturally follow will surely benefit from those experiences, while those who read the written versions of the papers included here should find much useful information.

On behalf of the Program Committee, I would like to extend my thanks to all authors/speakers who contributed their expertise to this critical review, and to those who attended the sessions as well as to those who will add this book to their libraries. We hope our collective efforts prove helpful throughout the optomechanical design community.

Paul R. Yoder, Jr.
Consultant in Optical Engineering

Optomechanical System Design

Chair
Paul R. Yoder Jr.
Consultant in Optical Engineering

Mechanical design of optical systems for space operation

Michael Krim

Hughes Danbury Optical Systems, Inc.
100 Wooster Heights Road, Danbury, CT 06810-7589

ABSTRACT

The age of large space optical systems began in the early 1960s. With very little precedent but driven by necessity, there emerged a series of new mechanical and structural technologies including dramatically lightweighted mirrors, ultra-stable low thermal expansion structures, large-scale testing thermal and vacuum environment testing, and robust alignment sensing and control systems. We will review in this monograph the growth of these and other technologies, beginning with a heritage that had its origins in amateur telescope making. We will also address a newer design technology, design-to-cost, which has become a dominant system discriminator in the past decade. This is not intended as a how-to technical paper. Instead it is intended to surface some of the questions and considerations that need to be answered in the development of any new systems concept, rather like a checklist, to ensure that low cost and a high probability of initial success are achieved.

1. TELESCOPE BUILDING IS NO LONGER AN AMATEUR AVOCATION

Until the advent of artificial satellites, the design and manufacture of large telescopes had been largely an amateur or at least non-commercial, albeit highly sophisticated, enterprise. The design of the 5-m Palomar telescope was actually developed in Springfield, Vermont by a group led by Russell Porter, an architect by training, and the chief engineer of the Jones and Lamson Corporation. This work was initiated in the late 1930s and completed after the war. Porter was also the founder of the renowned Stellafane Society, the oldest amateur telescope-making group in existence and which continues to meet every summer on Society Hill in Springfield. Attending some of the early 5-m telescope planning and design meetings was another non-professional telescope maker, Richard S. Perkin who with Charles Elmer founded the Perkin-Elmer Corporation. That corporation went on to design and build the Orbiting Astronomical Telescope, the OAO-C, in the early 1960s which was the pathfinder for larger and more powerful space-based astronomical systems.

What was different then from now? Those telescopes were largely financed by private funds, usually endowments from interested foundations. Without a high level of technical and financial oversight, their design often reflected a wider degree of experimentation than exists today. And sometimes these designs didn't perform as originally intended. Sometimes significant fabrication problems were encountered necessitating major redesigns. The Palomar primary mirror, for example, was originally to have been made from fused quartz produced in an electric furnace in Pittsfield, Massachusetts by General Electric. After nearly "browning out" the western part of that state an alternative material, Pyrex, made in gas fired furnaces by Corning in upstate New York, was substituted.

This anecdote is intended to provide a contrast to today's more structured technical culture, at least as far as space-based optical systems are concerned. Unlike a ground-based telescope where:

a) design flaws can be fixed and/or the system can be upgraded as new and desirable technologies become available; and

b) the means to transport the telescope to its operational site are always there, (i.e., roads),

space-based systems must rely on expensive, schedule driven launch vehicles and yet-to-be-proven manned repair missions. Space-based telescopes need to be "right" and they need to be on time. Pre-flight verification is a necessity as well as a very significant part of the cost of these systems which are between ten and fifty times the cost of their earth-based counterparts. The turn-key cost for an 8-m ground-based telescope is estimated to be about 100 million dollars, including the dome and mount structures. Space-based telescopes of that aperture have yet to be developed beyond the stage of speculative studies but costs are estimated in the 2 to 4 billion dollar category, exclusive of launch expenses. That's forty times as much as the ground-based version. And that kind of money doesn't come from foundations. It is Government money and with it of course must come insightful technical and programmatic management, both from the sponsoring organization as well as the performing one, i.e., the contractor.

The transition from the largely benign environment of ground-based telescope design to the unforgiving demands of space operation began about 30 years ago. To some extent it is still continuing both in the technical as well as in the programmatic sectors. Let us examine some of the principal factors that comprise this transition from the vantage point of a mechanical designer.

2. THE CHALLENGE OF SPACE OPTICS

Clearly, an objective of space-based operation is improved performance where limitations imposed by atmospheric distortion or "seeing" are absent. Free of these limitations, the theoretical resolving power of a telescope can be realized if the surface tolerances on the optics (mirrors) and the alignment between them is maintained to near-ideal conditions. This is often better than a fiftieth of a wave rms or about a tenth wave peak-to-peak. In mechanical engineering units this is about two millionths of an inch or less than a thousandth of the thickness of this page! The theoretical resolving power of a telescope, to first order, is directly proportional to the diameter of the aperture which in most cases is defined by the size of the primary mirror. But the tolerances on surface accuracy or alignment, being functions of the operating wavelength, remain constant or may even decrease. So as the quest for higher and higher performance continues and systems increase in size, the ingenuity and resourcefulness of the optical manufacturing teams and the mechanical designers become more and more challenged to achieve surface accuracies and element-to-element alignment tolerances that are no more forgiving than if they were for a 0.5-m system. And to do this without jeopardizing schedule or cost.

What are the specific challenges confronting mechanical design? Following is a partial list, not arranged in any particular order of priority or importance; different people would probably have different views of that, based on their own experiences.

- Cost
- Weight
- Mirror Design and Mounting
- Mirror Manufacturing and Gravity Release
- Structural Alignment and Gravity Release
- Mirror Accuracy and Temperature Change
- Fear of Failure
- Alignment Accuracy and Temperature Change
- Launch Survival and Verification
- Dynamics and Controls Interaction
- Alignment Sensing and Control
- Producibility
- Integration, Assembly, and Test
- Verification, beyond what is Testable
- Structural Modeling, the Glue that Ties it Together
- Schedule Credibility
- Cost.

These topics will be discussed in a qualitative sense and for each, a checklist of factors that need to be considered will be presented and explained. It was not the intent to address detailed technical issues but to alert the reader to the right questions to ask in planning an attack on them.

To those of you new to the space optics field, there is another way of looking at the mechanical engineering challenge of space optics. It is to recognize that the success criteria for the structural system is almost invariably not an easily isolated and well defined event such as collapse of a wing under static test, as might be the case in the aircraft industry. It is a subtle loss of a fraction of a micrometer worth of alignment between elements often many meters apart. Some of the ways available to the designer to meet this criteria, in a deterministic and affordable manner, is the subject of this paper.

3. COST

No program will be funded if it isn't affordable and no contractor will win that program if his costs are not credibly low. In fact programs can and have been terminated for cost growth reasons. Historically, "cost" was not considered to be in the technical sphere and yet in today's environment it must be treated as a design parameter, just like weight budgets or wavefront error allocations. One must price to win and then ensure that the costs to design and produce the hardware are within that price. Very much like the automobile business! This of course had not been the case in the beginnings of space optics. Then, there was indeed a race with the Soviets to demonstrate who the technical leader of the world was and rocketry and space was a visible and spectacular area of competition. Risking simplification, the order of priority then was performance, then schedule, and then cost. Today that order is reversed. However, there remains a heritage, and it may be indigenous to the engineering and scientific professions, to do better than what is required. Surpassing the specifications is still today regarded as a demonstration of one's technical virtuosity. It is the unusual case where this doesn't add cost, either in design or in test. Today's mantra is to meet the specified requirements with the least cost and with the smallest technical and schedule risk. In most cases the requirements are difficult enough, don't try to do better.

How does one know that the least-cost approach to a particular design has been achieved? Simple: estimate its cost, analyze the results, and look for the drivers. Are there alternative solutions that can reduce them, what is the specific cost "delta" for these alternatives? This needs to be done during the concept definition and trade phase along with other candidate approaches. Presupposing the existence of a design concept, the cost estimate needs to reflect at least the following elements:

a) *A Program Plan.* What needs to get done and when needs to be defined in as much detail as practical. This is the programmatic equivalent of the system layout drawing; it establishes the direction that the individual operating groups will take. And by the way, just like a layout drawing, it's never really finished and may be changed or updated to reflect program maturation over the life of the program. Without a program plan, which clearly identifies the program deliverables and the paths, milestones and submilestones by which these deliverables will be implemented, the cost estimate would have no credibility.

b) *Requirements.* Now that one knows what needs to be accomplished and when, how well they need to be done has to be defined. In today's complex, technically interdependent high performance systems, the requirements for the various subsystems must be developed from a top-level systems approach. Piecing together various subsystems, sometimes developed by highly competent or "star" engineers but with the attitude of prima donnas, simply will not be successful. The technical management of a program and the role of the chief engineer and systems engineering are beyond the scope of this paper. It is a subject which merits it's own discussion. Suffice to say that to design to cost, which is what this section is all about, demands that the requirements for the design be in-place and expressed in the language of the specific discipline responsible for its implementation. A structural engineer doesn't usually relate to fractional encircled energy or the power spectral density (PSD) of an optical surface. Not to demean those who operate in this area, but the structural and mechanical designer needs to have these terms translated into linear units of deformation or misalignment between elements will occur.

c) *Confidence in the design approach.* Is it readily analyzable and does it have a high probability of initial success? How dependent is the analysis on assumptions and idealizations, or what I sometimes refer to as the dilemma of "interlocking ifs." If the moments applied to a mirror are..., or if the CTE homogeneity of the material is..., or if the hysteresis in a joint was..., well that's what I mean by interlocking ifs. The assembled hardware has no way of knowing what assumptions the designer made. Discovering design faults is expensive, especially if they are discovered late in the development stage. Design the system to be verifiable, either by end-item test or by analysis augmented with subsystem or developmental test data. The verification process must be treated as an inherent element of the design process. It must be reflected in the program plan and usually expanded via an MIT or detailed Manufacturing, Integration, and Test plan which defines every step of the assembly and test process, including the facilities that are needed, and should dovetail with the system verification plan which defines how each element of the requirements specification and error budgets will be verified.

d) *A detailed drawing count or estimate.* It sounds trite but every part that comprises the design needs to be drawn, all drawings need to be engineered, all parts need to be made, bagged, tagged, inspected, and assembled. And all of those operations take time and time of course, multiplied by rates, is cost. Drawing counts and design labor estimates need to be made in concert between the program office and the various operating departments, neither alone is adequate. Don't omit any of the special test equipment or other factory equipment needed to verify the hardware. It needs to be designed with the same care and professionalism as the flight equipment itself.

e) *Real hardware fabrication estimates.* No substitute for cost quotes from vendors either, although in most instances (at least for one-off systems) the actual structural fabrication costs are small compared with the engineering and test labor that was needed to design and specify them.

f) *Historical precedence or heritage.* Lastly, innovation for its own sake needs to be minimized or better yet, avoided. Wherever possible, design trades should be confined to configuration variations and material selection. The latter is based on proven materials in similar applications. The development of new materials or the incorporation of non-traditional materials is best left for a pre-flight hardware development program which is independent of flight hardware schedule constraints.

g) *Sufficiency.* Know when you are done, know when the design is good enough, declare completion and move on to the next piece of work. It isn't required to do better than the specifications require, albeit one needs to provide for some margin. In only rare circumstances does doing better not add cost and as proclaimed earlier, cost is indeed a design parameter in today's competitive climate. That of course demands a clear definition of the requirements as stated at the outset plus sufficient judgment, insight, analysis, and possibly test data to confirm that the requirements have indeed been satisfied.

h) *Clearly defined work definitions.* "I thought you meant this, and now you're telling me that the customer wants a stand-alone flight safety analysis report. Where'n the heck do I have a charge number for that?" Not an uncommon complaint. A clearly defined and documented and agreed to statement of work between the program office and the customer and between the program office and the various operating groups supporting the project is absolutely essential. No further comments are necessary.

There is, of course, far more to the area of design-to-cost than what is briefly mentioned here. That, however, is a subject unto itself.

4. WEIGHT AND MIRROR MECHANICAL DESIGN

Progress in lightweighting has possibly been more dramatic in the field of space optics than in any other discipline. Whereas the aircraft industry provided the heritage for lightweight spacecraft structures, there was no such parallel for large space optics systems. Weight was often considered an asset for ground-based telescopes insofar as it provided rigidity and thermal stability. Some lightweighting was employed for mirrors, however, to reduce their thermal time constant in an effort to prevent localized seeing errors.

Mirrors in particular, over the last 35 years, have progressed from solid fused silica disks whose diameter-to-thickness ratio was 6, to early sandwich designs that achieved weight reductions on the order of 70%, to a variety of design approaches in both glass and metals that are no more than 10 to 12% of their equivalent-size 1960 predecessors in the 1-m class. And they perform as well or better. Where did the 6:1 ratio come from? Ritchey, of Ritchey-Chretièn fame, observed in 1903 that when the aspect ratio of a mirror was 6:1 there was no sensible difference in its figure when measured on edge to when it was measured with gravity normal to its surface. He described these conclusions, based on 12-inch mirrors, in his logbook at the Harvard Observatory. For constant self-weight deflection, the ratio of diameter squared to thickness is the relevant metric so that as the size doubles, the thickness needs to be increased by four. The 6:1 rule, despite not being rigorously correct, persisted into the 1970s and for some commercial telescopes, is still adhered to.

Presuming that the design and programmatic (cost/schedule) requirements have been defined, what are the "knobs" to turn for achieving a lightweight mirror and conversely what are the impediments to lightweighting. Some of these knobs are:

- Materials
- Configuration
- Substrate producibility
- Optical fabrication and test methods
- Mounting.

There is usually more than one design solution that will satisfy a specific set of requirements, a sample set of which is listed below:

- Diameter or geometry, f/no, prescription
- Fabrication and operational wavefront error allocations
- Type, i.e., passive, active, or correctable
- Thermal time constants
- Natural frequency
- Launch and operational dynamics loads
- Schedule, or lead time
- Weight
- Cost bogey, sometimes.

Producibility and cost tend to remain as key design and requirements factors. A brief discussion of some of the degrees of freedom, or knobs, that the mirror designer controls to achieve a lightweight design is presented below.

a) *Different materials are lightweighted differently.* Lightweighting may be achieved by selecting low density, high modulus materials like beryllium or by configuring the substrate to achieve high structural efficiency and thus not requiring much material in the first place. As with any structure, putting the right amount of material in the right places can lead to lightweight, efficient designs. However, not all materials used for mirrors are compatible with all lightweighting approaches. Today the designer has a variety of material options to select from (Corning's ULE and fused silica, Zerodur from Schott, fused quartz from Heraeus, beryllium both HIP'd and VHP'd, silicon carbide both CVD'd and reaction bonded and several others which are still developmental or otherwise not generally used in space applications such as borosilicate) but which generally are not lightweighted in the same way or to the same extent. ULE may be produced in a structurally efficient sandwich form by frit bonding thin faceplates onto a square or triangular grid core. Core depth to overall thickness ratios of 90 to 93% are achievable, which is near optimum from a stiffness to weight aspect. Zerodur on the other hand is usually lightweighted by machining recesses in the back to form a rib stiffened structure or is simply utilized as a thin meniscus, often in conjunction with shape control actuators. HIP'd beryllium can be produced as a sandwich by consolidating the powder over a sacrificial copper mandrel which is chemically removed through small holes in the back after the HIP'ing process is completed. Before selecting a final material/configuration it is important to confirm that the design is indeed producible. One must also be aware of materials and/or lightweighting processes that are developmental or have been accomplished in small sizes only. Early liaison with optical contractors and substrate manufacturers cannot be overemphasized.

b) *Is the material polishable?* Fully densified surfaces in the case of HIP'd beryllium and SiC mirrors and suitably small grain size are necessary to produce highly specular surfaces where microroughness needs to be limited to less than 5 or 10Å. The amorphous nature of glass, with no grain boundaries or other micro-structures, facilitates polishing. In certain instances, beryllium being most notable, electroless nickel plating several mils thick provides the hard, amorphous surface necessary for polishing. Unfortunately there is a mismatch in the coefficient of expansion of these two materials and especially with lightweight substrates, unexpected and certainly unwanted thermal bending occurred.

c) *Is the material available in the sizes needed?* This isn't much of an issue up to 1.5 meters but it is an important question for larger sizes. The size limit for HIP'd beryllium is 1.5 m, a facility issue. For frit bonded ULE, the limit is 2.5 m, also a facility limitation. Zerodur mirror blanks up to 8.2 m have been produced with thicknesses ranging from 0.2 to 1 m. Don't neglect the extra material needed at the edges to accommodate tool roll-off. On large mirrors the added weight can be significant as well.

d) *Optical fabrication and testing is another potential limit to lightweighting.* Two factors enter here: 1) the ability to grind and polish the mirror without causing the rib or grid structure to "print through" and 2) the ability to confidently support the mirror in a simulated zero-g environment. In the first instance this is largely dependent on faceplate thickness and rib spacing, (s^4t^3 to be specific) and other factors like tool size and pressure. This factor is typically between 250 and 1000 for glass and four times that for beryllium where the specific stiffness is greater by that amount. As the size increases and weight and stiffness decrease, the effect of gravity release needs to be accounted for with greater and greater accuracy. The Hubble Space Telescope (HST) 2.4-m primary mirror has 15 λ p-p of self-weight deflection when supported on three points at the rim. The error allocation for gravity release was only 0.004 λ rms, representing approximately a factor of 1000 on an rss basis. Essential to successfully predicting and/or supporting a mirror with an array of forces such that it approximates its weightless shape are highly accurate and verifiable finite element models. That in turn demands accurate knowledge of the mass distribution within the mirror.

5. MIRROR SUPPORT SYSTEMS

Mirror support systems for rocket launched, orbiting systems need to satisfy two widely divergent criteria, namely strength and stiffness, to survive launch and be compatible with the vehicle dynamic environment on one hand, and support the mirror with the least number of constraints necessary to ensure alignment stability. The latter is of course a restatement of the familiar phrase "statically determinate." Mirror support is included as a topic because of its potential (sometimes justified) as a schedule or performance risk. Unfortunately the success of a mirror mount is often not demonstrated until late in the program during thermal vacuum test or even worse, during flight. Because mirror mount design has not yet, and may never, become a standardized technology, we have included here some reminders of things to consider when developing the concept.

There was no precedent in the beginning. In its infancy, mirror mount designs were often seat-of-the pants concepts, necessitated by a lack of real experience and the absence of advanced finite element modeling codes such as are available today which enable detailed analyses of the stresses at the "glass-to-metal" interface or of the effect on surface figure of secondary loads imparted to the mirror by friction or flexure stiffness to be calculated. A wide variety of support system designs flourished, at least on paper. Over the years different people and different organizations developed mount designs which solved the problem in many different ways. It shouldn't be surprising that there is no "universal" approach to mirror attachment and support. Different applications as well as different mirror constructions, not to mention different requirements, dictate different solutions. Spherical ball bushings imbedded in recesses so their lines of action pass through the midplane of the mirror have been used in some instances (the OAO-C for example). The HST used athermalized rods passing through the mirror which provided a compressive interface between the metal and the glass regardless of the axial load direction. Other applications employed Invar pads which were bonded onto the substrate. Flexures and tension-compression rods with spherical bearing ends have been used to couple the mirror to the support structure. Flexures of course take many forms and it seems to this writer that they've all been used at one time or another.

Regardless of which specific design approach one uses, experience has shown that there are certain characteristics common to all successful designs. They are described below. Not every one of the factors included below applies to every mirror application; however, collectively they serve as a design checklist. Please presume that each of them is preceded with the silent phrase, "The mirror mount design should be..."

- *Based on a Clear Definition of the Allowable Constraint Forces.* Idealized "statically determinate" support systems exist only in theory. In the "real world," one needs to contend with friction in pivots and joints, non-zero flexure stiffness in what one would like to be the "soft" directions, and other similar issues which can and will produce secondary or redundant loads. The effect that these loads will have on figure need to be determined and accounted for in configuring and proportioning the mount hardware. Thus a figure error/loads sensitivity analysis needs to be performed for the mirror design in question from which the allowable magnitude of these secondary loads is determined in conjunction with the error budget.

- *Free of Gimmickry.* How the mount design works ought to be explainable, at least to 1st order, without having to resort to complicated or extensive analysis of a series of interconnected "ifs." No amount of analysis can make an ill-conceived architecture successful.

- *Able to Interface With Uncomplicated Mating Glass Interfaces.* Avoid complicated machined glass configurations like holes with reentrant pockets for "blind" clamps, sometimes referred to as overgrown Molybolts. These pockets in general will exhibit extensive subsurface damage which if etched out (maybe) might destroy precision. Don't depend on tension across the seal-plane or frit-line of a sandwich mirror, because of potentially high stress concentrations.

- *Configured to Avoid High Stress Gradients in the Glass.* High stress gradients such as those caused by pins bearing on holes for example, even if compressive, will be accompanied by shear stresses which ultimately are resolvable into tension and that's what initiates spalling failures. Pins in holes or shoulder bolts in holes, while I'm on the topic, may become cocked and produce much higher direct bearing loads than one might expect. Glass doesn't yield like metals to redistribute these stresses, it just breaks. Gluing the pin in can ameliorate this problem but permanent bonds introduce other problems. Flat mounting pads also can present hard interfaces between glass and metal (which might have the potential for relative motion) and can result in "micro-cocking" which in turn produces higher direct bearing stresses than ordinarily expected, too. In these cases, half mil "mylar" shims have been used to provide a cushion. If extreme temperature ranges are to be encountered where plastic is unacceptable, try 3000 series aluminum foil or something like it which can flow. Low intensity compressive glass-to-metal interface loads are always preferred followed by low intensity shears.

- *Deterministic*. Particularly in a "micro" sense. How accurately is the load distribution at the mount/glass interface known? Is the finite element simulation of that interface of sufficient fidelity. If a design can't be analyzed with a moderate degree of confidence and without too many "ifs" in series, a search for an alternative ought to be initiated, pronto. Areas with direct bearing loads need to be especially scrutinized particularly where misalignments or fabrication tolerances might cause cocking and so on. Do modeling idealizations possibly mask potential problem areas? Sensitivity analyses to test these idealizations, sometimes referred to as the "Principle of Overlapping Assumptions," need to be performed. Knowledge of the interface loads distribution is necessary to know where the lines of action of forces are and that in turn is necessary to evaluate the effect of constraint loads on the mirror.

- *Based on Actual Material Properties, Not Handbook Values*. Athermalized mount designs need to be based on CTE properties from actual lot data. The design needs to be tailorable to accommodate these properties variations, especially if Invar is employed. It's even more important if graphite epoxy is employed. The latter needs to be "mellowed in" by thermal cycling until the measured CTE stabilizes.

- *Able to Survive Large Temperature Changes Without Damage*. In the majority of cases, the operating temperature range of a mirror system is relatively small. However, that temperature may be hundreds of degrees away from the temperature that the mirror and mount was assembled at. Or large temperature excursions might occur during transportation or during a non-operating mode on orbit. Not only does the CTE of a material change with temperature, this change in CTE with temperature is different for different materials. Therefore, a specific CTE difference between two different materials in an assembly at one temperature, say the operating temperature, may be vastly different at another temperature. More to the point, the dimensional changes of different parts of an (athermalized) mirror mount system, $\Delta L/L$, over one temperature interval may be quite different than over another. This could lead to inordinately large incompatibilities between the parts and in certain cases, break the mirror. Check the $\Delta L/L$ curves for when large temperature ranges will be encountered. Watch out also for epoxies and other organics in confined volumes since large hydrostatic loads can be developed. Provide relief slots or whatever.

- *Tolerant of Non-Isothermal Conditions Especially if Athermalized*. In addition to the non-linear $\Delta L/L$ considerations just mentioned, the mount system may not be isothermal or may not have the temperature distribution planned for by the designer. Why is it that we always tend to beat up on the thermal guys? Athermalized designs in general depend on isothermal conditions. In some cases athermalization can accommodate linear gradients but we'll ignore those for now. The effect of off-nominal temperature distributions may be important in some configurations and in those instances needs to be evaluated. Either the athermalization might not behave as expected and a few errant micrometers of displacement might sneak in or worse, thermal stresses might break the glass.

- *Verifiable Before Assembly to the Flight Glass.* This applies to all mount assemblies but is most germane to those configurations where the possibility exists for interferences or high local shear or direct bearing loads to occur. If the mount is designed to be athermalized by the subtractive effect of two relatively high (compared to the glass) expansion materials and if there's a (credible) possibility that the ΔL/L or the temperature issues identified above might be a problem, verify the assembly in a test piece whose possible loss can be afforded before committing to the "real thing."

- *Insensitive to Machining and Assembly Tolerances.* All parts in the real world only approximate the dimensions that the designer wishes they might have. The smaller the part, in general, the larger these tolerances are as a percentage of the nominal dimension and therefore, the more sensitive that assembly might be to these variances. Worst case tolerance buildups need to be considered in evaluating the interface loads. Where applicable, the parts should be able to be adjusted by shims or machining or whatever to compensate for tolerance buildup.

- *Disassemblable, Just In Case.* Bonding is in many instances a convenient and uncomplicated way to join the mirror to its mount. But it doesn't come apart too easily if recoating or refiguring or, heaven forbid, redesign of the mount is required.

- *Not Dependent on Bond-line Flexibility.* Don't rely on the "...3-mil bondline .... giving a little in compression" to accommodate the relative expansion between Invar and quartz. Even over a relatively small 15°C temperature change, it doesn't! Bondlines for certain adhesives can be made deterministically flexible in shear by dicing the surface into a multitude of discrete pads whose dimensions (thickness and size) are selected on the basis of structural mechanics.

- *Manufacturable and Assemblable.* Don't expect that manufacturing engineering necessarily understands all of the subtleties of the design that you as the designer (hopefully) had in mind, especially when developing a new concept. The design of any assembly tooling or fixtures needs to be an inherent element of the mount design process.

- *Able to Accommodate Small Clearances or Able to Accommodate Small Interferences as an Alternative.* Unless it's an all-bonded design, there is the possibility that some clearances will exist at the glass-to-metal interface, and at other places too if cylindrical or spherical pivots are used. In fact, some designs deliberately provide for clearance to avoid the interference nightmares that I've been writing about. There's three things at least that one needs to consider relative to clearance in a design.

  1. Is it small enough so that excessive shocks will not be developed? Is the clearance sufficiently small that a "free-fall" at 5 g's for example doesn't introduce excessive shock. The velocity at impact resulting from a 1 mil "drop" in a 5 g acceleration field is about 2 inches per second, a lot slower than most of us would put down a coffee cup.

2. Are the parts protected from impact shock during launch (regardless of the gap size) by some absorbant material, such as the aluminum shims mentioned earlier?

3. Does the design provide for a preload to ensure that the mirror is constrained against a registration surface subsequent to launch? Will the preload spring overcome friction or anything else that might "hang" the mirror up and prevent it from being nudged towards the registration surface.

- *Based on Some Demonstrated Heritage, If at all Possible.* Always a nice attribute! Sometimes though a precedent can't be found. Which brings us to the final element in this checklist...

- *Verified by Test, Before Committing to Flight Hardware.* It is often not practical to test a complete mirror/mount assembly in a simulated operational environment where the success criteria is typically a figure error change of 0.01 waves rms or so. To a large extent we depend on accurate finite element modeling for that. But the results of the model need to have some "anchor point" into reality and that can be obtained from subscale tests of a mount and a mirror test piece which represents only the region where the mount is located. Cryogenic systems such as SIRTF are possibly more dependent on engineering development tests in as much as the expansion properties of materials over the 500°F range from room temperature to a 2K operating temperature are far from constant or even linear.

6. OTHER CONSIDERATIONS

6.1. Drawings

One of my pet peeves, especially today, is that the mechanical architecture of a system is often left unexplored until too late in the design process. Insufficient priority in the initial architectural trades can result in structures which are sometimes awkward, difficult to align and integrate, and overly complicated which of course means risk and cost. There are three drawings of extreme importance to the mechanical engineering of any space optical system and they need to be developed right up front. They are the:

a) optical drawing
b) functional layout
c) mechanical or physical layout.

A brief description of each of these drawings is in order.

The optical layout needs to be a large scale, dimensioned, and accurate drawing depicting all of the optical surfaces, ray bundles (in the case of scanning systems the swept volume of the scanned bundles needs to be included) and baffling. It needs to be done in at least two views, unless the system is perfectly axisymmetric. The clear aperture of all elements should be readily discernable. It does not necessarily need to include folds should the system require them for packaging or space allocation reasons. These can be depicted on the functional layout.

The functional layout is a very important drawing. It defines the location and volumes of all equipment that must be positioned in a specific way to the optical surfaces. This includes the thickness of the substrates, focal plane devices and radiators, alignment sensing instrumentation, calibration sources, and so on. If the system needs to be folded, this is the place to show that. The interfaces with adjoining equipment need to be explicitly defined. Where alignment adjustments will be made should be identified. If there are specific access requirements or installation and removal paths that are required, they need to be shown. The HST science instrument replacement paths are an example of one such requirement. This drawing ideally should depict the system as if the structure was "invisible." Secondly, those components that don't have any specific location requirement with respect to the optics are not shown. Electronics boxes are one such example.

The mechanical layout builds on this drawing by adding the structure and everything else. It needs to be consistent with interface envelopes and other space allocation requirements. It is at this time that any potential problems insofar as insufficient volume or excessively contorted load paths would be identified and presumably it would be early enough in the program to resolve any difficulties. This drawing provides the first real-world visualization of the system and as stated earlier, establishes the direction that the opto-mechanical aspects of the program will follow. The importance of this drawing, often the culmination of significant concept trades, to the overall success of the project cannot be overemphasized.

6.2. Modeling and Analysis

In the beginning there was Roark, Timoshenko, the moment distribution method, free-body diagrams, and Friden calculators. Accurate analysis of a point design was difficult enough, not to mention analysis and trades between competing design concepts. Optical performance criteria, as they flowed down to a mirror for example, were often specified as a "single valued" number, the rms wavefront. The emergence of finite element codes and their application to optical structures began in the late 1960s, with simple branched beam models whose principal use was in estimating dynamic characteristics. Today we have at our disposal sophisticated finite element codes which may be invoked both during the design phase for architectural and concept trades and later for performance predictions of the detailed and matured final design. Today we have the ability to model with unprecedented accuracy the behavior of mirrors under thermal, gravity, and mount constraint force loadings. The CTE distribution in the substrate can be modeled with measured data supplied by the manufacturer to determine precisely what the surface errors will be as a function of temperature change. We can use the resulting displacement data in optical performance codes to determine the specific nature of the image defects, whether they be in the geometric or diffractive regime.

So why do anomalies still occur? What happens when test and analysis don't agree or more acutely, when orbital performance is significantly different from what had been predicted? Unfortunately, no amount of analysis can make a poorly conceived design perform any better than it is intrinsically capable of. Secondly, the mere existence of a detailed finite element model does not necessarily pre-ordain the behavior of the structural systems. When these anomalies occur, costs increase due to program delays, "tiger team" investigations, diminished incentive fees, and the like. In concluding these remarks, I would like to offer a few cogent observations regarding the role of modeling and how to maximize agreement between prediction and test.

a) *Carefully review the design.* Does the as-built system look like what the analyst and the designer anticipated? If it doesn't, it's unlikely to provide accurate results. Some things to look for are:

   - Has the FEM been updated to reflect the geometry and supported loads of the as-built hardware or is it a relic from the CDR?

   - How are cable bundles accounted for? They may have a significant effect on the performance of low expansion structures.

   - Is the modeled weight the same as the actual weight? What about the CG?

b) *Carefully review the models.* Are the structural models behaving as expected? Over the years, we have become more and more confident about our FEMs and have begun to accept their results as being absolutely accurate. In many instances this confidence is justified but not in the world of large optical structures, at least not yet, where success or failure is measured in micrometers or even less of deformation. Some questions to ask yourself are:

   - Is there anything in the loadpath that might lead to anomalous performance? In composite structures, for example, are there any regions where the high expansion, through-the-thickness direction of the laminate is not accounted for. Three millimeter thick shear webs, if not specifically accounted for in the design of the HST metering truss, would have produced 3 µm of despace, twice the allowable for that 5-m long structure.

   - Have lower level assembly tests been performed and compared with predictions before subsystems are combined, often masking anomalies that might have been discovered earlier?

   - Joints are often a major source of inaccuracy, especially with composite structures because of complex geometries coupled with non-orthotropic material properties. Test joints at an engineering development test level and develop an accurate FEM representation of it.

   - How well are material properties in the models represented? This is especially true of composites where not only must part-to-part CTE variations be expected and accounted for, repeated thermal cycling can alter the CTE of certain laminate designs. In critical applications, measure the individual parts before assembly and account for it in the models. This applies to composites, mirror substrates, and certain metals, notably Invar.

- The CTE of most engineering materials change with temperature. This needs to be accounted for in the models. In the case of systems where the operating temperature range is far from assembly temperature, the effect of the different ΔL/L characteristics of the different materials comprising the structure need to be accounted for. Simply using the RT or even average CTE over the range may be insufficient to avoid unanticipated interferences at operating temperature. This is especially important in the area of mirror mounts for cryogenic systems.

c) *Carefully review the test.* Particularly in thermal/vacuum testing, the impressed temperatures may not be what had been used for the pre-test analysis. Ensure that sufficient instrumentation exists to determine the actual temperature distributions.

- Are the test conditions identical to those used for analysis? Is the system at equilibrium? Is it outgassed in the case of composites?

- Have the test fixtures been included in the thermal and structural models, i.e., does the model represent the test?

A more complete description of these and additional subtle issues associated with optical structures and the factors that set them apart from more conventional applications is contained in "Athermalization of Optical Structures" by Krim, dated 1/23/92, and published as Short Course Notes by the SPIE.

It is hoped that what is presented here will be a valuable adjunct to the other more technically focused papers contained in this compendium. As stated at the outset, this was not intended as a design manual or even as a summary definition of the state-of-the-art in designing for optical systems for space operation. Rather it is offered as a guide or checklist to help round out the total perspective of design in today's cost-driven environment and to that end, I hope that I have been successful.

Rugged Yet Lightweight: How Can We Achieve Both in Optical Instruments

Daniel Vukobratovich

Optical Sciences Center
University of Arizona
Tucson, Arizona 85721

Today's optical instruments must be lightweight, rugged, and economical. In the past, these three requirements were usually considered to be mutually exclusive. These demanding requirements are now being satisfied in a new class of optomechanical instrument designs. Creating rugged, yet lightweight optical instruments requires the use of new design techniques. New optomechanical design techniques include new materials, structural optimization, insensitive optical configurations, isolation of systems from environmental effects, and active control.

Rugged design implies a severe service environment. Typical severe service environments include mechanical shock, high inertial loading, mechanical vibration, temperature extremes, nuclear radiation, environmental contamination, and user abuse. Effective design requires precise knowledge of both the actual service environment and the effects of the service environment on the system.

Lightweight is a controversial word, but is usually taken to mean systems that are lower in mass than conventional systems. For example, a lightweight telescope primary mirror is defined as being lighter than a solid mirror of the same diameter and stiffness. A key parameter in lightweight systems is the ratio of stiffness to weight. This parameter affects response to both inertial and dynamic loading.

New materials include both newly developed materials, such as the metal matrix composites, and non-traditional metallic materials, such as titanium. Structural optimization is applied to improve stiffness to weight. Telescope primary mirrors are optimized by contouring the back of the mirror; support structures are optimized so that parts of the structure serve more than one purpose. Insensitive design includes traditional structural practice and clever optical design. An Airy point supported beam is highly resistant to dynamic and inertial loading. Tube bending is compensated by passive optical design in relay systems. Isolation from environmental effects includes the use of enclosures and vibration isolation systems. Finally, active control is used to reduce weight and stiffness of systems, while still maintaining optical alignment through the use of position sensors and actuators.

Design examples of rugged and lightweight systems include cryogenic space optics, ultra-lightweight telescopes, and large astronomical astrographic lenses. A 0.5 m, 15 kg, fused silica mirror for the Space Infrared Astronomical Facility operated successfully at 9 K, demonstrating the use of materials selection, structural optimization, and isolation from temperature effects. A series of ultra-lightweight telescopes, up to 0.4 m aperture, employs new metal matrix composite materials, and structural optimization. Passive optical compensation for temperature effects, non-traditional metallic materials, and isolation are effectively employed in a 2 m focal length, f/10 astrometric objective.

INTRODUCTION

Lightweight and rugged are controversial words when discussing optical systems. Lightweight mirrors are examples of the controversy in defining the word lightweight. Both a 300 mm diameter mirror weighing 2 kg, and a 1.8 m diameter mirror weighing 550 kg, are described as lightweight mirrors. This example suggests that the definition of lightweight involves more than just absolute weight. To avoid controversy, any definition of lightweight and rugged should involve comparisons with normal practice, and perhaps some form of scaling relation to allow for the great variation in physical size in optical systems.

Lightweight is simply defined as describing systems reduced in weight when compared with normal practice. The difficulty with the definition is in classifying normal practice. Optical systems vary considerably in both size and weight. Scaling relationships are one way of allowing for this variation.

An example of scaling relationships is the weight versus size relationship of the moving tube assembly of a conventional terrestrial astronomical telescope. In 1971, Rule suggested that the weight of the moving tube of a terrestrial astronomical telescope is related to the primary mirror diameter by the following relationship[1]:

$$W \simeq 2656d^{2.4} \tag{1}$$

where: W is the total tube moving weight in kg
d is the primary mirror diameter in m

Any telescope with a weight less than that calculated from Eq. 1 could be called a "lightweight." This is an acceptable approach provided that a scaling relationship exists. Frequently, the optical system in question is produced in small quantities, or is the only example of its type. In both cases, deciding if the system is lightweight on the basis of scaling information is difficult, since that information is not available.

Another approach to defining "lightweight" has been suggested by Schwesinger[2]. Although this definition was originally intended to apply to lightweight mirrors, it is equally valid for other types of structures. Schwesinger suggested that a lightweight mirror is "lightweight" if the self-weight deflection of the mirror is less than that of a solid mirror of the same diameter and weight.

Since the fundamental frequency of a structure is determined by both stiffness and weight, another possible definition for lightweight structures is to use a fundamental frequency criteria. This is an extension of Schwesinger's definition. Fundamental frequency of a self-weight loaded beam is given by[3]:

$$f_n = \frac{\lambda_i^2}{2\pi L^2} \left[\frac{EI}{m} \right]^{1/2} \tag{2}$$

where: f_n is the fundamental frequency
λ_i is a dimensionless constant depending on the type of beam support
L is the beam length
E is the elastic modulus of the beam material
I is the area moment of inertia of the beam about the neutral axis
m is the mass per unit length of the beam

Equation 2 can be rewritten in a more useful form by noting that the mass per unit length of the beam is equivalent to the beam cross section multiplied by the mass density of the beam material:

$$f_n = \frac{\lambda_i^2}{2\pi L^2} \left[\frac{E}{\rho} \right]^{1/2} \left[\frac{I}{A} \right]^{1/2} \tag{3}$$

where: A is the area of the beam cross section
I is the mass density of the beam material

Equation 3 indicates that for a given beam length and support condition, the two most important parameters controlling fundamental frequency are the ratios of elastic modulus to density, and moment of inertia to area. For a given material choice, a lightweight structure may be defined as one in which the ratio of moment of inertia to cross section is higher than that of a conventional structure. For example, a thin walled circular tube has about twice the value of this ratio when compared with a solid of the same diameter. The hollow tube is "lightweight" in comparison with a solid cross section. This definition is similar in concept to frequency scaling relationships for large telescopes suggested by Davison[4] and Schneermann[5].

Conventional measures of structural efficiency used in aerospace engineering are based on ratios of strength to weight. In many cases, failure through plastic yielding or elastic buckling is used in determining strength[6]. Conventional measures of structural efficiency are not suitable for evaluating opto-mechanical support structures since these structures are used at stresses well below material yield or buckling.

Rugged is a word used to describe the service environment. A system is said to be rugged if it is capable of operating in an environment that is more severe than normal. The difficulty with this definition is in understanding what is meant by a normal operating environment.

Traditional operating environments for optical systems are relatively benign, often approaching controlled laboratory environments. Systems designed to operate in industrial environments face more severe conditions. Military systems face very severe conditions. Extreme conditions are encountered in a wide variety of environments, including factories, satellites, and underwater systems.

Table 1 shows values for normal, severe, and extreme operating environments. A laboratory, or laboratory similar operating environment is assumed to be normal. This type of environment is described in standards on environmental control[7] such as ASHRAE Standard 55-81, and on vibration control[8]. Information for the severe portion of the table is taken from the current U.S. DOD standard on military operating environments[9]. Extreme service environments are taken from a variety of sources, and are intended to be representative. (The extreme column is not intended to be an opto-mechanical equivalent of the Guinness Book of Records!) In this paper, severe environments are assumed to be those given in Table 1.

Table 1. Service Environments.

	Normal	Severe	Extreme	Example of Extreme
Low Temperature	293 K	222 K	2.4 K	Infrared Astronomical Satellite (IRAS) Telescope
High Temperature	300 K	344 K	423 K	White Cell for combustion studies
Low Pressure	88 KPa	57 KPa	0	Infrared Astronomical Satellite (IRAS) Telescope
High Pressure	108 KPa	1 MPa	138 MPa	150° spherical sector Germanium dome on submersible
Humidity	25 - 75%	100% RH	(underwater)	150° spherical sector Germanium dome on submersible
Acceleration	2 g	12 g	11×10^3 g	107 mm Mortar shell ZnS dome
Vibration	200×10^{-6} m/s RMS $f > 8$ Hz	0.04 g^2/Hz $20 < f < 100$ Hz	0.13 g^2/Hz $30 < f < 1500$ Hz	Titan launch vehicle

MATERIALS FOR RUGGED, LIGHTWEIGHT SYSTEMS

The most important properties of materials used in lightweight, rugged optical systems are those determining structural efficiency, temporal stability, and thermal stability. Materials used in optical components must also have good optical performance properties. These properties are so important that if a material offers significant advantages in these areas, serious disadvantages in other areas, such as high cost and difficulty of fabrication are often overlooked. Beryllium, a very expensive and difficult to produce material, is an example.

Of primary importance are those material properties controlling the weight of the system. Stiffness and density strongly influence the weight of an opto-mechanical structure. For example, the self-weight deflection of a solid axisymmetric mirror is given by:

$$\delta = C \frac{\rho}{E} \left[\frac{d}{h}\right]^2 d^2 \tag{4}$$

where: δ is the self-weight deflection
C is the dimensionless support condition constant
ρ is the mirror material density
E is the mirror material elastic modulus
d is the mirror diameter
h is the mirror thickness

Equation 4 is very similar to Eq. 2. Both equations indicate that the ratio of material density to elastic modulus is an important material property influencing self-weight deflection. Many common structural materials have about the same ratio of density to elastic modulus, about 386 x 10^{-9} m^{-1}. Materials with a ratio of density to elastic modulus lower than this value are considered lightweight materials. Figure 1 shows the relationship between density and elastic modulus for a variety of materials.

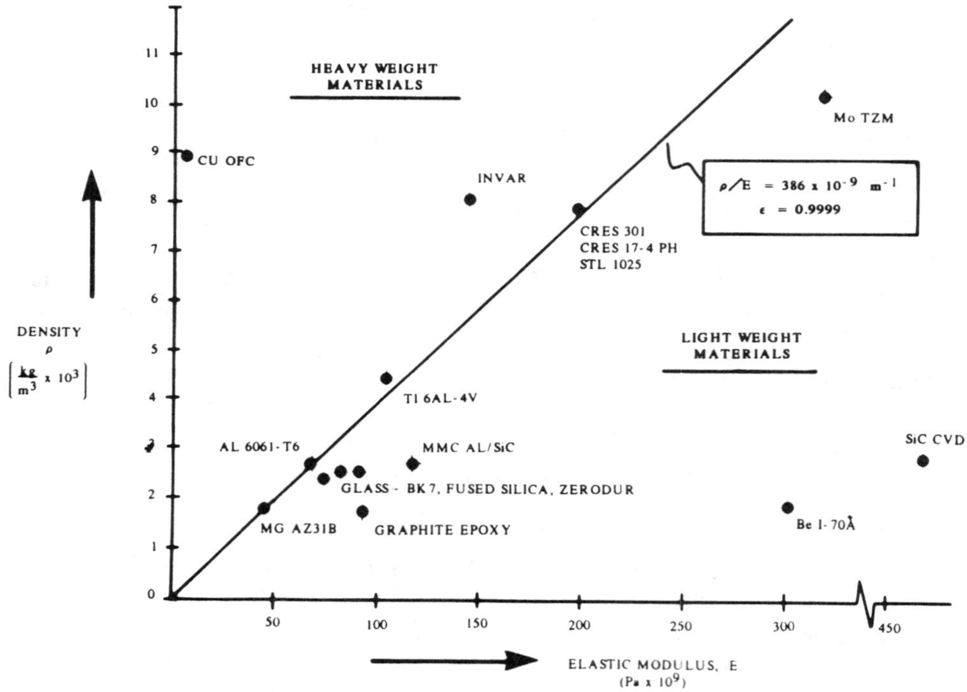

Figure 1: Material Specific Stiffness

The ratio between density and elastic modulus is a commonly suggested parameter for defining lightweight materials in a variety of applications, many of which are not concerned with optics[10]. This ratio is the inverse of the material parameter called specific stiffness. As is shown below, specific stiffness (or the inverse specific stiffness) of materials does not determine the final weight of a system. The weight of an axisymmetric mirror is given by:

$$W = \frac{\pi}{4} \rho h d^2 \tag{5}$$

where: W is the mirror weight
h is the mirror thickness
d is the mirror diameter
ρ is the mirror material density

Solving Eq. 4 for the mirror thickness, and substituting into Eq. 5, the mirror weight becomes:

$$W = \frac{\pi}{4} d^4 \left[\frac{\rho^3}{E}\right]^{1/2} \left[\frac{C}{\delta}\right]^{1/2} \tag{6}$$

where: W is the mirror weight
d is the mirror diameter
ρ is the mirror material density
E is the mirror material elastic modulus
δ is the specified maximum mirror surface deflection
C is the dimensionless mirror support constant

From Eq. 6, the material parameter controlling mirror weight is not the ratio of density to elastic modulus, but is the square root of the ratio of the cube of the density to the elastic modulus. This material parameter is identical to the dynamics parameter suggested by Krim[11].

Other properties controlling the structural efficiency include fracture toughness and damping capacity. Fracture toughness determines the ability of a material to resist failure by brittle fracture. This property is of particular importance for thin, highly stressed structural members. Fracture toughness defines the maximum acceptable stress to avoid brittle fracture for a given section thickness[12]. Fracture toughness decreases with temperature and is inversely proportional to the yield strength of the material[13].

Damping capacity is the material property that determines the dissipation of energy in the material when excited by vibration. This property is important in determining response of an optical support structure to excitation by vibration. For common structural materials, damping capacity increases with applied stress[14]. Of considerable interest for use in severe environments are materials with high damping capacity. Damping capacity of various materials are given in Schetky and Perkins[15], James[16], Adams[17] and Lee[18].

Long term stability of optical support structures is determined by the material dimensional stability properties. Paquin defines four categories of dimensional instability: temporal instability, thermal/mechanical instability, thermal instability, and hysteresis[19]. These dimensional instabilities are determined by the microstructural stability, the microyield strength, and homogeneity of the material. Microstructural stability is the inherent stability of the material in an unstressed condition. Microyield is defined as that amount of stress necessary to produce a permanent deformation in the material of 10^{-6}. Compilations of both of these properties are given in Marschall and Maringer[20].

For many service environments, the material thermal properties are important in determining the optical system response to temperature changes. Material thermal properties of interest include the thermal coefficient of expansion, the homogeneity of the thermal coefficient of expansion, the thermal distortion index, and the thermal diffusivity. The variation in index of refraction with temperature, or the therm-optic coefficient, is also of importance for materials used in transmission. Material properties control the ability of the system to maintain focus, optical alignment, and aberration control as temperatures are changed.

An extremely small thermal coefficient of expansion is usually not necessary in selecting a material for an opto-mechanical system. More important is that the thermal coefficient of expansion, or CTE, of the structural material should match the thermal response of the optical materials. In mirror systems, the CTE of the structure should match the CTE of the mirror material. In lens systems, the CTE of the structural materials should match the temperature induced change in focus of the system.

In the case of metal mirrors, a mismatch in the thermal coefficients of expansion of the mirror plating and mirror substrate materials can cause bi-metallic bending effects with changes in temperature. Aluminum and beryllium have different thermal coefficients of expansion than electroless nickel, the most common plating material. Plating both sides of the mirror is a partial solution to bi-metallic bending. Significant stress may be induced in the substrate due to the bi-metallic effect even when both sides of the mirror are plated. If this stress is greater than the micro-yield of the material, the optical surface figure changes because of dimensional instability when the mirror is thermally cycled.

Uniformity of the thermal coefficient of expansion of the material controls the optical surface figure of a mirror when subjected to a temperature change. Uniformity of the thermal coefficient of expansion is also important in maintaining alignment and focus in the support structure as the temperature changes. If the thermal coefficient of expansion varies spatially, a change in temperature causes different parts of the mirror to expand or contract by different amounts. This variation in property causes distortion in the optical surface figure. As a rule of thumb, most materials exhibit a uniformity of 3 to 5 percent in their thermal coefficient of expansion[21,22].

Four categories of materials are of particular interest in lightweight, rugged systems: non-traditional metals, metal matrix composites (MMCs), graphite epoxy composites, and plastics. There is no such thing as a perfect material; all of these categories of materials have both advantages and disadvantages. Selection of material must involve a careful analysis of the actual performance advantages of the materials, against possible drawbacks.

One of the most interesting, and controversial non-traditional metallic materials for use in lightweight, rugged optical systems is beryllium. Beryllium offers exceptional specific stiffness, about 16.8 x 10^6 m, when compared to the specific stiffness of 2.5 x 10^6 m for most common structural materials. In addition, the thermal distortion index of beryllium is less than half that of aluminum, 50.9 x 10^{-9} m/w versus 136 x 10^{-9} m/w. Beryllium has high nuclear hardness, which is important for certain military applications.

Beryllium is a very expensive material, more so than just about any other optical material. This expense is due largely to the difficulty of processing the raw material and in fabricating parts. Toxicity of beryllium is a complication leading to increased cost during fabrication. Toxicity is not, however, an issue for normal use of fabricated beryllium components. Temporal and thermal stability of beryllium is controversial. Low microyield strength and anisotropy of beryllium are two of the material properties leading to the controversy. Fracture toughness of beryllium is low, requiring extreme care in design for fracture control.

The exceptional properties of beryllium, coupled with the very real problems associated with the use of this material, has led to polarization of the optical community. People are of two opinions concerning the use of beryllium: use anything but beryllium, and beryllium is wonderful. In reality, beryllium is a material like any other and should be evaluated on the merits of known performance rather than hearsay. Unfortunately, until very recently it has been extremely difficult to obtain information concerning performance of beryllium systems, making unbiased evaluation of the material very difficult.

A relatively new category of materials are the metal matrix composites. These materials consist of a metal matrix with a non-metallic material imbedded in the matrix. Although a variety of such materials exist, the aluminum/silicon carbide metal matrix composites have been the most commonly used[23]. In particular, the aluminum/silicon carbide composite material developed by the Advanced Composite Materials Corp., trade named SXA, has seen considerable use in lightweight, rugged optical systems. SXA consists of an aluminum matrix, typically a 20 series aluminum alloy, reinforced with between 20 and 30 per cent by volume with silicon carbide. For optimum strength, silicon carbide whisker reinforcement is used; for improved uniformity of properties a silicon carbide particulate reinforcement is used. SXA has a specific stiffness of 4.1 x 10^6 m, which is 1.6 times higher than that of aluminum. Microyield strength of SXA is higher than that of most common structural materials used in optomechanics; for the whisker reinforced material microyield is about 346 MPa. In mirror substrates, the use of SXA eliminates bi-metallic bending with temperature when the mirror is plated with electroless nickel. Bi-metallic bending is eliminated by tailoring the CTE of the SXA substrate to exactly match the CTE of the electroless nickel. Adjustment of the thermal coefficient of expansion is possible through controlling the amount of silicon carbide in the matrix. About 30 volume per cent of silicon carbide is used to match the thermal coefficient of expansion with that of electroless nickel. SXA is not toxic, and can be machined in ordinary shops, leading to reduced fabrication costs.

An example of the use of SXA is a lightweight telescope designed by the Optical Sciences Center, University of Arizona, for Aerojet Electro-Systems. This telescope,shown in Fig. 2, is an all SXA off-axis 3 mirror system of 0.2 m aperture. A fundamental frequency of 94 Hz is obtained through the use of the MMC material. Maximum self-weight stress in a terrestrial environment is only 3.2 MPa.

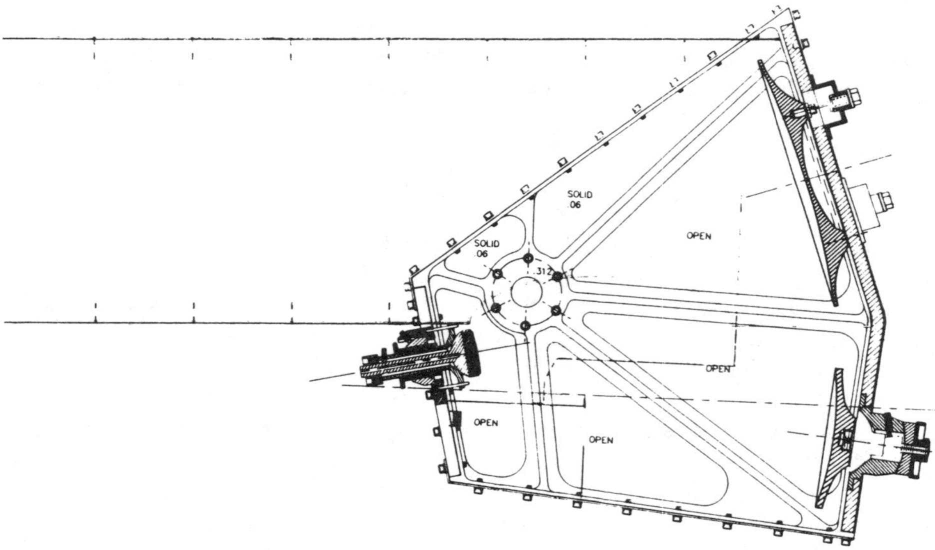

Figure 2: Advanced ULW Telescope Employing SXA MMC Materials
(Figure courtesy Aerojet Electro-Systems)

Graphite epoxy composite materials are used in a variety of high performance space optical systems, including the Hubble Space Telescope and Mars Observer Camera[24]. Graphite epoxy materials have a high specific stiffness. For example, the specific stiffness of isotropic G70 graphite epoxy is about 5.2×10^6 m. The thermal coefficient of expansion of a graphite epoxy composite can be tailored to be near zero, or even negative. Thermal conductivity of the graphite epoxy composites is about twice that of stainless steel. The combination of good thermal conductivity and low thermal coefficient of expansion gives the graphite epoxy composites a thermal distortion index which is comparable or better to that of a "zero expansion" material such as Schott Zerodur.

Graphite epoxy composites absorb moisture, and as the moisture is adsorbed the material changes in physical size. One relation for the change in length with per cent absorbed moisture is[25]:

$$CME = \frac{163 \times 10^{-6}}{\%M} \tag{7}$$

where: CME coefficient of moisture expansion
%M percent absorbed moisture

Graphite epoxy materials microcrack when the temperature is lowered, leading to dimensional instability[26]. Hard radiation causes a deterioration of the material. Graphite epoxy materials are usually anisotropic in their properties. In many cases, material properties may be controlled in a single direction, or a single plane. Currently, fabrication of graphite epoxy components for optical application is extremely labor intensive, leading to high costs.

Plastics are gaining in importance as materials for use in both the optics and structures of lightweight, rugged optical systems. Plastic materials are lightweight, low in fabrication cost, and resistant to corrosion. Low fabrication cost has been particularly important in the use of plastic materials in applications such as binocular structures[27].

Most plastics have high thermal coefficients of expansion which complicates the structural use of these materials in combination with more conventional optical materials. The therm-optic coefficients of the optical plastics are high, leading to difficulties in athermalizing the performance of plastic optical systems. Temporal stability of plastics is poor. Stability with time has been improved through the use of glass filled polycarbonate plastic materials. Plastics are soft in comparison with ordinary optical glasses, and are thus more vulnerable to surface scratching.

Table 2 gives the representative properties of some materials of interest for lightweight, rugged optical systems.

STRUCTURAL DESIGN

Structural design of lightweight, rugged systems is determined by the service environment and by the optical tolerances of the system. Typically, the most important characteristics of the service environment are acceleration, vibration, temperature change, and thermal gradients. Optical system characteristics of importance are the size and weight of the optics, and the optical tolerances.

Design of an effective optical support structure requires considerable interaction between optical and mechanical design. In most cases, it is necessary to know both the optical tolerances of the system and the sensitivities of the tolerances. Difficult mechanical problems can sometimes be solved by modification of the optical configuration if such interaction occurs.

Acceleration is a service environmental specification which is universal. Even space systems, which must operate in a "zero gravity" environment, are fabricated on the ground, in a one g environment. In addition, space systems are exposed to high acceleration during launch into space. The effect of gravity must therefore always be considered in the design of optical support structures.

Acceleration causes the optical support structure to deform in the direction of the acceleration. This deformation induces mis-alignment of the optics. In large terrestrial optical components, gravity induces an unacceptable optical surface deformation. In extreme cases, large stresses may be induced by high accelerations.

Conventional practice is to limit acceleration effects through the use of sufficiently stiff structures. Unfortunately, stiff structures are relatively heavy, particularly as the size of the system increases. Reducing system weight may require abandoning the use of a stiff structure.

In most optical systems it is not necessary to have a universally stiff structure. Only deformation of the optics relative to each other affects optical system performance. It

Table 2. Material Properties.

	Density (ρ) Kg - M^{-3} ($x10^3$)	Elastic Modulus (E) (GPa)	(MYS) (MPa)	(ρ/E) (m^{-1}) ($x10^{-9}$)	$(\rho^3/E)^{1/2}$ ($N\text{-}m^{7/2}$)	$(E/\rho)^{1/2}$ m ($x10^3$)	Damping (η_s)	Yield Stress (GPa)	Yield Stress/ MYS	Specific Heat (Cp) $J\cdot(Kg\text{-}k)^{-1}$	Conductivity (K) $W\cdot(m\cdot K)^{-1}$	Expansion (α) ($x10^{-6}/K$)	$\Delta\alpha$ ($x10^{-9}/K$)	Thermal Distortion (α/K) ($x10^{-9}$ m/w)	Diffusivity [$K/C_p\rho$] ($x10^{-6}$ m^2/sec)
2024-T4 Al	2.77	73.1	262	372	16.6	1.64	$5x10^{-5}$	290	1.11	877	121	23		190	49.8
2024-T6 Al	2.77	73.1	172	372	16.6	1.64		345	2.00	877	150	23		153	61.7
5456-H34 Al	2.66	71.0	121	368	15.8	1.65		214	1.77	962	117	24		205	45.7
6061-T6 Al	2.71	68.9	127	386	16.5	1.61		276	2.17	879	167	23	60	138	70.1
A356-T6 Al	2.68	70.3	22	374	16.1	1.64	10^{-3}	165	7.50	962	152	21		138	59.0
I-70 Be	1.85	303	17	59.9	4.44	4.09		172	10.1	1925	180	11.5	130	63.9	50.5
I-220 Be	1.85	303	34	59.9	4.44	4.09		241	7.09	1925	180				
I-400 Be	1.84	303	34	59.6	4.41	4.10				1925	180	11.5	400	63.9	50.8
HP-21	1.85	290	31	62.6	4.54	4.00				1925	180				
HP-40	1.85	290	55	62.6	4.54	4.00				1925	180				
HP-81	1.85	290	17	62.6	4.54	4.00				1925	180		220		
Cu-2Be	8.23	117	64	690	67.1	1.20		1140	17.8	418	105	17.4	835	166	30.5
OFHC Cu	8.89	117	15	745	75.3	1.16	5.10^{-3}	69.0	4.60	385	391	16.9	880	43.2	114
Az-31	1.77	44.8	22	388	10.8	1.61	2.10^{-3}	152	6.89	1004	76.1	25.2		331	42.8
TZM Mo	10.2	324	165	309	55.6	1.80	$1.6x10^{-4}$	621	3.76	272	146	5.0	270	34.2	52.6
ARMCO IRON	7.81	208	75	368	46.5	1.65	$6.4x10^{-3}$								
1020 STL	7.86	200	166 ($2x10^{-6}$)	386	47.9	1.61	10^{-3}	2.12	352						
1040 STL	7.84	200	138 ($2x10^{-6}$)	385	47.7	1.61		490	3.55						

Table 2. Material Properties (cont.).

	Density (ρ) Kg - M^{-3} (x10^3)	Elastic Modulus (E) (GPa)	(MYS) (MPa)	(ρ/E) (m^{-1}) (x10^{-9})	(ρ^3/E)$^{1/2}$ (N-m$^{7/2}$)	(E/ρ)$^{1/2}$ m (x10^3)	Damping (η_s)	Yield Stress (GPa)	Yield Stress/ MYS	Specific Heat (Cp) J-(Kg-k)$^{-1}$	Conductivity (K) W-(m-K)$^{-1}$	Expansion (α) (x10^{-6}/K)	$\Delta\alpha$ (x10^{-9}/K)	Thermal Distortion (α/K) (x10^{-9} m/w)	Diffusivity [K/C$_p\rho$] (x10^{-6} m^2/sec)
1080 STL	7.84	200	205 (2x10^{-6})	385	47.7	1.61		1240	5.49						
4340 STL	7.86	200	270	386	47.9	1.61		1480	5.49	477	38.1	11.3		297	10.2
52100 STL	7.86	200	276	386	47.9	1.61									
304 CRES STL	7.9	200	50	387	48.2	1.61		241	4.83	502	16.3	17.3		1061	4.11
310 CRES STL	7.9	200	345	387	48.2	1.61		310	(.9)	502	14.2	15.8		1113	3.58
440 CRES STL	7.7	200	522	378	46.4	1.63		448	(.86)	460	24.1	10.1		419	6.80
300 MARAGING STL	7.92	183	1840	425	50.6	1.54		1860	1.01	293	38.1	9.4		247	16.4
400 MARAGING STL	7.92	183	2210 (2x10^{-6})	425	50.6	1.54									
6AL-4V Ti	4.43	114	524	381	26.8	1.62		827	1.58	565	7.2	8.8	900	1222	2.88
5AL-2.5 SnTi	4.46	110	552	398	27.6	1.59		793	1.44	523	7.8	9.4		1205	3.34
INCONEL-X	8.30	214	448	380	50.2	1.62	10^{-3}	689	1.54	439	11.2	12.6		1125	3.07
INVAR 36	8.0	145	41	541	57.7	1.36		276	6.73	502	11.1	1.3	50	117	2.76
SXA 24-T6 30 V/0 Sic	2.91	117	117	244	14.1	2.02		523	4.47	795	125	12.4	800	99.2	54.0
GRAPHITE EPOXY G70/x30 UNi	1.8	293	138	60.3	4.34	4.07		550	3.99	920	51.9	1.0		19.3	31.3
GRAPHITE EPOXY G70/30 ISo	1.8	97	100	182	7.53	2.34		193	1.93	920	37.5	0.1	100	2.67	22.6

is therefore possible to use structures which have relatively poor overall stiffness, but which nevertheless maintain optical alignment.

The simplest example of a low stiffness structure which maintains optical alignment is the Airy point supported beam. Deformation of a self-weight loaded beam is controlled through the careful selection of the support locations for the beam[28]. The Airy point support solution is applicable to a uniform cross section, self-weight loaded beam. The "Airy points" are the locations of two simple supports for the beam such that the beam ends have equal and parallel deflections. For a simply supported uniform cross section beam, the use of the Airy point supports produces a zero slope condition at the ends of the beam. Since both beam ends have the same deflection, and are parallel to each other (zero slope), optics mounted at the beam ends remain in alignment. The two Airy points are located at a distance of 0.2113 times the overall length of the beam from either beam end.

Although the original Airy point solution is for a uniform beam on two simple supports, the solution can be extended to a multiple point supported beam. For the beam ends to have a zero slope condition and to have equal deflections, the n points of support must be spaced apart by the amount given by the following equation:

$$L_S = \frac{L_B}{(n^2-1)^{1/2}} \qquad \textbf{(8)}$$

where: L_S is the spacing between two adjacent support points
L_B is the overall beam length
n is the number of support points

Equation 8 indicates that the Airy point condition is independent of the stiffness of the beam. Equation 7 is also valid regardless of the orientation of the beam. Since stiffness is not important in the performance of an Airy point supported beam, the beam may be made very light in weight. The Airy point supported beam is a possible lightweight optical support structure.

In reality, the weight of the optics at the beam ends influences the beam deformation. The effect of this additional deformation is eliminated by adjusting the location of the simple supports of the Airy point supported beam. The relation between the support locations, beam weight, and weight of the optic is given by:

$$L_S = \frac{L_B}{2}\left\{1-\left[1-\frac{W_B}{6\left(\frac{W_O}{2}+\frac{W_B}{4}\right)}\right]^{1/2}\right\} \qquad \textbf{(9)}$$

where: L_S is the location of the beam supports
W_B is the beam weight
W_O is the weight of the optics at one of the beam ends

Equation 9 requires that the optics at either end of the beam have the same weight. In some circumstances this may require the use of additional ballast weight at one end of the beam. Although the use of ballast may seem inappropriate in a lightweight structure, the very good performance of an Airy point supported beam usually overrides this objection.

Larger structures benefit from the principle of equal and parallel deformation used in the Airy point supported beam. The classic example of a structure using equal and parallel deflections is the Serrurier truss. The Serrurier truss, shown in Fig. 3, is a pin jointed, three dimensional space truss used to control deflection in terrestrial telescopes. A center of gravity support is provided for the truss; and both sides of the truss are designed to have equal and parallel deflections[29].

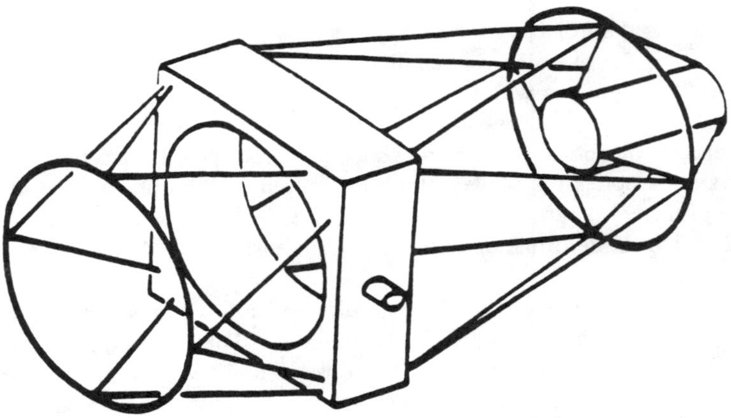

Figure 3: Serrurier Truss

For equal and parallel deflections in the end rings of a Serrurier truss, the following condition must be satisfied:

$$\frac{W_1}{E_1A_1}\left(\frac{4L_1}{d^2}+1\right)^{3/2}=\frac{W_2}{E_2A_2}\left(\frac{4L_2}{d_2}+1\right)^{3/2} \tag{10}$$

where: W_1,W_2 are the end ring loads
L_1,L_2 are the distances from the center of gravity support to the end rings
E_1,E_2 are the modulus of elasticity values of the truss member materials
A_1,A_2 are the truss member cross sections
d is the end ring diameter

The Serrurier truss is an important class of lightweight structure which has been heavily used in terrestrial telescopes. It is not an exaggeration to say that the performance of this type of structure was critical in the development of large telescope technology. The principal of parallel deflection in the Serrurier truss has been exploited in other types of lightweight structures[30] derived from this type of truss. Despite its very good deflection performance, the Serrurier truss has some significant disadvantages.

Stiffness to weight of the Serrurier truss is low, which implies a low fundamental frequency. This low fundamental frequency can be a problem in some applications. When the overall length to width ratio of a Serrurier truss is greater than about 1.4, a multiple bay truss has a better stiffness to weight ratio. In general, the number of bays required to provide optimum stiffness to weight in an end point loaded truss is given by the following relationship[31]:

$$[n(n-1)]^{1/2} < \frac{L}{d} < [n(n+1)]^{1/2} \qquad (11)$$

where: n is the number of truss bays
L is the overall truss length
d is the truss width

Certain classes of optical instruments, such as periscopes, bore scopes, and passive range finders have optics distributed along the length of the optical support structure. Deformation of the entire structure is important in such cases. One approach to reducing the effect of structural deformation is to employ insensitive optical elements.

The optical invariant implies that long tubular optical instruments must have a series of relay elements to obtain realistic fields of view. Tube bending causes the relay elements to tilt and move away from the optical axis, which disturbs the optical performance of the system. The effect of tube bending can be reduced by an optical solution. Each relay element is replaced by a positive and negative set of elements. The location of the rear principal point of the combination of positive and negative elements is placed at the intersection of the normal to the focal plane and the optical axis of the objective at the end of the tube. Tube deformation causes the combination to rotate about its rear principal point. Rotation of a lens about its rear principal point does not induce any image motion, so the system is insensitive to tube deformation. Optical compensation of flexure using this principal is employed in submarine periscopes and range finders[32]. Meinel has combined the principal of optical compensation with the Serrurier truss structure to develop a class of lightweight telescope structures that are insensitive to both gravity and wind induced bending[33].

Vibration effects are another consideration in the design of lightweight, rugged optical support structures. The effects of ordinary periodic vibration can be reduced by changing the fundamental frequency of the system, or providing a vibration isolation system[34]. The most important characteristic of a system exposed to vibration is the fundamental frequency of this system. Often it is not possible to measure the fundamental frequency of the system. Computation of the fundamental frequency may be difficult due to the complexity of the system. In such cases, the fundamental frequency may be approximated from the known self-weight deflection of the system using the following equation:

$$f_n \simeq \frac{1}{2\pi}\left[\frac{g}{\delta}\right]^{1/2} \tag{12}$$

where: F_n is the fundamental frequency
δ is the self-weight induced deflection
g is the acceleration due to the Earth's gravity (9.81 m-sec^{-2})

An important parameter in describing system response to vibration is the transmissibility. Transmissibility is defined as the ratio of the amplitude of the exciting vibration to the amplitude of the system response. A value of transmissibility greater than one implies that the system amplifies the amplitude of the exciting vibration, while a value of the transmissibility of less than one implies attenuation of the exciting vibration. The system response is given by:

$$T = \frac{X_S}{X_E} = \left\{\frac{1 + \left[2\frac{f}{f_n} C_R\right]^2}{\left[1 - \left[\frac{f}{f_n}\right]^2\right]^2 + \left[2\frac{f}{f_n}C_R\right]^2}\right\}^{1/2} \tag{13}$$

where: T is the transmissibility
X_E is the amplitude of the exciting vibration
X_S is the amplitude of the system response vibration
C_R is the system critical damping ratio
f_n is the system fundamental frequency
f is the exciting frequency

For real optical support structures, the critical damping ratio is usually below 0.10, and in many case below 0.01. If the critical damping ratio is small, and if the system frequency is much less than the exciting frequency, Eq. 11 simplifies to:

$$T \simeq \left[1 - \left[\frac{f}{f_n}\right]^2\right]^{-1} = \left[1 - \frac{4\pi^2 f^2 \delta}{g}\right]^{-1} \tag{14}$$

where: T is the transmissibility
f_n is the fundamental frequency
f is the exciting frequency
δ is the system self-weight induced deflection
g is the acceleration due to Earth's gravity

Both Eqs. 13 and 14 assume that the system is excited by a single frequency. In many service environments, the system is exposed to random vibration. A random vibration spectrum contains all frequencies between its frequency limits. If the system fundamental frequency lies between the frequency limits of the random vibration spectrum, the system is excited at its fundamental frequency. A system excited at its fundamental frequency is in resonance, and Eq. 11 predicts that transmissibility at resonance is high for lightly damped systems. Random vibration may induce very high inertial loads into an optical support structure.

Random vibration is normally quantified by a power spectral density function (PSD). The response of a simple, single degree of freedom system to a specified power spectral density function is given by:

$$g_{rms} \simeq \left[\frac{\pi}{2} \, f_n \, Q \, (PSD) \right]^{1/2} \qquad \textbf{(15)}$$

where: g_{rms} is the system root mean square (RMS) response, in units of dimensionless "g"
f_n is the system fundamental frequency
Q is the system transmissibility at resonance or at its fundamental frequency
PSD is the power spectral density at the system fundamental frequency, in units of $g^2\text{-hz}^{-1}$

It is common in vibration engineering to assume that most structural damage is done by the "3-sigma" peak acceleration. This is found by multiplying the RMS acceleration given by Eq. 15 by 3. The displacement response of the system is given by:

$$\delta = \frac{g_{rms}}{(2\pi f_n)^2} \qquad \textbf{(16)}$$

where: δ is the displacement response
f_n is the system fundamental frequency
g_{rms} is the root mean square acceleration response

In evaluating system response to random vibration, Eq. 16 requires accurate knowledge of the power spectral density of the service environment, system damping, and system fundamental frequency. Fundamental frequency is readily calculated, but damping must be measured by experiment. The transmissibility of a system at resonance is called the "Q" of a system, and is an important figure of merit in judging response of the system. Values of Q for optical support systems may range from as low as 25 to over 160. A value of 100 is a often used for estimating system response. Table 3 gives representative power spectral densities for a variety of service environments.

Random vibration usually forces the designer to increase system stiffness, or damping. Most structural materials are very poorly damped, as indicated by the values of Q given in table 1. Material damping is proportional to stress, with high damping values corresponding to high stresses. Very stiff optical support structures are lightly stressed, leading to low damping. A second source of damping is coulomb friction. Coulomb friction damping requires relative motion between the parts of a support

Table 3. Power Spectral Densities (PSDs).

	F_n	PSD
Ariane	5-150	+6 dB/Octave
	150-700	0.04 G^2/Hz
	700-2000	-3 dB/Octave
Shuttle	15-100	+6 dB/Octave
Orbiter Keel	100-400	0.10 G^2Hz
	400-2000	-6 dB/Octave
STS PAM-D	10	0.0020 G^2/Hz
	10-80	+4 dB/Octave
	80-170	0.033 G^2/Hz
	170-2000	-4dB/Octave
	2000	0.0012 G^2/Hz
Apollo	20-80	+3 dB/Octave
	80-400	0.040 G^2/Hz
	400-2000	-3 dB/Octave
Thor-Delta	20-200	0.07 g^2/Hz
Thor-Able		
Mars Observer	20-100	+6 dB/Octave
Camera	100-900	0.2 g^2/Hz
	900-2000	-6 dB/Octave
Titan	10-30	+6 dB/Octave
	30-1500	0.13 g^2/Hz
	1500-2000	-6 dB/Octave
Navy warships	1-50	0.001 g^2/Hz
Minimum integrity	20-1000	0.04 g^2/Hz
test	1000-2000	-6 dB/Octave
(MIL-STD-810E)		
Typical aircraft	15-100	0.03 g^2/Hz
	100-300	+4 dB/Octave
	300-1000	0.17 g^2/Hz
		-3 dB/Octave

structure. This type of relative motion is unlikely to occur in a properly designed optical support structure, again suggesting that damping values will be low.

The need for high stiffness to weight for service in severe random vibration environments leads to development of new types of optical support structures. Conventional optical support structures are made in the form of tubes or trusses. Thin tube sections, although efficient in bending, may be limited in performance due to concerns about buckling. Truss structures provide good control of deformation due to acceleration, but have poor stiffness to weight.

Two alternate structures used in lightweight, rugged Cassegrain telescopes are the Dall and Meinel type structures[35]. A 400 mm aperture Dall type telescope is shown in Fig. 4 and a 200 mm Meinel type telescope is shown in Fig. 5. Both structures reduce the structural weight to approximately the same weight as the primary mirror.

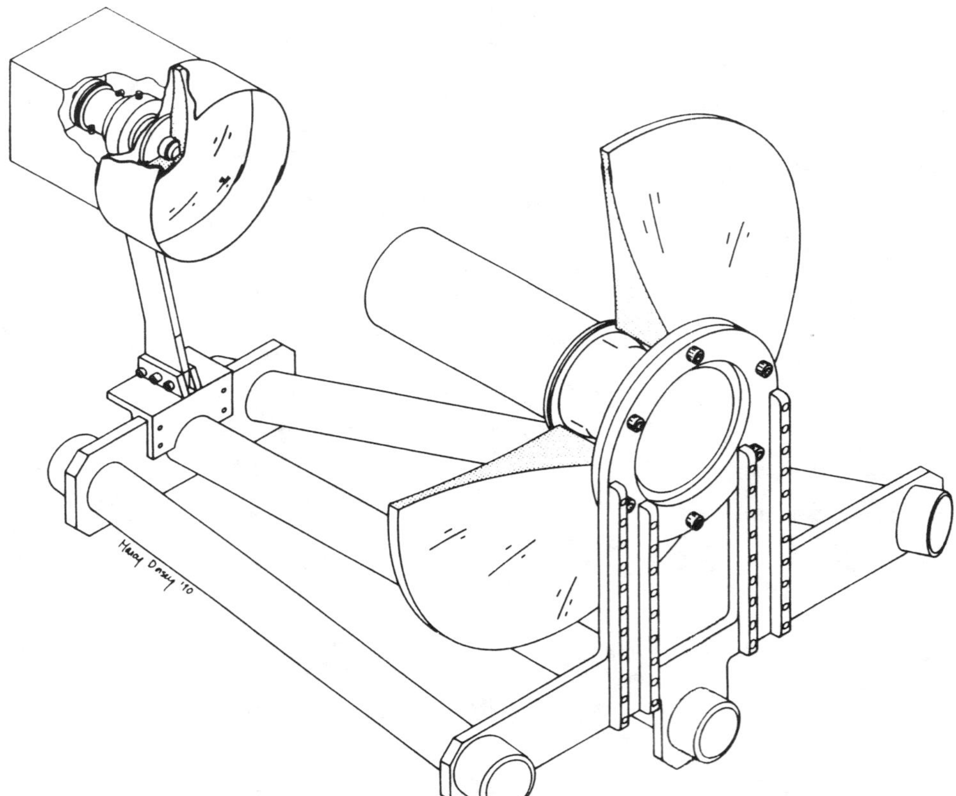

Figure 4: 400 mm Dall Type Telescope, with SXA MMC Optics and Structure

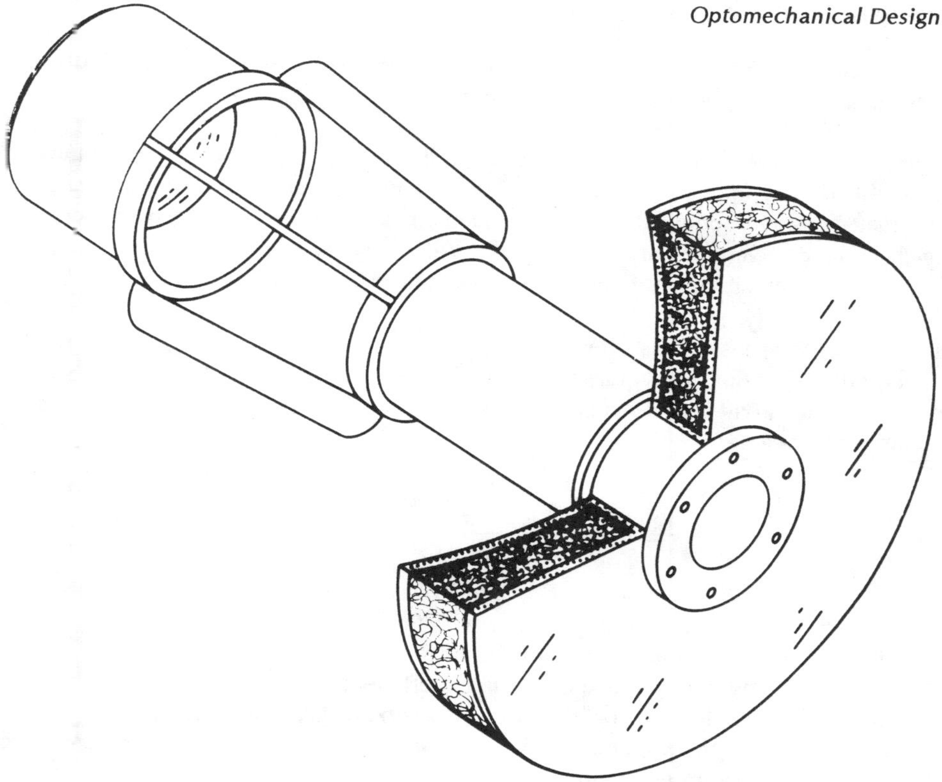

Figure 5: 200 mm Meinel Type
Telescope with SXA Foam Sandwich Mirror

In the Dall structure, three bars connect the back plate, which carries the primary mirror to the secondary mirror support strut. Both the secondary mirror and primary mirror are supported by cantilever beams, limiting overall system stiffness. Supporting the Dall structure at the center of gravity of the system causes both primary and secondary to deflect relative to the support. This self-weight deflection is exploited in the Dall structure by adjusting the proportions of the structure to produce near parallel and equal deflections of the primary and secondary. This is very similar to the principle of the Serrurier truss. A 0.3 m aperture f/5 Cassegrain using a Dall type structure had an overall weight of 4.5 kg, including optics. The measured fundamental frequency of this telescope was 34 Hz, with a Q of 25.

Efficiency of the Dall type structure falls off rapidly above system apertures of about 0.4 m. In comparison, the Meinel type structure has good structural efficiency over a range of apertures from 0.07 to 4 m. Stray light control baffles are required in Cassegrain telescopes, and are used as part of the Meinel structure. Because the baffles are compact, a relatively thick tube wall thickness is used to eliminate buckling without compromising weight. The remaining portion of the Meinel design is the vane structure used to connect the two baffles. These vanes have the same obscuring effect on the telescope aperture as a conventional spider. Remaining concerns are possible lack of torsional stiffness in the vanes, and the structural connections between vanes and baffles. For system apertures up to 0.4 m, these are not serious problems.

Many severe service environments are characterized by large changes in temperature. Lightweight, rugged optical systems must therefore be athermalized. Athermalization implies that focus and optical alignment are maintained when the system temperature is changed. Three techniques are used to athermalize lightweight systems: bi-metallic compensators, common material athermalization, and metering rods.

Bi-metallic compensation is often used when an extremely low or negative thermal coefficient of expansion is required. In particular, bi-metallic schemes are often employed in lightweight lens systems. For a simple lens, the change in lens power with temperature is given by:

$$B = K\frac{dK}{dT} = \alpha - \frac{1}{n-n_{AIR}}\left[\frac{dn}{dT} - n\frac{dn_{AIR}}{dT}\right] \qquad (17)$$

where:
- dk/dT is the change in lens power with temperature
- K is the lens power (reciprocal of focal length)
- B is the opto-thermal expansion coefficient
- α is the thermal coefficient of expansion of the lens material
- dn/dT is the therm-optic coefficient of the lens material (change in index of refraction with temperature)
- n is the refractive index of the material
- n_{AIR} is the refractive index of air
- $\frac{dn_{AIR}}{dT}$ is the change in the refractive index of air with temperature

Neglecting air temperature effects, an approximate version of Eq. 17 is:

$$B \simeq \alpha - \frac{1}{n-1}\frac{dn}{dT} \qquad (18)$$

For a system of thin lenses, Eq. 17 is used to find the change in system power with temperature:

$$\frac{dK_S}{dT} \simeq K_S(-B_S) = -K_S\sum_{i=0}^{n}\frac{K_i}{K_S}B_i \qquad (19)$$

where:
- $\frac{dK_S}{dT}$ is the change in system power with temperature
- B_S is the opto-thermal coefficient of expansion of the system
- K_S is the system power
- K_i is the individual element power
- B_i is the opto-thermal coefficient of expansion of the individual element
- n is the number of elements

Passive athermalization requires that the system opto-thermal coefficient of expansion must equal the thermal coefficient of expansion of the structural material used in the system. It is difficult to satisfy this requirement with a single type of material. By combining high and low thermal coefficients of expansion materials in series, the effective CTE of the structure is adjusted to match the opto-thermal coefficient of expansion of the system. The following equations give the material lengths necessary for athermalization in a bi-metallic compensator:

$$B_S f = \alpha_1 L_1 + \alpha_2 L_2 \tag{20}$$

$$L_2 = \frac{f\left(\frac{B_S}{\alpha_1} - 1\right)}{\left(\frac{\alpha_2}{\alpha_1} - 1\right)} \tag{21}$$

$$L_1 = f - L_2 \tag{22}$$

where:
- f is the system focal length
- B_S is the opto-thermal coefficient of expansion of the system
- L_2 is the length of the high thermal coefficient of expansion material
- α_2 is the thermal coefficient of expansion of the high thermal coefficient of expansion material
- L_1 is the length of the low thermal coefficient of expansion material
- α_1 is the thermal coefficient of expansion of the low thermal coefficient of expansion material

The bi-metallic compensator is a very old idea, first used to athermalize the length of the pendulum in Harrison's marine chronometer two centuries ago. The graphite epoxy composite materials have given this old idea a new range of applications for lightweight, rugged systems. Graphite epoxy composite tubes are used in some athermalization schemes to act as the low thermal coefficient of expansion material. Such a technique is employed in the Hubble Space Telescope truss[36].

Same material athermalization is used with reflective optical systems. When both structure and optics are made of the same material, an athermal structure is produced. A uniform change in temperature of a reflective optic changes both the focal length and the physical size of the mirror. If the mirror is mounted using a material that is the same as is used in the mirror, the same change in temperature produces a corresponding change in the physical size of the mount. This corresponding size change in the mount eliminates mounting stress due to the temperature change.

Change in focus with temperature is proportional to the thermal coefficient of expansion of the mirror. If the focus is located using a structure made of the same type of material as the mirror, the length of the structure changes with temperature an amount corresponding to the change in focus. Focus is maintained if both mirror and structure are made of the same type of material.

Reflective systems are constructed using the same metal for both mirror and structure to obtain this type of same material athermalization. Common materials used in same material athermalized systems include aluminum, beryllium, and the metal matrix composites. The all beryllium cryogenic IRAS (Infrared Astronomical Satellite) system is a successful example of the use of the principle of same material athermalization[37]. Although this technique has been employed in all-glass systems, such systems are expensive and fragile[38].

A third approach to producing a lightweight, rugged athermal structure is the use of metering rods. In a metering rod structure, the optics are supported using a lightweight structure made of efficient structural materials. The structural materials are selected on the basis of structural efficiency and do not have to have a thermal coefficient of expansion which matches that of the optics. The optical mounts are attached to the structure by a compliant connection. Motion of each mount along the optical axis is possible. Motion of the mounts along the optical axis is controlled by attaching the mounts to metering rods. The metering rod material is selected to have the thermal coefficient of expansion necessary to match the system change in focus with temperature. Metering rods are typically made of Invar, or Super Invar, although low thermal coefficient of expansion glasses or glass ceramics such as Schott Zerodur have been used.

Ideally, the metering rods are equally spaced around the circumference of the optics. If the metering rods are symmetric about the opticsal axis, and have a low thermal coefficient of expansion, the system is protected against changes in optical alignment or focus due to transverse temperature gradients.

In a dynamic environment, the stability of the optics along the optical axis may be an issue. The axial stiffness of a long thin rod is low, leading to relatively low axial stiffness for the optics. Transverse bending of the metering rods is usually not a problem. A very large transverse deformation of the metering rod is required to produce a small change in axial location. Transverse deformation effects are greatly reduced by mounting the metering rods at their Airy points. Structural joints may be necessary when metering rods are used in structures that employ the principle of equal and parallel deflection, such as a Serrurier truss.

Connection of the metering rods to the optics is important; and the thermal coefficient of expansion of the mount material between metering rods and optic must be considered. Ideally, the metering rods should be in physical contact with the optics. This may require that the optic be oversize to provide a mounting surface. Figure 6 shows a lightweight telescope built by RMR Design Group Inc. for the University of Utah that employs fused silica metering rods. Glass metering rods are not unusual, and have been employed on systems exposed to very severe dynamic loading. For example, fused silica metering rods were used in the OAO-2 telescope[39].

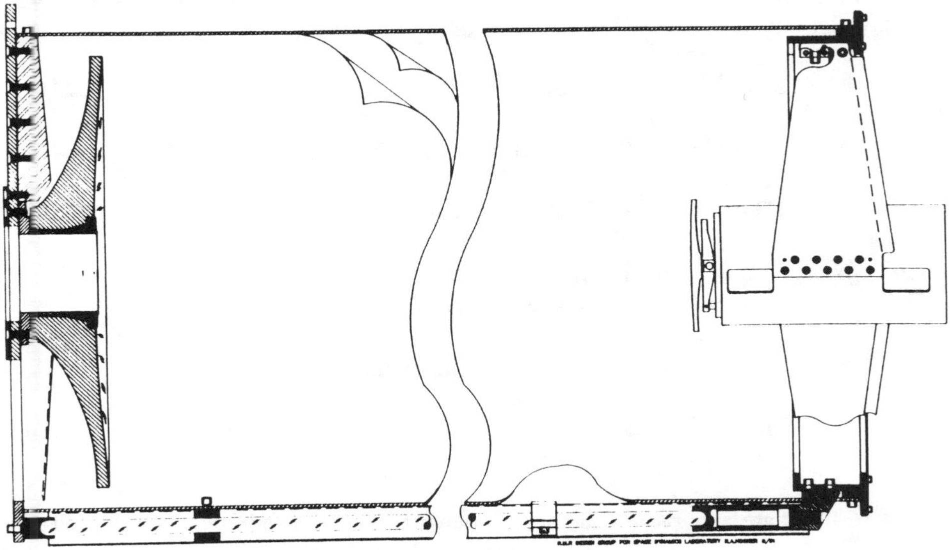

Figure 6: Cassegrain Telescope with Fused Silica Metering Rods for Athermalization
(Courtesy RMR Inc. and University of Utah)

LIGHTWEIGHT OPTICS

In most optical systems, the weight of the optical elements is a small fraction of the overall system weight. This would suggest that the weight of the optics is not of great importance in designing a lightweight, rugged optical system. To the contrary, optical weight is very important in determining the overall weight of a system. The larger the optics weight, the stiffer that the structure must be to hold the deflection (or fundamental frequency) to an acceptable level. Structural weight is proportional to stiffness; the stiffer the structure, the greater the weight. If the system must be pointed using a gimbal system, increased structural weight will cause an increase in gimbal weight. Reducing optical weight is an important means of reducing overall system weight.

Since the radii of curvature, thickness, and spacing of lens elements are set by the optical performance of the system, there is not much that can be done to reduce the weight of transmissive elements. "Lightweight" glasses are employed during optical design to reduce weight. In ordinary optical glass, high index and high dispersion usually imply high density. Lightweight glasses are glasses with high index and dispersion and low density. A variety of such glasses are now available. Poor weathering characteristics, and low strength are possible drawbacks in the use of some of the lightweight glasses. Optical plastics offer significant weight reduction in lens systems. Current optical plastics have a limited range of indices of refraction and dispersion, which limits the design possibilities of these materials. Optical plastics are subject to the problems of instability with temperature and time discussed earlier.

Considerable scope exists to reduce mirror weight in reflective systems, with potential large system weight reduction. Based on empirical study of existing lightweight mirrors,

Valente suggests that the following scaling laws can be applied to determine the weight of lightweight mirrors[40]:

$$W = 246D^{2.92} \quad \text{Solid} \quad \textbf{(23)}$$

$$W = 120D^{2.82} \quad \text{Conventional lightweight} \quad \textbf{(24)}$$

$$W = 53D^{2.67} \quad \text{Ultra-lightweight} \quad \textbf{(25)}$$

where: W is the mirror weight in kg
D is the mirror diameter in m

Five technologies are used to produce lightweight mirrors:

1. Contoured back mirrors where the back of the mirror is contoured to improve stiffness to weight.

2. Plate and radial rib mirrors consist of a thin face plate reinforced by a series of radial ribs on the mirror back.

3. Open back mirrors consist of a thin face plate reinforced by a regular array of intersecting ribs forming a repetitive pattern of open pockets or cells in the back of the mirror.

4. Sandwich mirrors consist of thin face and back plates connected by a regular array of intersecting ribs forming a repetitive pattern of pockets or cell inside the mirror.

5. Sandwich mirrors where the thin face and back plates are connected by a structural foam.

For optimum weight reduction, a lightweight mirror should have the best possible stiffness to weight. The structural efficiencies of different types of lightweight mirrors were evaluated by Valente and Vukobratovich[41]. The self-weight deflection of the open back, symmetric sandwich, and contoured back mirror shown in Fig. 7 were studied, and compared with the mirror weights. Figure 8 shows the resulting self-weight deflection versus mirror weight for single arch, open-back, solid, double arch, and sandwich mirrors. In all cases, the best mirror in terms of deflection for a given weight, was considered. According to Fig. 8, the best stiffness to weight is obtained in sandwich mirrors, followed by the double arch type of contoured mirror. Open back mirrors show virtually the same deflection as a solid mirror of the same weight.

Other considerations apart from self-weight deflection and mirror weight are often important in selecting a mirror technology. Cost of fabrication, complexity of the mirror mount, and the thermal response of the mirror can influence selection of lightweight mirror technology. Technological risk must be considered, particularly when exotic lightweight materials are used.

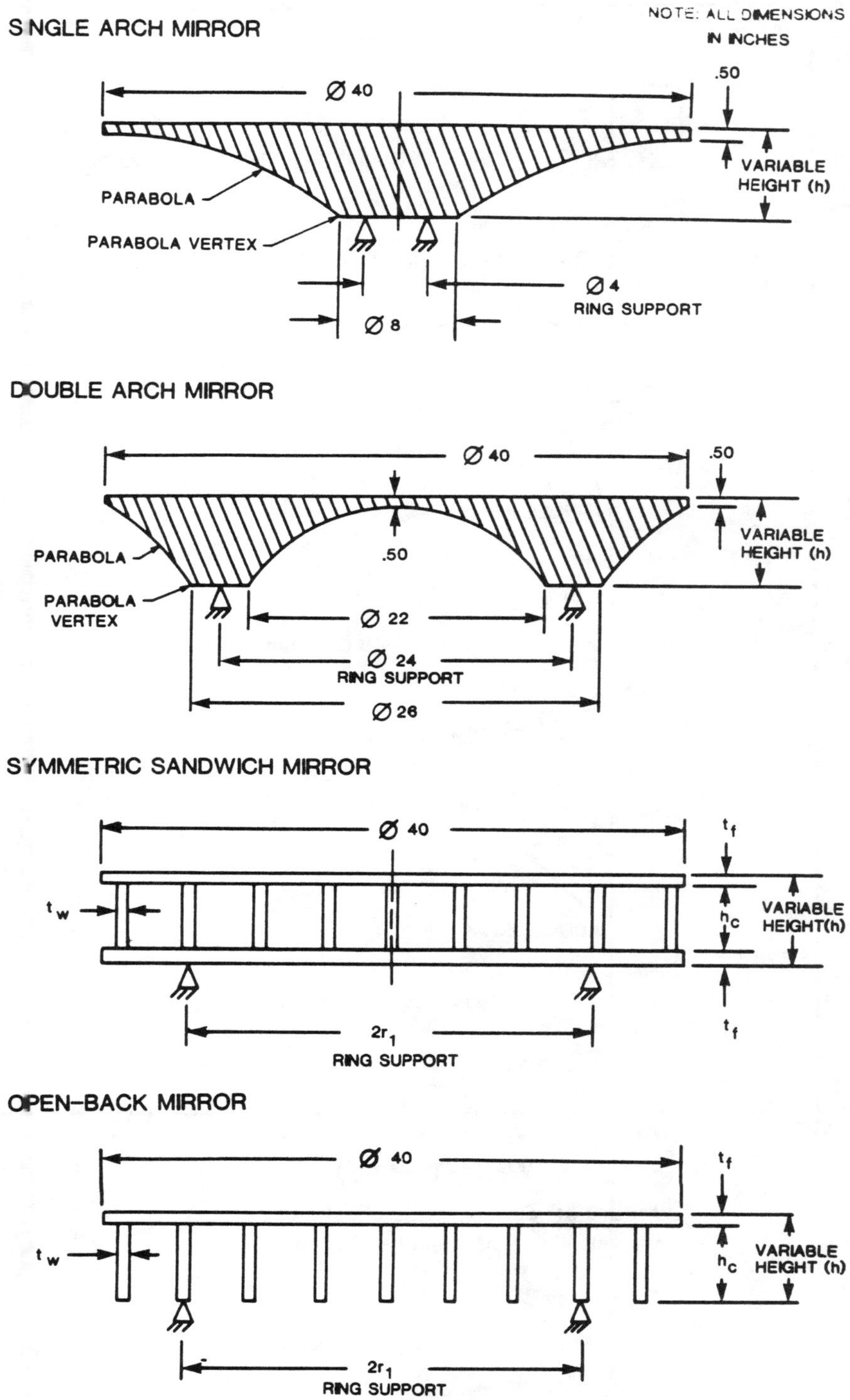

Figure 7: Lightweight Mirror Configurations

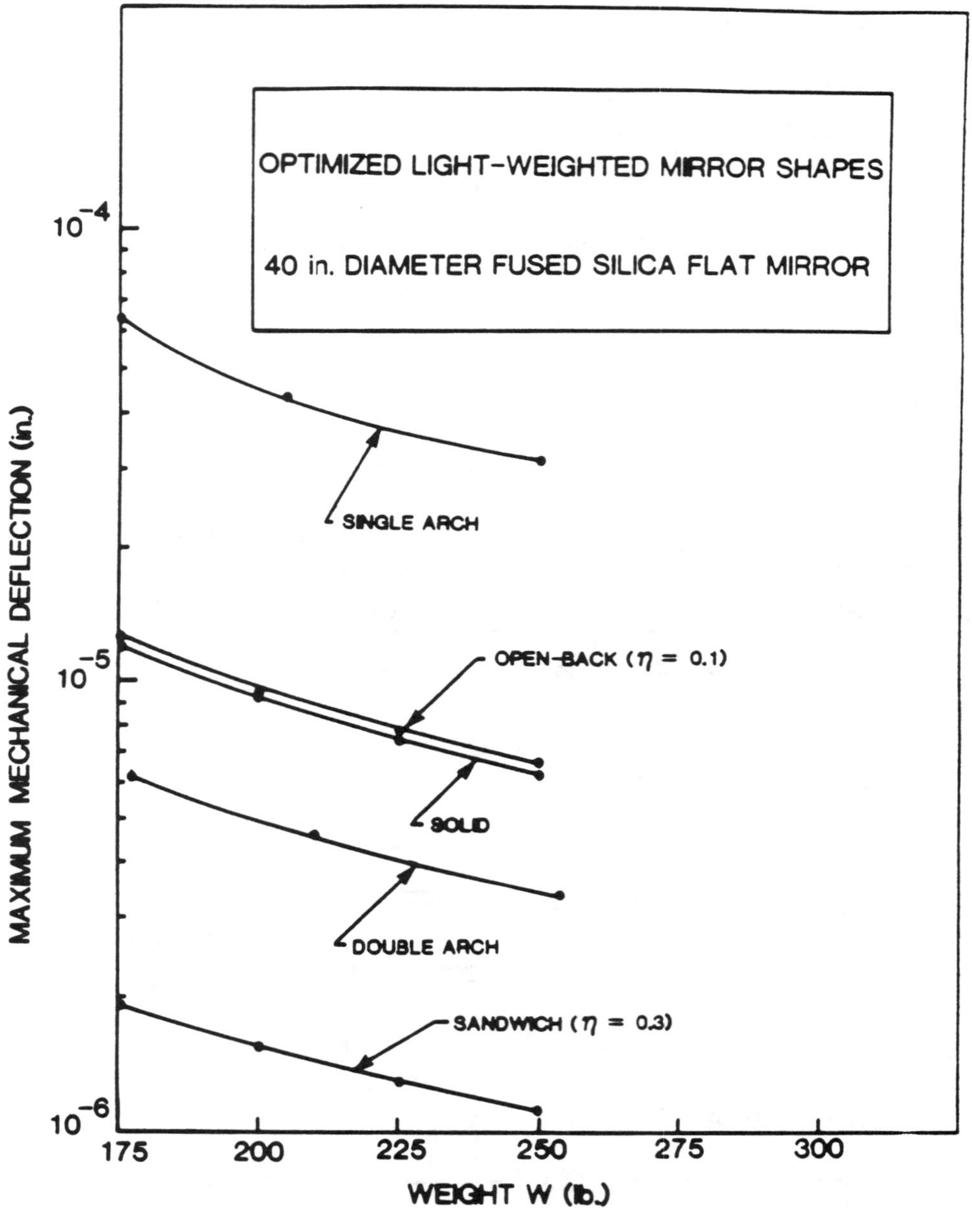

Figure 8: Self-Weight Deflection of Various Types of Lightweight Mirrors

Contoured back mirrors are a relatively low cost solution to the lightweight mirror problem. Figure 9 shows the three most important types of contoured mirrors: the single arch, double arch, and double concave shapes. Optimum shapes for minimum self-weight deflection have been developed by Cho[42]. A contoured back mirror of optimum design can have a deflection ratio of 0.5 or less when compared to a solid mirror of the same weight. Such a mirror easily satisfies Schwesinger's lightweight criteria mentioned earlier.

The simplest type of contoured back mirror is the single arch mirror, which has a parabolic taper from a thick center to a thin edge. Deflection efficiency of the single arch mirror is low, with a deflection ratio of 2 when compared with a solid mirror of the same weight. Simplicity of the mirror and mount is the main virtue of this mirror configuration. A single arch mirror is normally mounted using a central hub mount. Since the weight of the mirror mount is very low, good efficiency is possible when the weight of the single arch and its mount are compared to the weight of a conventional mirror and mount. A 0.35 m fused silica single arch mirror is used as the primary mirror of the Mars Observer Camera[43].

A double parabolic taper, from a thick intermediate ring towards both center and edge, is used in the double arch mirror. Self-weight deflection of a double arch mirror is about 0.25 of a solid mirror of the same diameter to thickness aspect ratio, and about 0.16 that of a solid mirror of the same weight. Double arch mirrors are normally mounted using three or more support sockets in the back of the mirror.

If the mirror is intended for use in the axis horizontal position (gravity acting perpendicular to the optical axis), a symmetric shape has the advantage of minimum self-weight deflection. Symmetric double concave or double convex mirrors have very small radial deflections when in this position. When used in the optical axis vertical position, the symmetric shape produces excessive self-weight deflection. Symmetric mirrors are normally mounted using discrete point supports around the circumference of the mirror. The 0.5 m fused silica TEAL RUBY primary mirror is a double concave, edge mounted mirror[44].

Sandwich mirrors have the best stiffness to weight of current mirror technologies. Mehta has shown that there is an optimum geometry for a sandwich mirror to produce the best possible stiffness to weight[45]. For a given overall height or weight, the optimum face plate thickness can be found which produces a mirror with the greatest possible flexural rigidity. The optimum face plate thickness for a given mirror weight is found from:

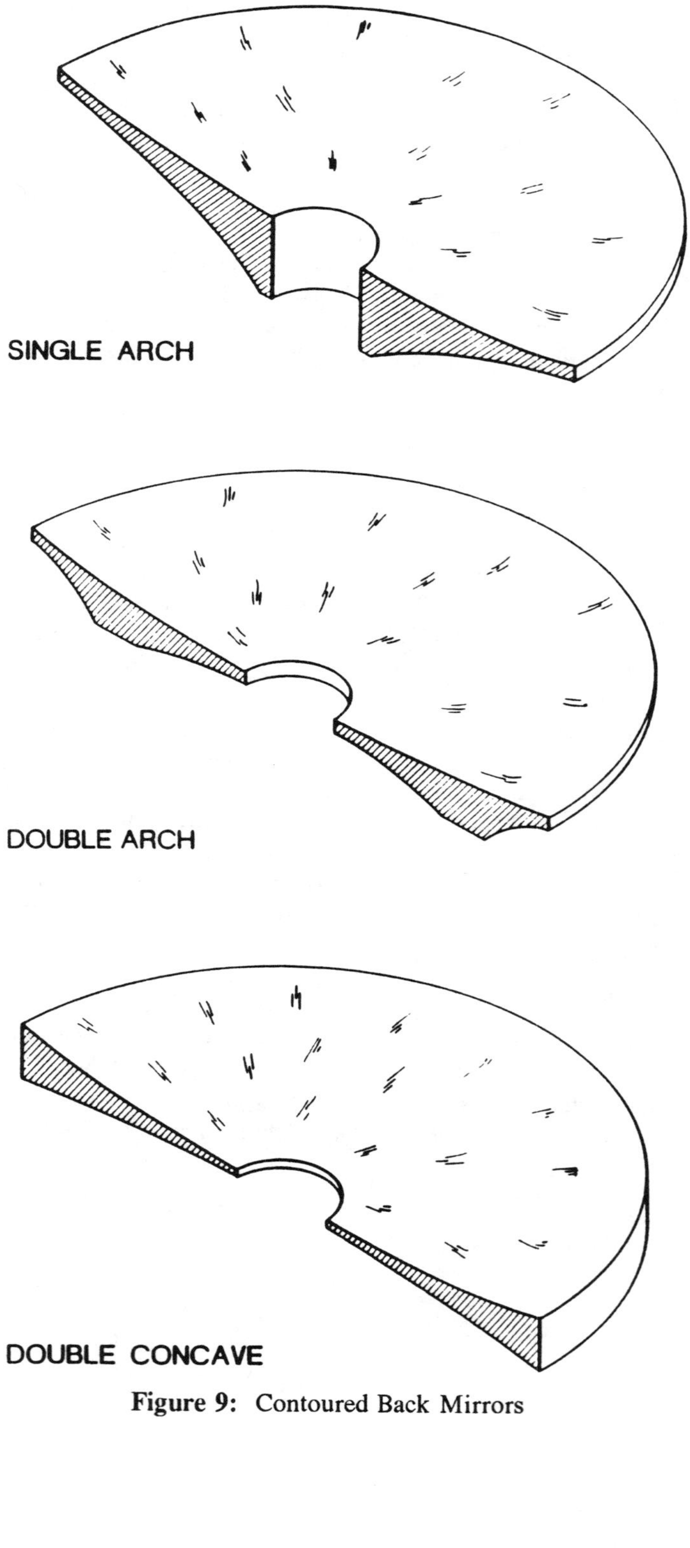

Figure 9: Contoured Back Mirrors

$$h_f = \frac{W}{\rho A} \left\{ \frac{\left[\left(1 - \frac{\eta}{2}\right)^{1/2} - (1 - \eta)^{1/2}\right]}{2\left[\left(1 - \frac{\eta}{2}\right)^{1/2} - (1 - \eta)^{3/2}\right]} \right\} \quad (26)$$

$$\eta = \frac{(2B + t_r)t_r}{(B + t_r)^2} \quad (27)$$

where: h_f is the optimum sandwich mirror face plate thickness
- W is the mirror weight
- ρ is the mirror material density
- A is the mirror projected area
- η is the rib solidity ratio
- B is the diameter of a circle inscribed in the mirror rib cells
- t_R is the rib thickness

After the optimum mirror parameters are found, the flexural rigidity and mirror weight can be computed from:

$$D = \frac{Eh_B^3}{12(1 - \nu^2)} \quad (28)$$

$$W = \rho A(2h_f + \eta h_c) \quad (29)$$

$$h_B^3 = (2h_f + h_c)^3 - \left(1 - \frac{\eta}{2}\right) h_c^3 \quad (30)$$

where: D is the flexural rigidity
- E is the mirror material elastic modulus
- hB is the mirror equivalent bending thickness
- ν is the Poisson's ratio of the mirror material
- ρ is the mirror material density
- A is the mirror projected area
- h_c is the mirror rib height
- η is the rib solidity ratio
- h_f is the face plate thickness

Figure 10 shows the relationship between flexural rigidity and face plate thickness, while Fig. 11 shows cell wall thickness as a function of rib solidity ratio.

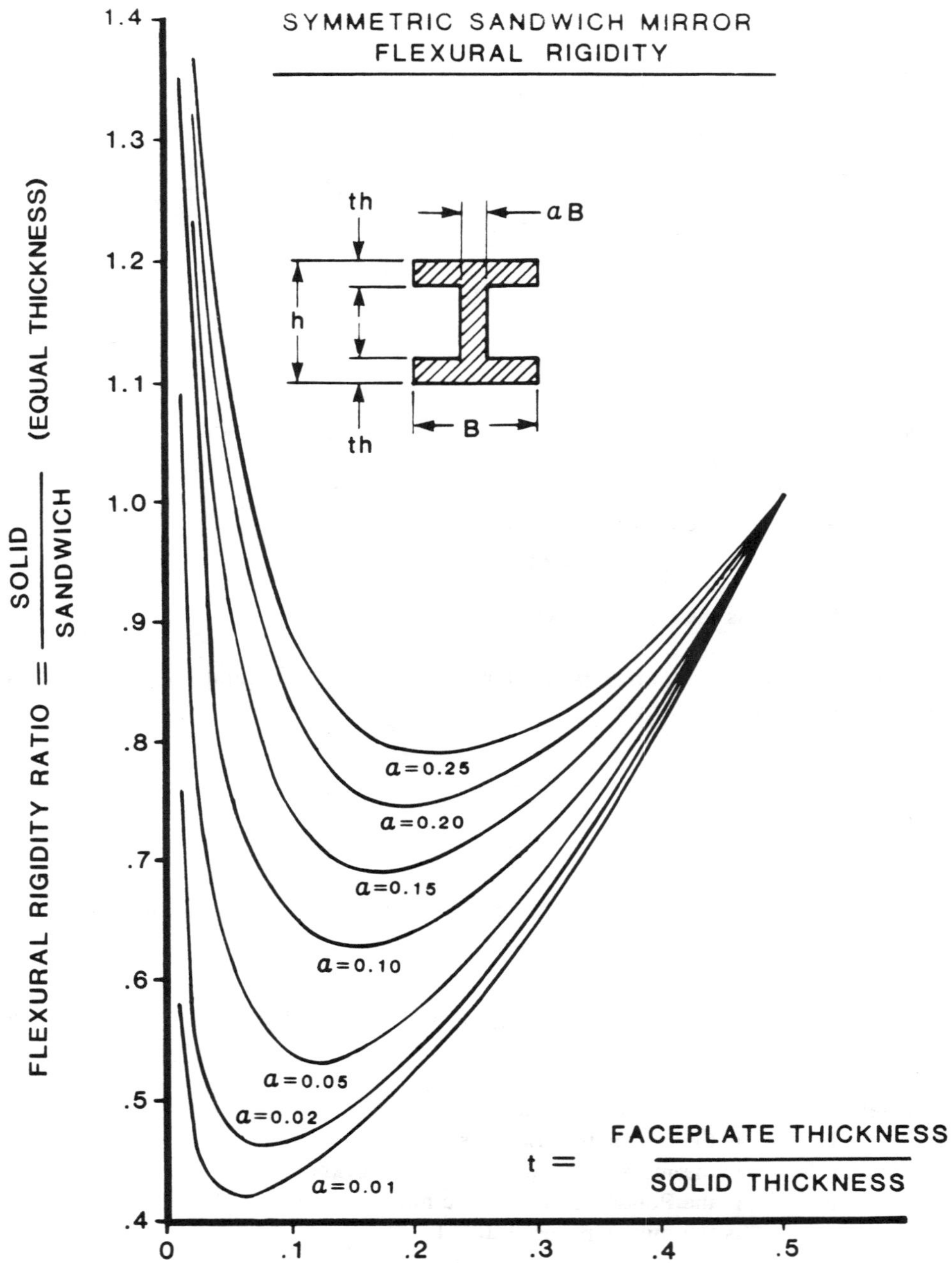

Figure 10: Flexural Rigidity of Sandwich Mirrors

During polishing, the face plate of the mirror deflects under polishing pressure. This deflection causes a permanent periodic deformation of the mirror surface, corresponding

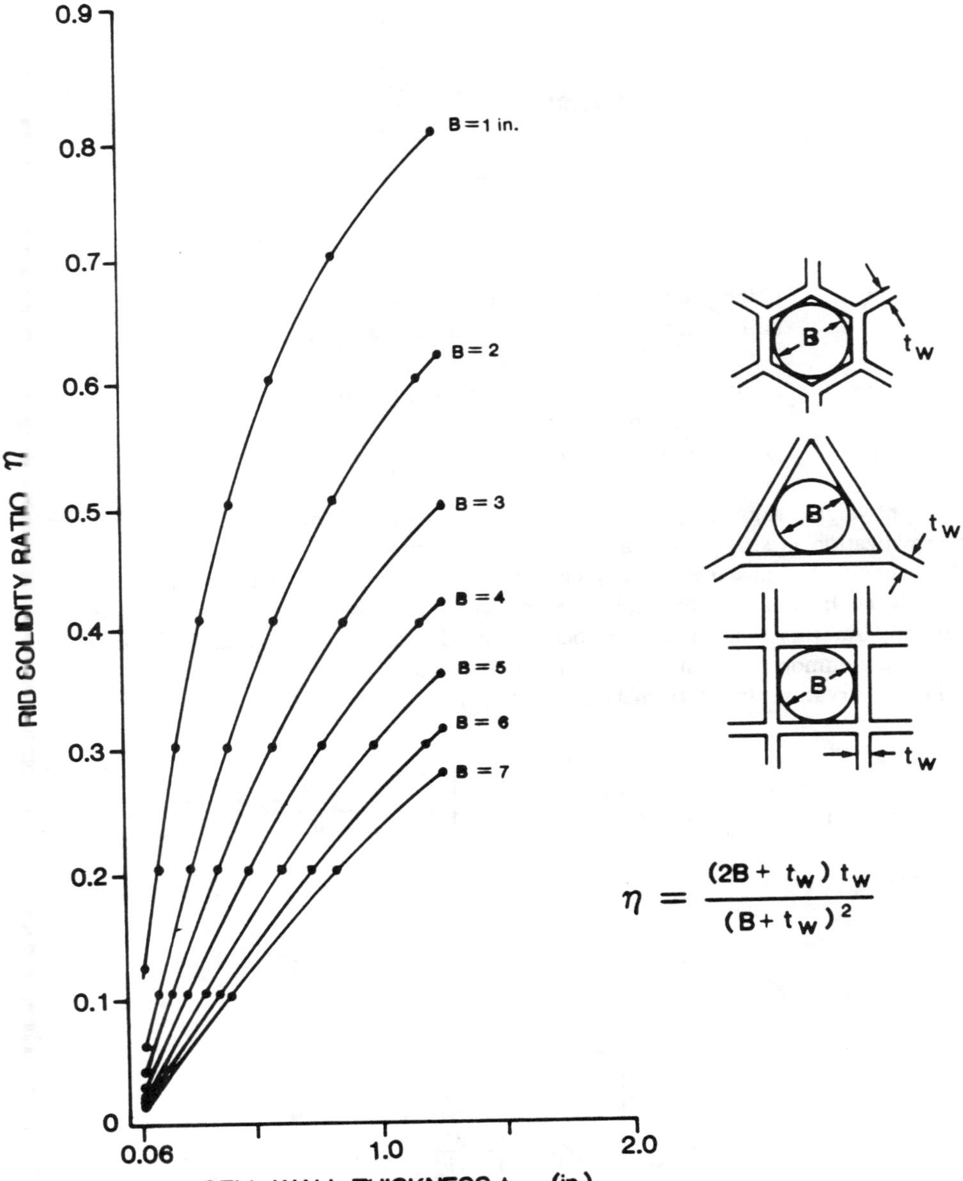

Figure 11: Rib Solidity Ratio for Lightweight Sandwich Mirrors

with the cell location. This permanent deflection, called quilting, causes a reduction in energy in the central maximum of the diffraction disk, and should be reduced to as low a level as is possible. Mirror quilting amplitude is estimated using an expression developed by Barnes[46]:

$$\delta_c = \psi \left[\frac{E h_f^3}{12(1 - \nu^2)} \right]^{-1} P B^4$$
$$\psi_{SQUARE} = 0.00126$$
$$\psi_{TRIANGLE} = 0.00151$$
$$\psi_{HEXAGONAL} = 0.00111 \qquad (31)$$

where: δ_c is the peak to peak quilting amplitude
E is the mirror material elastic modulus
h_f is the face plate thickness
ν is Poisson's ratio for the mirror material
P is the polishing pressure
B is the size of the circle inscribed in the mirror rib cells

Lightweight, rugged optical systems often employ metal mirrors. Same material athermalization is simple in systems with metal mirrors. Mounting metal mirrors is relatively simple, since metal mirrors can be drilled and tapped. Thermal gradient effects are reduced by the high thermal conductivity of the metals typically used in mirrors. Temporal stability, and bi-metallic bending of electroless nickel plated mirrors are the two most common problems encountered in using metal mirrors. The change in mirror radius of curvature due to bi-metallic bending is given by[47]:

$$\frac{1}{R_0} - \frac{1}{R} \simeq \frac{6}{h} \frac{\psi_p}{\psi_m} (\alpha_p - \alpha_m) \Delta T \left(\frac{h_{p1} - h_{p2}}{h} \right) \left[1 - \left(\frac{h_{p1} - h_{p2}}{h} \right) \left(4 \frac{\psi_p}{\psi_m} - 1 \right) \right] \qquad (32)$$

$$\psi_m = \frac{E_m}{1 - \nu_m} \qquad (33)$$

$$\psi_p = \frac{E_p}{1 - \nu_p} \qquad (34)$$

where: R_o is the original mirror radius of curvature
R is the mirror radius of curvature after the temperature change
h is the mirror axial thickness
h_{p1}, h_{p2} are the plating thicknesses on the front and back of the mirror
α_p is the plating material thermal coefficient of expansion
α_m is the mirror material thermal coefficient of expansion

E_p	is the plating material elastic modulus
E_m	is the mirror material elastic modulus
ν_m	is Poisson's ratio for the mirror material
ν_p	is Poisson's ratio for the plating material
ΔT	is the temperature change
ψ_p	is the plating material parameter
ψ_m	is the mirror material parameter

Equation 32 is used to determine the mirror thickness necessary to limit the change in radius for a given change in temperature and plating thickness variation. It is very difficult to predict final plating thickness variation, since plating is removed in an unpredictable way from the mirror surface during fabrication of the mirror optical surface.

OPTICAL COMPONENT MOUNTING FOR SEVERE ENVIRONMENTS

Optical components for use in severe environments must be mounted in a low stress condition. Any stress induced during assembly is likely to be increased in the service environment. In transmissive components such as lenses, induced stress will cause an optical path difference due to stress birefringence of the glass. This optical path difference is given by:

$$OPD = Kh\sigma \tag{35}$$

where:

OPD	is the optical path difference (or stress birefringence)
h	is the lens thickness
σ	is the applied stress
K	is the stress-optical coefficient of the glass

Table 6 gives representative values of the stress optical coefficient for various glasses.

Mechanical stress induced in a metal mirror can cause a permanent change in optical figure. For certain metals in the range of strain from 10^{-7} to 10^{-4} the following equation is used to predict microyield behavior[48]:

$$\sigma = K'\epsilon^{n'} \tag{36}$$

where:

σ	is the applied stress
K'	is the microstrain hardening coefficient
n'	is the microstrain hardening exponent
ϵ	is the permanent strain

Table 4. Weibull Parameters of Glasses.

	Material Type	m	σ_o (MPa)
1.	Schott Zerodur	14	11.3
2.	ULE Corning 7971	6.8	11.6
3.	Silicon Single Crystal	4.54	346.5
4.	Al_2O_3	5-15	30-50
5.	CaF_2	3	5
6.	SrF_2	3	4
7.	ZnSe	6-9	23-27
8.	ZnS	4-9	15-29
9.	Soda Lime Kimble R-6	6.18	128.8
10.	Pyrex Corning 7740	4.99	257.2

Table 5 gives typical microstrain parameters for representative materials.

Table 5. Microstrain Parameters

Material	n′	K′ (MPa)	MYS (MPa)
2024-Tr AL	0.093	915.0	253.0
6061-T4 AL	0.097	668.8	175.0
5456-H34 AL	0.029	20.52	13.8
TZM Mo	0.109	1662.0	369.0
I-4000 Be	0.338	6895.0	64.6
AZ-31 Mg	0.177	382.0	33.1
Type 321 Stainless Steel	0.198	948.1	61.5

Under high acceleration, severe mechanical vibration, or rapid changes in temperature, stresses in glass optical materials may approach or exceed the failure stress of the material. Unlike metals, glasses do not have a well defined yield stress. Glass failure is due to surface fracture, and is predicted by statistical means. Failure probability of a glass material is given by the Weibull distribution, where:

$$P_f = 1 - \exp\left[- \left(\frac{\sigma_a}{\sigma_o} \right)^M \right] \quad (37)$$

where: P_f is the probability of failure
σ_a is the applied stress
M is the Weibull modulus
σ_o is the scale factor

Weibull failure data is usually taken on relatively small samples, and must be scaled to the actual surface area and flaw size using the following equation:

$$\frac{\sigma_s}{\sigma_c} = \left(\frac{S_c}{S_s} \right)^{1/2} \left(\frac{A_c}{A_s} \right)^{\frac{1}{m}} \quad (38)$$

where: σ_s is the failure stress on the test sample at a given probability of failure
σ_c is the scaled failure stress on the component at the same probability of failure
A_s is the sample surface area
A_c is the component surface area
m is the Weibull modulus
S_s is the maximum sample surface flaw depth
S_c is the maximum component surface flaw depth

Table 4 gives typical Weibull data for some representative optical materials.

Conventional mechanical seat and retainer lens cell designs will induce excessive stress in the lens during large temperature changes or under large accelerations. Stress in the lens is reduced through the use of adhesive bonding in the mount. A semi-flexible adhesive is placed around the circumference of the lens to isolate the lens from expansion or contraction of the cell with temperature, and from mechanical shock. An equation developed by Bayar is used to determine the optimum radial adhesive thickness necessary to isolate the lens from temperature effects[49]:

$$h_r = \frac{d_g (\alpha_m - \alpha_g)}{2 (\alpha_r - \alpha_m)} \quad (39)$$

where: h_r is the required adhesive thickness
d_g is the lens outer diameter
α_g is the lens thermal coefficient of expansion
α_m is the lens cell thermal coefficient of expansion

α_r is the adhesive thermal coefficient of expansion (for the silastic adhesives often used in this application, this about 200 x 10^{-6} m/m-K)

Table 6. Stress-Optical Coefficients.

	Material	Wavelength (nM)	K 10^{-12} Pa^{-1}
1.	F2	589.3	2.81
2.	BK7	589.3	2.77
3.	SF5	589.3	2.28
4.	SF4	589.3	1.36
5.	SF6	589.3	0.65
6.	SF57	589.3	0.02
7.	SF59	589.3	-1.36
8.	ZnSe	1150	-1.98
9.	CaF_2	1150	2.16
10.	BaF_2	1150	4.79

The adhesive thickness predicted by Eq. 37 is relatively large in comparison with normal adhesive bonding practice. Under the weight of the lens the very thick layer of adhesive may deflect, causing the lens to move with respect to the cell. The radial deflection of the center of a lens mounted in a semi-flexible adhesive mount is given by[50]:

$$\delta_L = \frac{W_L}{\frac{\pi}{2} d_g \frac{L_r}{h_r} \left(\frac{E_r}{1-\nu_r} + G_r \right)} \qquad (40)$$

where:
δ_L is the motion of the center of the lens with respect to the mount
W_L is the lens weight
d_g is the lens diameter
L_r is the length of adhesive in contact with the side of the lens
h_r is the adhesive radial thickness
E_r is the elastic modulus of the adhesive
ν_r is the Poisson's ratio of the adhesive
G_r is the shear modulus of the adhesive

An exceptionally rugged, lightweight lens system is developed by using the above adhesive bonding technique to locate individual lens elements in sub-cells. These subcells are then pressed into the main barrel. This assembly technique effectively isolates the optical components from severe changes in temperature and mechanical shock, while at the same time enabling centering tolerances as small as 2.5 microns to be attained. Figure 12 is a cross section of a 2 m focal length, f/10 astrometric astrograph lens built by the Optical Sciences Center, University of Arizona, for the United States Naval Observatory. Overall weight of this very large lens assembly is 50 kg; which is split equally between optics and lens barrel. The lens barrel is made from 6Al-4V titanium. This material was selected to reduce weight, and to obtain a low thermal coefficient of

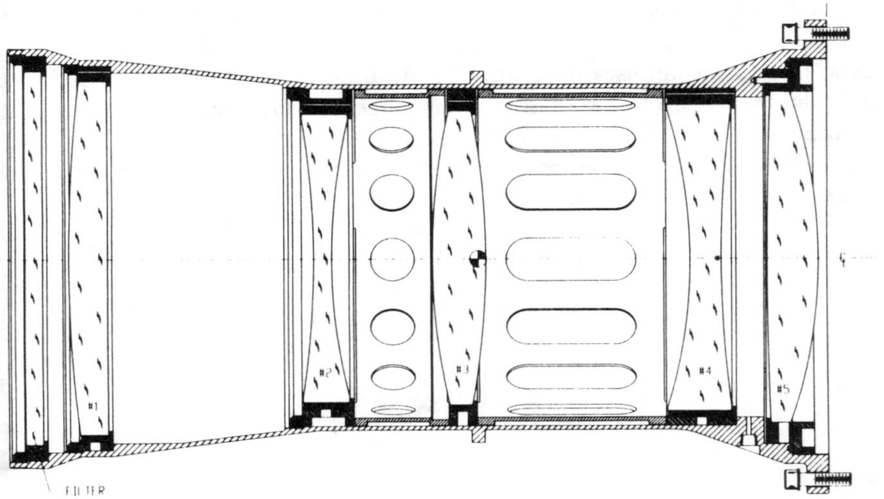

Figure 12: Astrometric Astrograph Lens with Elastomeric Subcell Mounts for the Individual Lens Elements

expansion. Each of the individual optical elements is adhesive bonded into a titanium sub-cell. The sub-cells are then pressed into the main cell. A Tiodize surface finish is used to prevent galling of the metal parts during assembly.

Mirrors are more sensitive to surface deformation than lenses, and require special care in mounting. Irregularity in the mirror mount can induce unacceptable optical surface deformation. The mirror mounting problem may be broken into two parts: the actual mirror to mount connection and the isolation of the mirror from deformation of the mount.

Adhesive bonding of the mirror into a mount requires special precautions to insure that unacceptable stress levels are not induced in the mirror. Bonded contacts induce stress in the mirror from shrinkage of the adhesive during curing, and from the thermal coefficient of expansion mis-match between mirror and adhesive. Both effects are reduced by using small areas of contact between the mirror and adhesive. If high bond strength is required, it is better to use many small areas of adhesive to obtain strength rather than a single large adhesive area. Isolation of the mirror surface figure is possible through modification of the mirror configuration. One such configuration uses a cylindrical "stalk" on the back of the mirror. This stalk is usually about one third the mirror diameter, and has a length equivalent to the mirror thickness. Adhesive is placed around the circumference of the stalk. This type of mirror mount is called a mushroom mount, from the resemblance of the mirror to a mushroom.

Large changes in temperature, such as are encountered in cryogenic optical systems, require more sophisticated mounting techniques than simple adhesive bonds. One successful approach to the cryogenic mirror mount problem is the use of flexural mounts. Flexural mounts are similar in principle to semi-kinematic mounts[51]. Kinematic principles are used to determine the stiffness of each flexure used to mount the mirror. The classic flexural mount uses three identical flexures, attached either to the edge or back of the mirror. Each flexure is stiff in the tangential direction and flexible in the radial direction. This combination of stiffness and compliance accurately locates the mirror, yet allows the mount to expand or contract radially with respect to the mirror. The stiffness of a flexure mounted mirror is given by:

$$K_R = \frac{3}{2} (K_{FR} + K_{FT}) \tag{41}$$

$$K_A = 3\, K_{FA} \tag{42}$$

where:
- K_R is the radial stiffness of the mirror mount
- K_{FR} is the radial stiffness of one mirror mount flexure
- K_{FT} is the tangential stiffness of one mirror mount flexure
- K_A is the axial stiffness of the mirror mount
- K_{FA} is the axial stiffness of one mirror mount flexure

Error in locating the flexures with respect to the mirror can induce bending forces into the mirror. This lack of assembly accuracy requires modification of the design of the mounting flexures. Additional directions of compliance and stiffness must be designed into each mirror mount flexure.

An example of the use of flexure mounting for isolating a mirror from the effect of temperature changes is a 0.5 m fused silica test mirror made as part of the Space Infrared Telescope Facility program[52]. The test mirror was a concave double arch design with three althermalized support sockets in the back of the mirror. Titanium parallel spring guide flexures were used to connect the support sockets to the aluminum mounting plate. Cryogenic distortion of the surface, or the change in optical surface figure from 300 K to 7 K, was 0.12 waves RMS (1 wave = 633 nm)[53].

REFERENCES

1. B. H. Rule, "Large telescope mounts," in *ESO/CERN Conference on Large Telescope Design*, edited by R. M. West, Geneva, March 1-5, 1971.

2. G. Schwesinger, "Elastostatischer Vergleich von Telesckopiegeln in Leichbauweise," *Messtechnik*, Vol.9, p. 229 (1969).

3. R. D. Blevins, *Formulas for Natural Frequency and Mode Shape*, Van Nostrand Reinhold, New York, NY, 1979.

4. W. Davison, Design strategies for very large telescopes," *Proc. SPIE 1236*, p. 878 (1990).

5. M. W. Schneermann, "Structural design concepts for the 8 meter unit telescopes of the ESO-VLT," *Proc. SPIE 628*, p. 412 (1986).

6. F. R. Shanley, *Weight-Strength Analysis of Aircraft Structures*, Dover, New York, NY, 1952.

7. F. C. McQuiston and J. D. Parker, *Heating, Ventilating and Air Conditioning*, 2nd. Ed., John Wiley & Sons, New York, NY, 1982.

8. E. E. Ungar, D. H. Sturz, and C. H. Amick, "Vibration control design of high technology facilities," *Sound & Vibration*, Vol. 24, No. 7, pp. 20 (July, 1990).

9. *Environmental Test Methods and Engineering Guidelines*, MIL-STD-810E, 14 July 1989.

10. W. Wrigley, W. M. Hollister, and W. G. Denhard, *Gyroscopic Theory, Design, and Instrumentation*, The M.I.T. Press, Cambridge, MA, 1969.

11. M. Krim and J. Russo, *LDR Segmented Mirror Technology Assessment Study*, NASA Contractor Report 166493, NAS2-11104, Ames Research Center, Moffett Field, CA 94035, March 1983.

12. A. Blake, *Practical Stress Analysis in Engineering Design*, Marcel Dekker, New York, NY, 1982.

13. R. L. Tobler and H. I. McHenry, "Fracture mechanics," in *Materials at Low Temperatures*, R. P. Reed and A. F. Clark ed., American Society for Metals, Metals Park, OH, 1983.

14. B. J. Lazan, *Damping of Materials and Members in Structural Mechanics*, Pergamon Press, New York, NY, 1968.

15. L. M. Schetky and J. Perkins, "The 'quiet' alloys," *Machine Design*, Vol. 50, No. 8, p. 202 (April 6, 1978).

16. D. W. James, "High damping materials for engineering applications," *Mater. Sci. Eng.*, Vol. 4, p. 1 (1969).

17. R. D. Adams, "The damping characteristics of certain steels, cast irons and other metals," *Journal of Sound and Vibration*, Vol. 23, No. 2, p. 199 (1972).

18. L. T. Lee, *A Graphical Compilation of Damping Properties of Both Metallic and Non-metallic Materials*, AFML-TR-66-169, Air Force Materials Laboratory, Research and Technology Division, Air Force Systems Command, Wright-Patterson Air Force Base, OH, May, 1966.

19. R. A. Paquin, "Dimensional stability: an overview," *Proc. SPIE 1335*, p. 2 (1990).

20. C. W. Marschall and R. E. Maringer, *Dimensional Instability*, Pergamon Press, Elmsford, NY, 1977.

21. S. F. Jacobs, D. Shough, and C. Connors, "Thermal expansion uniformity of material for large telescope mirrors," *Applied Optics*, Vol. 23, No. 23, p. 4237 (Dec. 1, 1984).

22. C. J. Pellerin, F. Ayer, Y. Mehrotra, and A. K. Hopkins, "New opportunities from materials selection trade-offs for high precision space mirrors," *Proc. SPIE 542*, p.5 (1985).

23. W. R. Mohn and D. Vukobratovich, "Recent applications of metal matrix composites in precision instruments and optical systems," *Opt. Eng.*, Vol. 27, No, 2, p. 90 (Feb., 1988).

24. J. E. Stumm, G. E. Pynchon, and G. C. Krumweide, "Graphite/epoxy material characteristics and design techniques for airborne instrument application," *Proc. SPIE 309*, p. 188 (1981).

25. A. R. Telkamp and E. A. Derby, "Design considerations for composite materials used in the Mars Observer Camera," *Proc. SPIE 1303*, p. 416 (1990).

26. G. C. Krumweide, "Attacking dimensional instability problems in graphite/epoxy structures," *Proc. SPIE 1533*, p. 252 (1991).

27. D. Vukobratovich, "Binocular performance and design," *Proc. SPIE 1168*, p. 338 (1989).

28. H. K. Baumeister and R. A. Sebrosky, "Locating simple beam supports," *Machine Design*, Vol. 54, No. 4, p. 106 (Feb. 25, 1982).

29. M. Serrurier, "Structural features of the 200-inch telescope for Mt. Palomar observatory," *Civil Engineering*, Vol. 8, No. 8, p. 524 (Aug., 1938).

30. R. H. Hammerschlag, "Construction outlines of the Utrecht Open Solar Telescope," in *Solar Instrumentation: What's next?*, R. B. Dunn ed., Sacramento Peak Solar Observatory, Sunspot, NM 88349, 1981.

31. J. Lubliner, *Preliminary design of tube for TMT*, TMT Report No. 50, Jan. 5, 1981.

32. G. W. Moffitt, "Compensation of flexure in range finders and sighting instruments," *J. Opt. Soc. Am.*, Vol. 37, No. 7, p. 582 (July, 1947).

33. A. B. Meinel and M. P. Meinel, "Wind deflection compensated, zero-coma telescope truss geometries," *Proc. SPIE 628*, p. 403 (1986).

34. D. Vukobratovich, "Principles of vibration isolation," *Proc. SPIE 732*, p. 27 (1987).

35. A. B. Meinel, "A 1.8-m lightweight doubly asymmetric equatorial telescope design," *Applied Optics*, Vol. 10, No. 2, p. 249 (Feb., 1971).

36. M. H. Krim, "Design of a highly stable optical support structure," *Opt. Eng.*, Vol. 14, No. 6, p. 552 (Nov./Dec., 1975).

37. P. Young, M. Schreibman, "Alignment design for a cryogenic telescope," *Proc. SPIE 251*, p. 171 (1980).

38. C. W. F. Everitt, D. E. Davison, and R. A. Van Patten, "Cryogenic star-tracking telescope for Gravity Probe B," *Proc. SPIE 619*, p. 89 (1986).

39. P. R. Yoder Jr., *Opto-mechanical Systems Design*, Marcel Dekker, New York, NY, 1986.

40. T. M. Valente, "Scaling laws for light-weigh optics," *Proc. SPIE 1340*, p. 47 (1990).

41. T. M. Valente and D. Vukobratovich, "A comparison of the merits of open-back, symmetric sandwich, and contoured back mirrors as light-weighted optics," *Proc. SPIE 1167*, p. 20 (1989).

42. M. K. Cho, R. M. Richard, and D. Vukobratovich, "Study on mirror shape for a lightweight mirror subject to self-weight," *Proc. SPIE 1167*, p. 2 (1989).

43. R. W. Applewhite and A. R. Telkamp, "Effects of thermal gradients on the Mars Observer Camera primary mirror," *Proc. SPIE 1690*, (in press) (1992).

44. J. W. Pepi and R. J. Wollensak, "Ultra-lightweight fused silica mirrors for cryogenic space optical system," *Proc. SPIE 183*, p. 131 (1979).

45. P. K. Mehta, "Flexural rigidity characteristics of light-weighted mirrors," *Proc. SPIE 748*, p. 158 (1987).

46. W. P. Barnes Jr., "Optimal design of cored mirror structures," *Applied Optics*, Vol. 8, No. 6, p. 1191 (June, 1969).

47. W. P. Barnes Jr., "Some effects of aerospace thermal environments of high-acuity optical systems," *Applied Optics*, Vol. 5, No. 5, p. 701 (1966).

48. C. W. Marshall, R. E. Maringer, and M. E. Hoskins, *Continuation of a Study of Stability of Structural Materials for Spacecraft Applications for the Orbiting Astronomical Observatory Project*, NASA NAS5-11195 (Battelle Memorial Institute, Columbus, OH, 1969).

49. M. Bayar, "Lens barrel optomechanical design principles," *Opt. Eng.*, Vol. 20, No. 2, p. 181 (Mar./Apr., 1981).

50. T. M. Valente and R. M. Richard, "Analysis of elastomer lens mountings," *Proc. SPIE 1533*, p. 21 (1991).

51. D. Vukobratovich and R. M. Richard, "Flexure mounts for high resolution optical elements," *Proc. SPIE 959*, p. 18 (1988).

52. B. Irnaninejad, D. Vukobratovich, R. M. Richard, and R. K. Melugin, "A mirror mount for cryogenic environments," *Proc. SPIE 450*, p. 34 (1983).

53. J. H. Miller et al, "Ames Research Center cryogenic mirror testing program. A comparison of the cryogenic performance of metal and glass mirrors with different types of mounts," *Proc. SPIE 973*, p. 62 (1988).

What's Different about Ultraviolet and Infrared Optics?

R. Barry Johnson
Center for Applied Optics
University of Alabama in Huntsville
Huntsville, Alabama 35899

Abstract

The design and fabrication of components comprising ultraviolet and infrared optical systems require the same general engineering considerations to be given. In actuality, a variety of significant differences exist in the magnitude of considerations for optical systems in these two spectral regions. Among these are the optical materials available for use, surface finish, mechanical and optical fabrication techniques, housing and mounting methods, alignment and test, and cost. This paper contrasts and compares the practical issues facing the designer, fabricator, and assembler dealing with optical systems for the ultraviolet and infrared spectrums.

1 Introduction

The design and fabrication of components comprising ultraviolet and infrared optical systems require the same general engineering considerations to be given. So what's different about UV and IR optics from a practical point of view? The objective of this paper is to address this question.

In actuality, a variety of significant differences exist in the magnitude of considerations for optical systems in these two spectral regions. Among these are the optical materials available for use, surface finish, mechanical and optical fabrication techniques, housing and mounting methods, alignment and test, and cost. In the following sections, these considerations will be explored.

2 Optical Materials

The UV spectrum can be divided into several portions, viz., 400–300 nm (near ultraviolet), 300–200 nm (far ultraviolet), and < 200 nm (deep ultraviolet). In a like manner, the infrared spectrum can be segmented into different parts, viz., 0.7–3 μm (near infrared), 3–6 μm (middle infrared), 6–15 μm (far infrared), and $> 15\ \mu m$ (extreme infrared). The most obvious point is the large difference in the wavelengths — a wavelength ratio of about 10 for the middle-infrared to the near-ultraviolet spectrums exemplifies this point.

The number of optical materials available for the different portions of the spectrum varies greatly with the smallest number being in the ultraviolet and the greatest being in the visible. The visible and near-infrared regions are dominated by glasses whereas crystalline materials dominate the other spectral regions. Figure 1 illustrates the approximate number of optical materials available as a function of wavelength.[1] It is evident that the richness in materials falls off rapidly for wavelengths < 300 nm and $> 10\ \mu m$. In this sense, infrared and ultraviolet optical systems share the same limitation. However, the range of refractive index and the maximum value are much lower for the ultraviolet than for the infrared. This is graphically illustrated in Fig. 2 which shows the approximate upper and lower boundaries for the refractive index as a function of wavelength.[1] The lesser upper values of refractive index for ultraviolet materials tend to require more lens elements than for an equivalent infrared lens system. On the other hand, the lower refractive index produces less surface reflection losses for the ultraviolet optics; but, coatings become more difficult as the wavelength shortens. A rather significant difference between ultraviolet and infrared materials is the dispersion. Figure 3 illustrates the approximate upper and lower boundaries of the dispersion ($\frac{dn}{d\lambda}$) as a function of wavelength.[1] It is readily apparent that ultraviolet materials are much more dispersive than infrared materials.

Fused or synthetic silica is the most commonly used glass for wavelengths greater than 170 nm. Other commonly-used UV-materials include calcium fluoride, lithium fluoride, and magnesium fluoride. Significant research and development of fluorocrown glasses with relatively low refractive indices and dispersions have occurred during the past few years. Examples of such glasses are the Schott Ultran series of ultraviolet glasses. These glasses provide an alternative to the crystalline materials typically used to chromatically correct lens systems incorporating silica.

A problem suffered by ultraviolet materials, and to a much lesser degree by infrared materials, is solarization. This is a serious problem faced by designers of lenses to be used in microlithography applications. In the infrared, internal absorption is often the limiting factor when refractive ma-

[1]Figure prepared by Dr. Chen Feng of the Center for Applied Optics at the University of Alabam in Huntsville.

terial is used for high-power infrared laser optics.

A modest variety of materials are available for use in the infrared spectrum. Most popular of the crystalline materials are germanium, silicon, sapphire, and CVD-processed zinc selenide and zinc sulfide. Several special glasses have been developed with the chalcogenides being somewhat more useful. Others include calcium aluminate glasses, germanate glasses, metal fluorides and oxides. Several of these materials can be used together to form color corrected lens systems.

3 Structural Materials

Generally speaking, the structural materials used for both ultraviolet and infrared optical systems are the same. The principal difference is that the precision and dimensional stability requirements for ultraviolet systems are much more exacting than for infrared systems. The reason is basically due to the large ratio in wavelengths between these two spectrums.

Selection of the structural materials is affected by the coefficient of thermal expansion (CTE) of the optical materials. The value of the CTE for the optical materials varies from almost zero to over $50 \cdot 10^{-6}/^{o}C$. CTE values for popular optical materials used in the ultraviolet or infrared spectrums are fused silica ($0.5 \cdot 10^{-6}/^{o}C$), silicon ($4.2 \cdot 10^{-6}/^{o}C$), germanium ($6.1 \cdot 10^{-6}/^{o}C$), sapphire ($6.7^{p} \cdot 10^{-6}/^{o}C$ and $5^{s} \cdot 10^{-6}/^{o}C$), zinc selenide ($7.8 \cdot 10^{-6}/^{o}C$), chalcogenides (e.g., AMTIR-1, $13 \cdot 10^{-6}/^{o}C$), calcium fluoride ($24 \cdot 10^{-6}/^{o}C$), and lithium fluoride ($37 \cdot 10^{-6}/^{o}C$)[2]. The CTE for structural materials ranges from about $2 \cdot 10^{-6}/^{o}C$ to over $25 \cdot 10^{-6}/^{o}C$. Four commonly used materials and their CTE values are 6061 aluminum ($23.6 \cdot 10^{-6}/^{o}C$), 17-4 PH stainless steel ($10.8 \cdot 10^{-6}/^{o}C$), Invar ($0.9 \cdot 10^{-6}/^{o}C$), and titanium ($8.6 \cdot 10^{-6}/^{o}C$).

In addition to the CTE of the structural materials, the refractive index of the optical materials varies as a function of temperature. For example, the temperature coefficient of refractive index (TCRI) is $6.7 \cdot 10^{-5}/^{o}C$ for germanium, $3.9 \cdot 10^{-5}/^{o}C$ for silicon, $-1.6 \cdot 10^{-5}/^{o}C$ for lithium fluoride, and $-9 \cdot 10^{-5}/^{o}C$ for calcium fluoride. As the temperature of an optical system varies, the optical designer faces a significant challenge to maintain the focal length and focus as a consequence of the differences in CTE and TCRI of the materials comprising the optical system. Generally, maintaining focus is the only objective although holding the value of the focal length can be of importance in certain applications.

[2] W. L. Wolfe, *Handbook of Military Infrared Technology*, Office of Naval Research, Washington (1965).

Over the years, much effort has been expended to create designs that achieve these objectives. In general, passive techniques are employed for infrared optics while active means are often needed for high precision ultraviolet optics. As can be readily shown[3] the combined effects of CTE (α) and TCRI ($\frac{\partial n}{\partial T}$) cause a defocus ($\delta$) for a single lens with a focal length of f in a housing of overall length L of

$$\delta = \left[-\frac{f}{n-1}\frac{\partial n}{\partial T} + \alpha f + \alpha L \right] \Delta T.$$

Passive techniques usually involve the use of materials with high thermal expansion coefficients arranged in a manner such that the focal shift is eliminated. In practice, this is difficult to attain in the infrared and almost impossible in the ultraviolet. Recalling that for a lens with a numerical aperture of NA the Rayleigh criterion for resolution ($\mathcal{R}$) is

$$\mathcal{R} = \frac{0.61\lambda}{NA}$$

and that the Rayleigh depth of focus ($\mathcal{D}$) is given by

$$\mathcal{D} = \frac{\pm\lambda}{2(NA)^2},$$

the strong influence of wavelength is evident. Since ultraviolet microlithographic lenses and infrared lenses often have F/#'s near unity, the resolution and depth of focus differences are due to wavelength. For example, if the $NA = 0.48$, then $\mathcal{R} = 0.61\ \mu m$ and $\mathcal{D} = \pm 0.71\ \mu m$ for an ultraviolet lens[4] operating at $0.365\ \mu m$ and $\mathcal{R} = 12.7\ \mu m$ and $\mathcal{D} = \pm 21.7\ \mu m$ for an infrared lens operating at $10\ \mu m$. Clearly, the focus stability requirement with temperature changes is about 30 times tighter for the ultraviolet optics than the infrared optics. Consequently, the engineering demands for a passive optomechanical design to maintain focus for an ultraviolet lens system is very difficult to achieve (if not impossible) over a reasonable temperature range. On the other hand, infrared optical systems with passive thermal compensation can and have been designed and constructed[5] although substantial attention to the CTE and TCRI of the materials must be given.

High-performance ultraviolet lenses require that they be operated within a very narrow temperature range or that some form of active mechanical

[3] J. M. Lloyd, *Thermal Imaging Systems*, Plenum Press, New York, pp. 257–267 (1975).

[4] The resolution for a microlithographic lens includes a multiplicative factor that expresses the process dependent aspect of the lithographic process. A typical value of 0.8 was used in this case.

[5] P. J. Rogers, "Athermalization of IR Optical Systems," Proc. SPIE, Vol. CR38, pp.69–94 (1991).

athermalization be used.[6] Active means include the use of temperature sensors that provide input for the computation of mechanical compensation. This is often implemented by motors, piezo-electric, etc., devices. It should be noted that infrared optics sometimes include active athermalization although the accuracy of the movements is much less demanding than for the ultraviolet case.

4 Lens Configurations

Both ultraviolet and infrared optical systems share several common features, viz., the use of all-reflective optics, catadioptric optics, and relatively few optical elements. Figure 4 shows the configuration of a catadioptric objective lens for an infrared scanner along with the "eyelens," scanner, and imaging lens.[7] The next figure illustrates an ultraviolet catadioptric microscope objective of the basic Schwarzschild class.[8] Figure 6 contains a rendering of a five-element ultraviolet lens used in the 0.2–0.4 μm spectrum.[9] The lens is made of CaF_2 and quartz, has a focal length of 5.0 inches, the F/# is 4.0, and the total field of view is 10 degrees. In general, infrared optics have a much larger field-of-view than ultraviolet optics whereas ultraviolet optics have a much tighter tolerance on field flatness and distortion.

An example of an athermalized infrared zoom lens with a magnification range of 3.5–20 power is shown in Fig. 7.[10] This lens is made of germanium, zinc selenide, and zinc sulfide. Athermalization is accomplished by combining lens groups that are each athermalized about the central movable elements that are used for the zoom function. The elements comprising the first and third lens groups shown in Fig. 7 are made of germanium, zinc selenide, zinc selenide, and zinc sulfide. Specific details of how to realize passive athemalization is taught in the patent. Unfortunately, this type of passive athermalization is not viable for ultraviolet optics due to the lack of appropriate optical materials.

A typical projection optical system for submicron lithography is illustrated in Fig. 8.[11] The complexity of the optical system is similar to that of the more sophisticated infrared optical systems. The fabrication and alignment precision required by the intermediate and wafer triplet assemblies is far bet-

[6] A. Ahmad, "High Resonance Adjustable Mirror Mounts," Proc. SPIE, Vol. 1167, pp. 313–317 (1989).

[7] U.S. Patent 4,432,596 (1984).

[8] J. R. Benford, "Microscope Objectives," in *Applied Optics and Optical Engineering*, R. Kingslake Ed., Academic Press, New York, pp. 172–174 (1965).

[9] Adapted from M. Laikin, *Lens Design*, Marcel Dekker, New York, pp. 198–199 (1991).

[10] U.S. Patent 4,679,891 (1987).

[11] Private communication from A. Ahmad.

ter than for infrared optical systems. The wafer triplet assembly is shown in its housing in Fig. 9. The tolerances are so exacting that the housing must be diamond turned and the alignment of the elements is accomplished on an air bearing table. Each optical element is carefully measured in order that the housing can be modified to compensate for optical fabrication errors. Tolerances can be as small as 0.5 μm. Infrared optical systems rarely have housing related mechanical tolerances tighter than 10 μm. Consequently, the effort to construct and align these optical systems is significantly greater for ultraviolet optics. Yoder[12] and Ahmad[6,13] have discussed many of the issues concerning design, fabrication, and alignment of mechanical systems for optics.

The use of binary or diffractive optical surfaces for infrared optical systems is gaining popularity.[14] Although the performance is adequate in the infrared spectrum, this surface form cannot be used in the ultraviolet due to material and fabrication limitations.

5 Coatings

Effective coatings for antireflection, bandpass, etc. require that appropriate materials be available from which coatings can be constructed. The number of materials available for coatings in the ultraviolet and infrared spectrums is limited to approximately the same number as can be used to construct lenses. Even with this limitation, coating designers and engineers have been able to produce excellent coatings throughout the infrared spectrum and over much of the ultraviolet spectrum. Although almost any functional coating type can be realized in the infrared, in the ultraviolet spectrum antireflective coatings are not available for wavelengths below about 250 nm. In this spectral region, only bandpass filters can be made, but with great difficulty and expense. Special fabrication techniques have been developed in order to make even these filters with adequate transmission and out-of-band rejection.

Reflective coatings in the infrared are relatively straightforward and typically use gold, aluminum, or enhanced silver. In the ultraviolet, the reflectivity of useful coating materials, such as UV-enhanced aluminum, is limited to wavelength greater than 150 nm.[15] Special coating stacks can be designed

---

[12] P. R. Yoder, Jr., *Opto-Mechanical Systems Design*, Marcel Dekker, New York (1986).

[13] A. Ahmad, "Fabrication techniques for high-resolution lens assemblies," Proc. SPIE, Vol. 1335, pp.194–198 (1990).

[14] M. J. Riedl and J. T. McCann, "Analysis and Performance Limits of Diamond Turned Diffractive Lenses for the 3- and 8–12 Micrometer Regions," Proc. SPIE, Vol. CR38, pp.153–163 (1991).

[15] S. Musikant, *Optical Materials*, Marcel Dekker, New York, pp. 194–197 (1985).

that yield a reflective surface over a very narrow spectrum in the far and deep ultraviolet spectrums.

6 Fabrication and Test

Fabrication of the optical elements used in the ultraviolet and infrared spectrums is typically accomplished using conventional grinding and polishing techniques. Infrared optics are being made on an increasing basis by diamond-turning methods; however, the as-turned surface quality is simply inadequate for ultraviolet applications. In some cases, diamond turning is used to prepare the base substrate for an ultraviolet mirror that will be subjected to additional processing. For example, an aluminum substrate might be diamond turned, then overcoated with nickel, diamond turned again, and then post polished to remove the residual machining marks and surface roughness. An appropriate reflective coating is then deposited. This process provides a means to attain surfaces that have low light-scattering properties. Figure 10 presents the theoretically predicted levels of scattered light as a function of surface roughness and wavelengths covering both the ultraviolet and infrared regions.[16] Typical types of surface polish are indicated by the dashed lines. As can be seen from the figure, the relationship between surface roughness and wavelength for a constant amount of scattered light is directly proportional. This implication is that ultraviolet optical surfaces require greater care and quality than is necessary for infrared optical surfaces. Cost and schedule seem to grow exponentially with decreasing surface roughness.

Because of the very high quality surfaces needed for ultraviolet optics, conventional polishing methods are essentially the only option. The fundamental surface figure requirements are about the same for both ultraviolet and infrared optics, e.g., $\frac{\lambda}{4}$; however, the physical scale is extremely different. At 365 nm, the physical displacement is 91 nm while at 10 μm the displacement is 2,500 nm! Testing of surface figure is more difficult for ultraviolet optics since the wavelength of the light used in a typical surface measuring interferometer is about two or so times greater. This means that the fringe sensitivity is reduced by the wavelength ratio. The converse is true for infrared optics, i.e., testing of the surface is much easier. In a like manner, the same is true for system level testing (resolution, MTF, etc.).

Measurement of scattered flux from the surface of the optics or from the entire lens assembly is significantly more challenging for ultraviolet optics than for infrared optics. One of the most basic problem in testing ultraviolet optics is that as the wavelength shortens, the atmospheric absorption

[16] H. Bennett, "Optical properties of optical materials," Opt. Eng. **17**(5) (1978).

dramatically increases to such a degree that measurements must be made in a vacuum environment. Although a number of facilities exist that can measure infrared optics, by comparison, very few are available for testing ultraviolet optics. For example, to the author's knowledge, quality measurement of the scattered flux in the deep ultraviolet from a surface can be measured only at NASA Marshall Space Flight Center.

The parts comprising mounts and housings for the optical systems typically are fabricated by standard machine shop equipment. In order to achieve the highest performance levels for ultraviolet optics, the parts must often be hand lapped and/or diamond turned. For example, improvements in microlithographic lens design and manufacturing methods have produced improvements in the number of pixels per image field of over an order of magnitude during the past decade.[17] Although the precision required for infrared optics is lower, diamond-turned mechanical components are used as a means to reduce overall fabrication and alignment costs.

7 Conclusions

Ultraviolet and infrared optics share many common features and characteristics while at the same time having a number of practical differences. For the most part, these differences can be directly related to the large difference in wavelength between the two optical spectrums and the environments in which they operate. The most exacting ultraviolet applications requires that the optical system be maintained within a very narrow temperature range in a clean room whereas a precision military FLIR *must* operate over a broad temperature range in often hostile environments. Another interesting difference is that polarization and partial coherence are of little importance in infrared optics, but are often of great significance in the ultraviolet. For example, the partial coherence specified for the Zeiss 10-78-65 lens for i-line microlithography is 0.61. Also, ultraviolet optical systems do not suffer the narcissus problem that many infrared system must address.[18]

Continued interest and application of the ultraviolet spectrum will result in further improvement of fabrication, alignment, and test methods and techniques. Optics for the infrared will reap the benefit of these developments — but the differences will remain.

---

[17]W. H. Arnold, A. Minvielle, K. Phan, B. Singh, and M. Templeton, "0.5 Micron Photolithography using High Numerical Aperture I-Line Wafer Steppers," Proc. SPIE, Vol. 1264, pp. 127–142 (1990).

[18]E. Ford and D. Hasenauer, "Narcissus in current generation FLIR systems," Proc. SPIE, Vol. CR38, pp. 95–119 (1991).

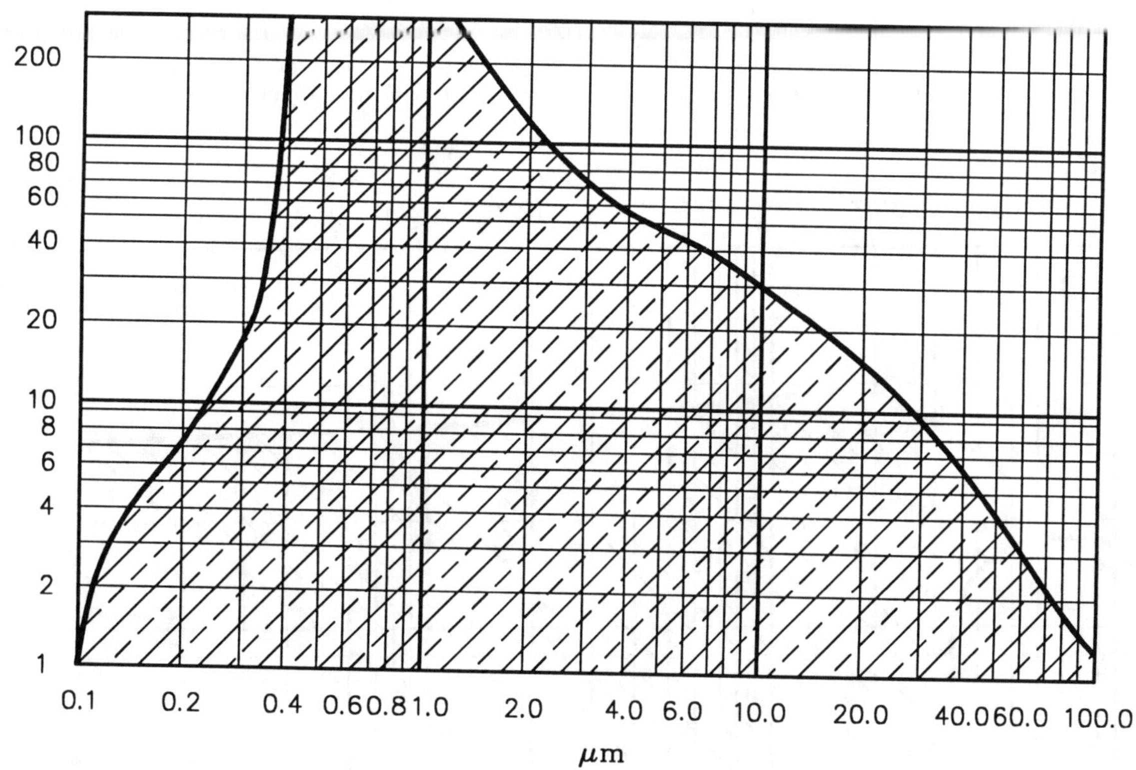

Figure 1. Approximate distribution of optical materials available at wavelengths from the ultraviolet to infrared spectrums.

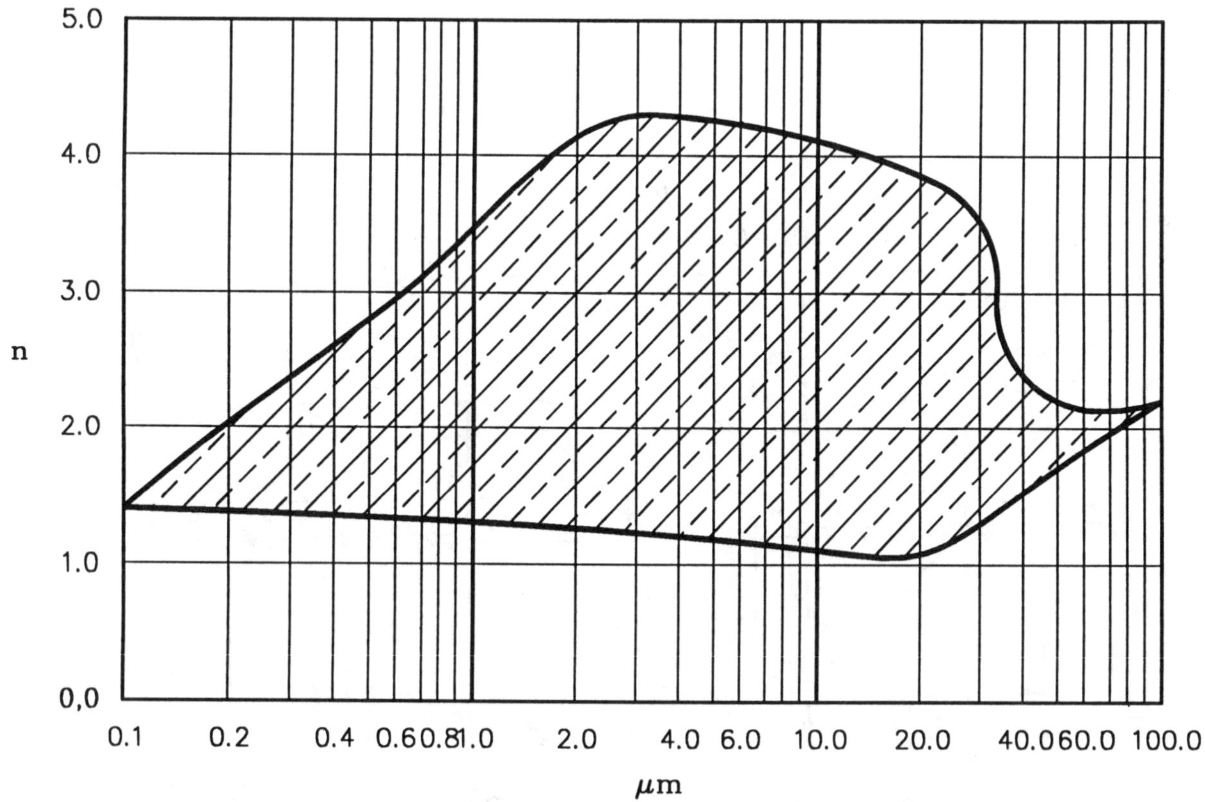

Figure 2. Approximate upper and lower boundaries of the refractive index of optical materials as a function of wavelength.

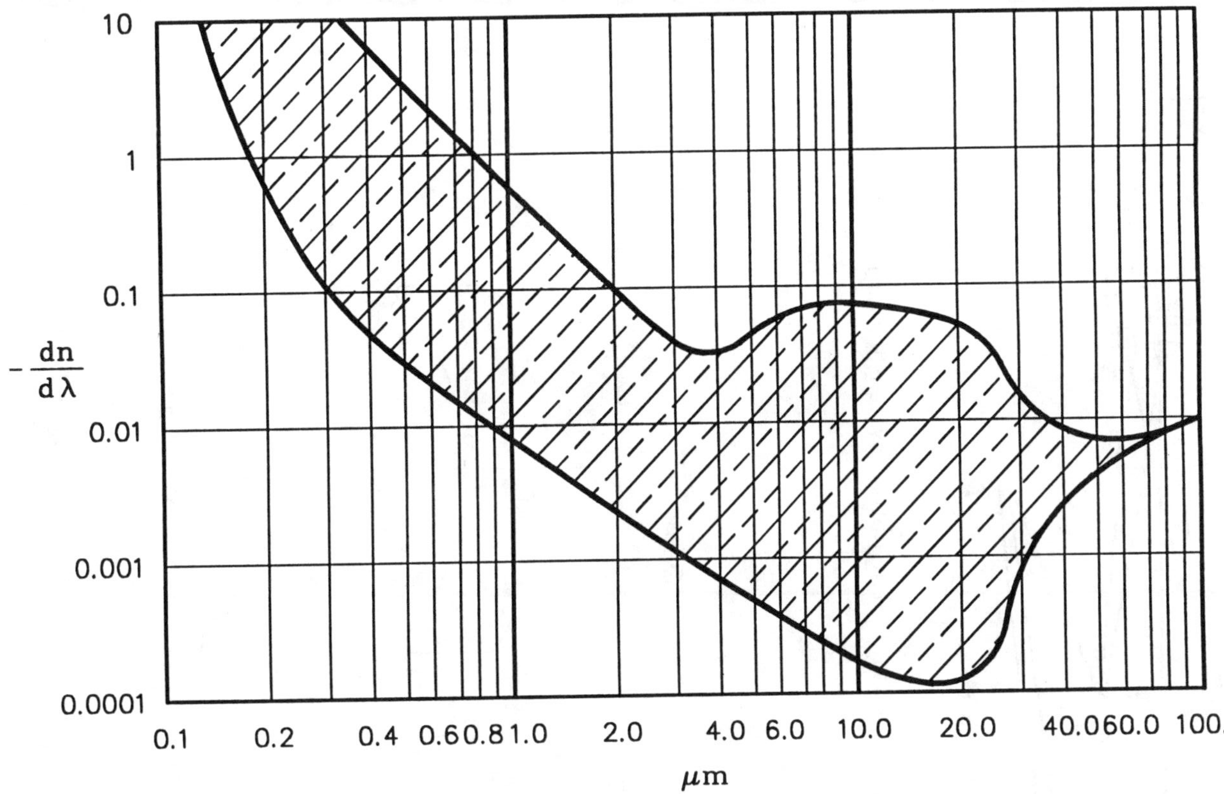

Figure 3. Approximate upper and lower boundaries of the dispersion of optical materials as a function of wavelength.

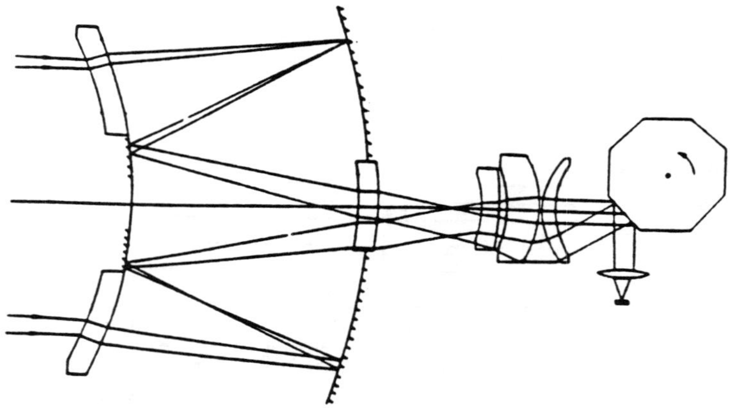

Figure 4. Catadioptric optical system for an infrared scanner.

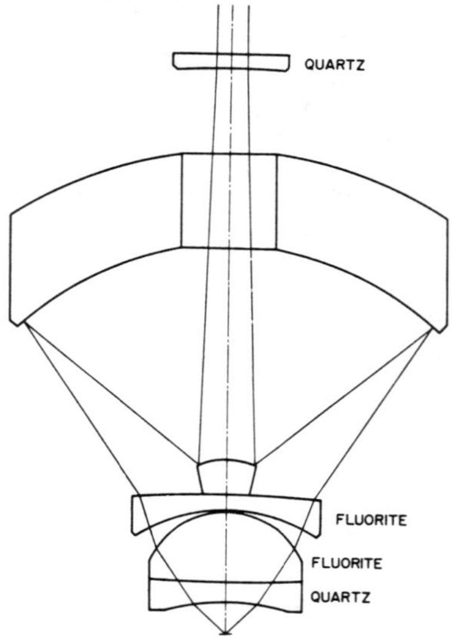

Figure 5. Ultraviolet catadioptric microscope objective.

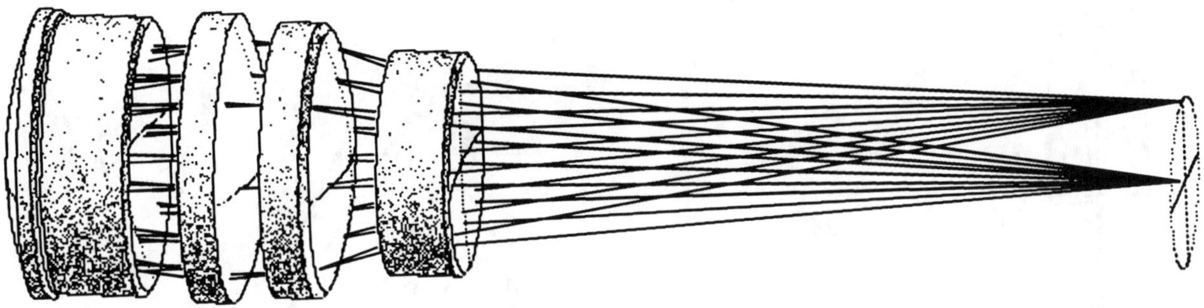

Figure 6. F/4 ultraviolet lens for use in the 0.2–0.4 μm spectrum.

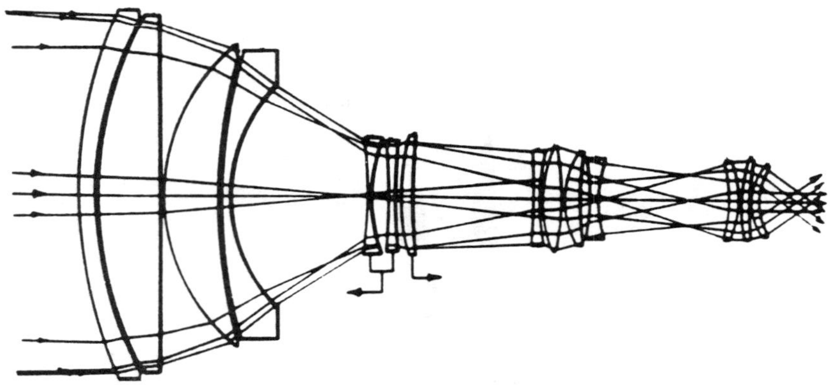

Figure 7. Athermalized infrared zoom lens.

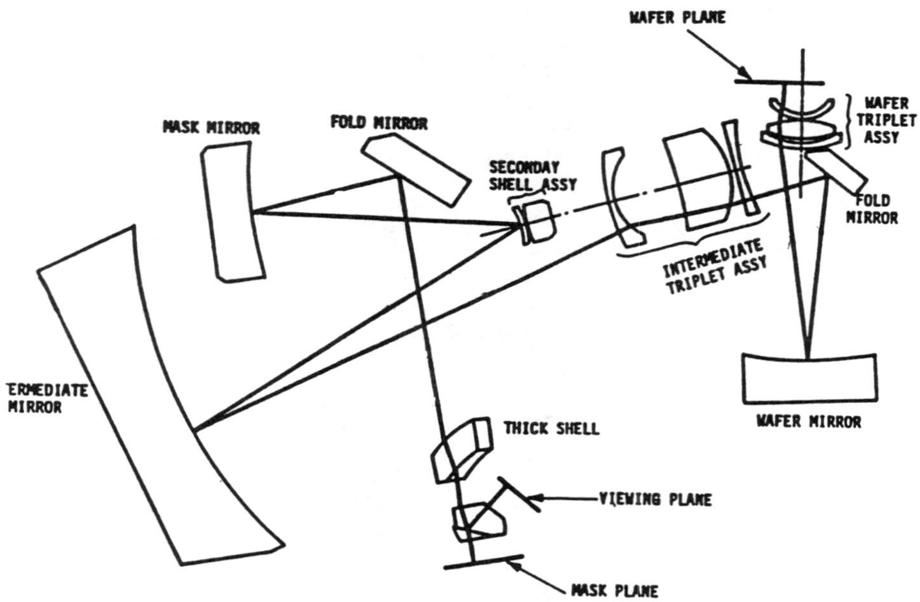

Figure 8. Projection optical system for submicron lithography.

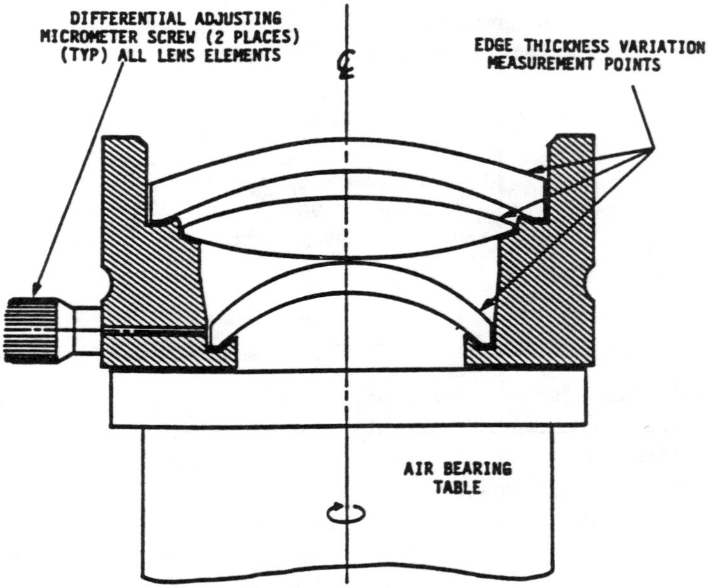

Figure 9. Wafer triplet assembly and housing for the projection optical system for submicron lithography shown in Fig. 8.

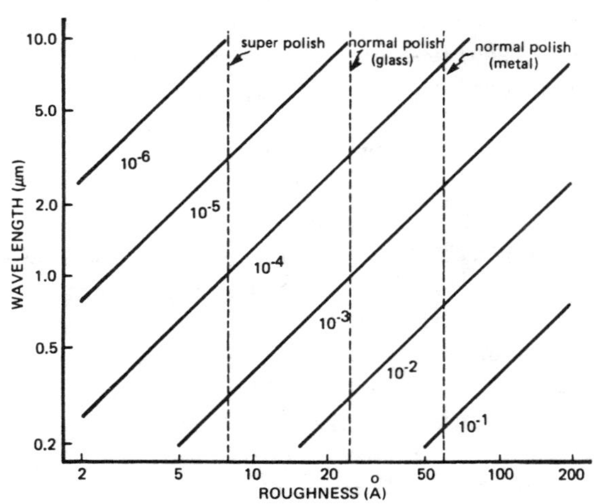

Figure 10. Theoretically predicted levels of scattered light as a function of surface roughness and wavelength covering both the ultraviolet and infrared regions.

Maximizing production yield and performance in optical instruments through effective design and tolerancing

Ronald R. Willey and Mark E. Durham

Opto Mechanik, Inc.
P.O.Box 361907
Melbourne, Florida 32935

ABSTRACT

The focus of this paper is concerned with the practical aspects of designing optical instruments which are intended for production in large or even small quantities. Certain aspects of performance are required of an instrument which is its reason to exist. It is usually expedient, plus fiscally and ecologically sound, to strive for an overall efficient use of resources in the life cycle of the instrument (i.e., keep the net cost down). This includes the processes from concept through design, prototypes, production, use, and disposal. The design and tolerancing aspects of the process have a major effect on the life cycle cost and efficiency of the system and that is the principal subject of this paper.

We discuss what makes up the cost of a lens and the effects of tolerances and other factors on that cost. This results in a new lens cost estimation formula. We describe the interactions of lenses and lens cells from the tolerance viewpoint. We then explain the principles whereby the system tolerances can be determined which will give the minimum cost system which meets the performance requirements. We conclude with an example from real life of the preliminary application of the principles.

1. INTRODUCTION

The assignment of tolerances to the various dimensions and parameters of an optical system is a CRITICAL element in determining the resulting performance and cost of the system. This has historically been complex and ill-defined and therefore a distasteful aspect which the designer has had to deal with in one way or another. Much of the tolerancing of systems in practice has been done by art and experience more than by scientific calculation. Here, we attempt to make the ENGINEERING principles as simple and clear as possible so that they may be applied in a straightforward manner. We use the term "engineering" to imply that practical approximations based on empirical data are used to reduce the problem to terms that can be handled in the real world. That is to further imply that we are dealing here with the application of optical design to produce a functional set of hardware as opposed to the theory and/or science of lens design as an exact and abstract study. In the production of an optical system, random errors in parameters occur. These cause the results to be statistically predictable, but not exactly calculable. Therefore, the use of reasonable engineering approximations is appropriate and justifiable.

It is sometimes possible to tolerance a system such that each of the components is fabricated to an accuracy which will insure that the instruments will be adequately precise and aligned to give the required performance by simple assembly with no alignment or adjustment. This may be the case in certain diamond turned optical assemblies. The other extreme is where almost every parameter of a system can be adjusted to bring the system to the desired performance. However, neither of these approaches is usually the least cost way to meet the performance requirements. We discuss philosophical principles and practical ways of approaching the least cost solutions and give an illustration of the application of the techniques.

This paper is the forth of a series (1,2,3) which started in 1982 where the authors invesitgated how to achieve the least-cost tolerancing of an optical system. Since that time, Wiese (4) has compiled a very useful collection of papers on tolerancing in its many aspects. That volume includes a paper by Plummer (5) which played a role in our earlier work and a paper by Adams (6) which judiciously utilized some of our findings. Fischer (7) did a more recent survey based on Plummer's work which we compare with some of our updated cost versus tolerance data below. Parks (8,9) and Smith (10) have papers in Wiese's collection that have many practical and helpful discussions on the subject. The works referenced form a good general background for this paper, but we will reiterate the salient points below for the convenience of the reader. We will also cite other specific references as they apply below.

It is the authors' experience and opinion that there has been a great deal of waste resources in the past due to poor tolerancing "art". A rigorous and all encompassing treatment of all but the simplest system can be VERY complex. It is our aim to move the practice of tolerancing from the art stage to the engineering stage with as much simplification as is reasonably justifiable. Warren Smith (10) has done a great deal to move the status of the practice in this direction, we are attempting to move another step along the way by adding the real influence of cost into the tolerancing process.

The optical and mechanical designers of instruments have by far the greatest influence on the ultimate cost and performance of an instrument. All others, including the manufacturing operations, cannot have as much power as the designers to change the potential satisfaction of the user and profitability of the producer. Figure 1 illustrates this graphically. Figure 2 outlines the detail design process and how the subjects of this paper interplay in the process.

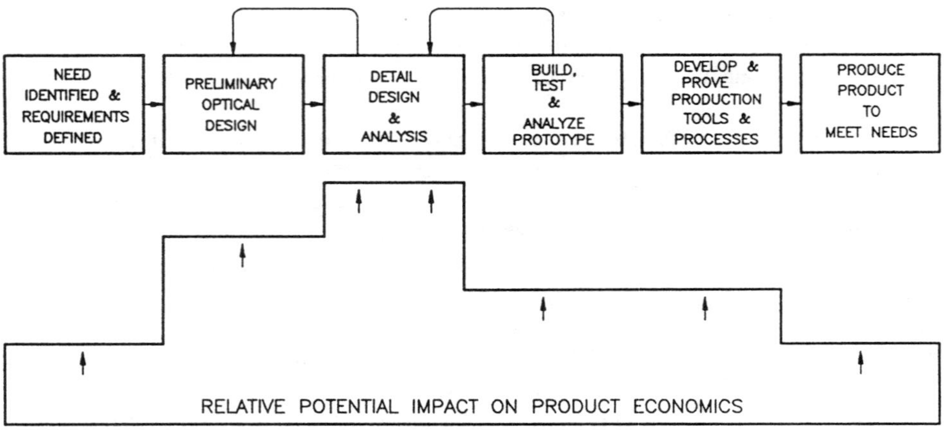

FIGURE 1. OVERALL PROCESS TO DEVELOP AND PRODUCE A NEW OPTICAL PRODUCT

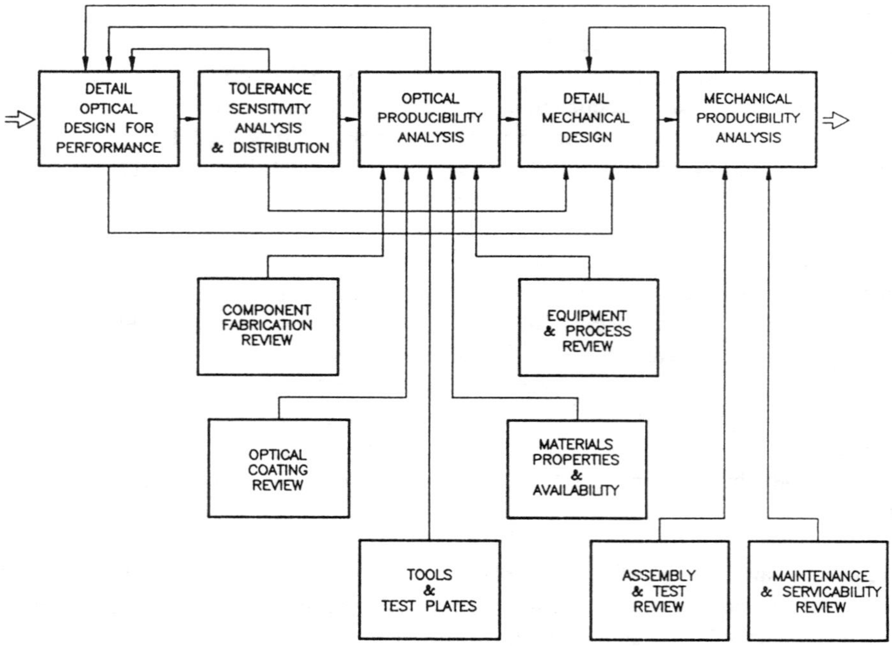

FIGURE 2. DETAIL DESIGN AND ANALYSIS PROCESS OF AN OPTICAL PRODUCT

2. BASE COSTS

We will first review the concepts of base costs from our previous work and then add the results of new work on the estimation of the base costs directly from the data on a drawing and/or the specifications of a given component.

Let us take the example of fabricating a single lens. If someone asked that we make a biconvex lens of glass, we would typically have to go through the following steps:

generate (or mill) a radius on the two sides
mount the lens on a spindle
grind and polish the first side of the lens
remount the other side of the lens on a spindle
grind and polish the second side
edge the lens

There are obviously a few other small steps such as obtaining materials and grinding and polishing tools, dismounting, cleaning, etc. We have said nothing to this point about further specifications such as diameter, radii, thickness, and tolerances. Without these there is still a cost in time and materials and equipment necessary to make the biconvex glass lens. This is what we call the base cost. As we get more specific about the lens and add more restrictive tolerances on the parameters, more care and time and equipment will probably be required to make the lens to the new specifications. Therefore the cost will increase with increasingly stringent requirements/tolerances/specifications.

Our previous work (1,2,3) went into some detail on the relationships of the increase above the base cost with changing tolerance values. However, neither we nor Plummer(5) quantified those base costs. We will quantify the base costs in this work and thereby add significantly to the usefulness of the technique and somewhat to the accuracy of the results. What was not made clear in the earlier work was that the base costs to be used in a given cost versus tolerance case are not the same for grinding and polishing as they are for centering and edging. As we shall show below, the centering and edging operation is not influenced by grinding and polishing costs or tolerences and vice versa. Not incorporating this concept can lead to some error in the application of our earlier work and Adams'(6) extension of it. There is a different base cost to which the increase with tolerance is to be applied for the two classes of operation. Figures 3 and 4 show the tolerance of lens diameter and eccentricity costs as a percentage of base cost for the centering and edging operations. This applies to the centering and edging base cost (CE). Figures 5 through 10 show the tolerance costs as a percentage of the base grinding and polishing costs (GP) for tolerances of radius of curvature, irregularity, diameter to thickness ratio, center thickness, scratch and dig, and the glass stain

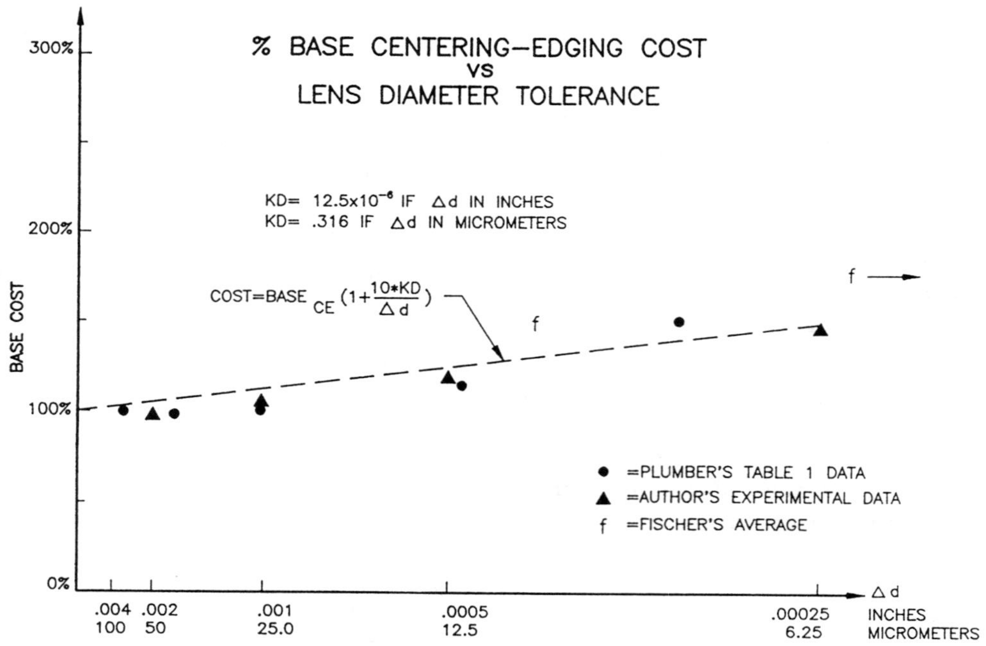

FIGURE 3: LENS DIAMETER TOLERANCE (PLOTTED ON RECIPROCAL SCALE)

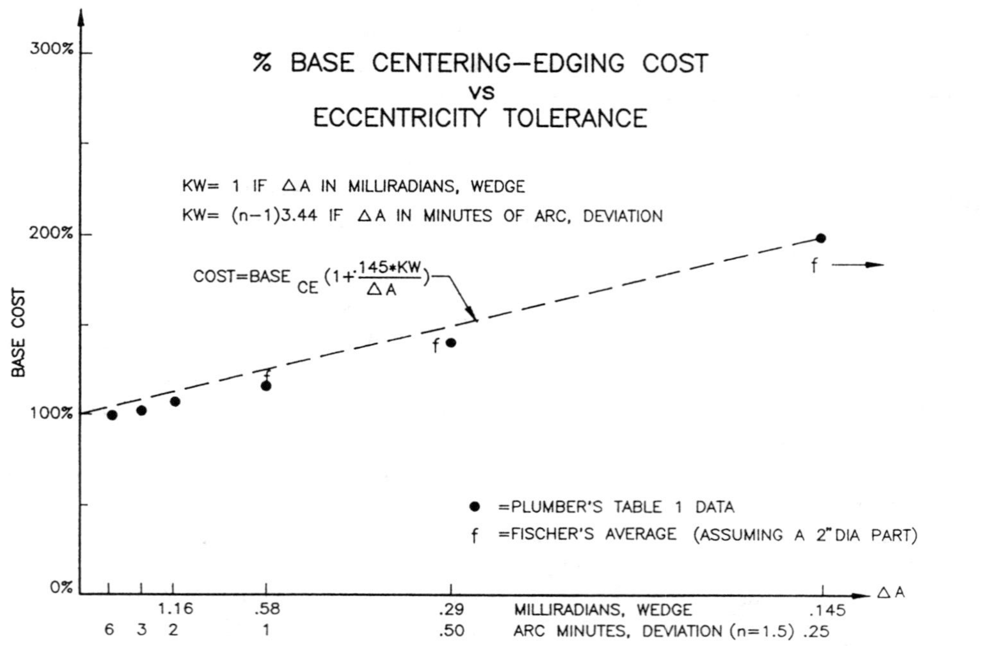

FIGURE 4: ECCENTRICITY TOLERANCE (PLOTED ON RECIPROCAL SCALE)

characteristics. It was also interesting to us to recognize that the milling or generating costs are not particularly affected by tolerances with today's equipment and they are not part of the base which the tolerance factors multiply. The milling costs are therefore part of the base, but only a function of material and dimensions, not tolerances.

Before we discuss the cost versus tolerances in detail in the next section, we will show the development of the base cost formula. The present view and cost estimating scheme is the authors best effort to date resulting from their experience and discussions drawing on the work and experience of Stephen Cupka, Manager of Estimating and Reinhard Seipp, Assistant Manager of Optical Manufacturing at Opto Mechanik, Inc., most of these persons also have experience from one or more other shops than OMI to draw upon.

Let us call the total base cost to make a lens MT, which is in either time or money which differ only by some multiplicative factor. If we call the milling or generating cost MG, we can represent the total base costs as the sum of milling, grind and polish, and centering and edging costs as given in Eqn. 1.

$$MT = MG + GP + CE \qquad (1)$$

2.1 Milling/generating costs

We find a reasonable fit with experience in milling to be given in Eqn. 2 where LM is the number per lot to be milled and d is the diameter in inches. The milling of both sides of the lens is included here.

$$MG = 4 + 90/LM + 0.1*d^2 \qquad (2)$$

This implies that there is some base cost to mill the optic plus some setup cost divided by the number of parts to be run from that setup plus a factor due to lens size. Since the material to be removed from a molded blank is usually about the same thickness independant of blank diameter, the generating time is only a function of a blank's area (d^2).

2.2 Centering and edging costs

We will deal with CE before GP because it is simpler. The cost is a function of the number in the lot LC to be centered on one setup, the diameter d, the number of chamfers C, and the number of flats F (planes perpendicular to the lens axis). Equation 3 represents our collective best estimate.

$$CE = (2 + d + C + F)/3 + (30 + 10*C + 15*F)/LC \qquad (3)$$

This accounts for a setup cost for the diameter, chamfers, and flats plus the edging of each lens.

2.3 Grinding and polishing costs

The grinding and polishing cost is a very strong function of the number of lenses which can be ground and/or polished at one time on a block. If the radius is short the number NS which can be blocked for that side is determined by the radius R and the lens diameter d. If the radius is long, the number which can be blocked is determined by the maximum block diameter G and the lens diameter d. The precise calculation of this number can be done when flats, chamfers, and center thicknesses are properly accounted for. That subject might be worthy of a paper on its own. For simplicity and practicality, we will use a conservative approximation without showing its derivation here. If the R/d is less than .87, only a single lens can be polished at one time, and at least 3 per block can be done if the ratio is greater. If the radius is short the number per block NS will be given by integer value in Eqn. 4.

$$NS = INT(2.9*(R/d)^2) \tag{4}$$

Here R is the radius and d is the diameter of the side in question, and N1 would be NS for side 1 and N2 would be NS for side 2.

If the radius of the side is long, the number per block will depend on the diameter G of the largest block which can be used as shown on Eqn. 5.

$$NS = INT(.64*(G/d)^2) \tag{5}$$

Whichever NS is smaller for a given side (for R or G) will be the number which can be ground and polished per block for that side. Our experience is that less than the maximum number will often be used when the radius is long. This is not of great cost consequence, however, because the change in cost with a few parts more or less when NS is large is seen in Eqn. 6 to be small relative to other costs.

Our collective experience with the grind and polish costs GP per lens and per side in a block of NS lenses is given in Eqn. 6.

$$GP = 7 + 14/NS \tag{6}$$

It is also usually appropriate to consider a yield related factor Y due to scrapage, etc. This is the factor times the number of lenses to be delivered which must be started to give the yield; it is actually inverse yield then. When this is applied and both sides are considered, the total GP base costs are given in Eqn. 7 where N1 and N2 are the numbers

$$GP = Y*(14 + 28*(1/N1 + 1/N2)) \tag{7}$$

of lenses per block for side 1 and side 2. We now have all of the components of the base cost (in relative units) for a lens based on a collective empirical history and in a fairly workable form. We only need to apply the effects of tolerances and other influences such as material properties and diameter to thickness ratio to these base costs in the next section and we can then predict the cost to produce a given lens to its specifications. The derivative of that cost with a change in any tolerance or parameter will be used to determine the distribution of tolerances which will minimize system cost and insure the desired yield.

3. THE EFFECT OF TOLERANCES AND OTHER FACTORS ON COSTS

We have generally updated and included here the cost versus tolerance data presented in our previous work (3) in Figs. 3 through 10. The dots were Plummer's work (5), the triangles were our previous experimental and or estimated values, the c's are factors from Cupka's earlier estimates, and the f's are the averages from Fischer's survey (7). The lines drawn are the functions which we currently choose to use as the best estimate from experience and the functional equations are indicated on each figure. On the figures where appropriate, we have included information from Smith(10). He indicated tolerances that he thought we "low cost" which we have plotted as "$". We show his "commercial" as "$$", "precision as "$$$", and "extra precise" is shown as "$$$$". These are in general accord with other data. For a more extensive discussion of the previous work, see the references.

Figure 3 shows the cost effect of lens diameter tolerances which is not a strong factor up to the limit of the capabilities of the edging process. The various authors' data are in good agreement. We have shown the tolerance scale in inches, and micrometers on this and some of the other figures for the readers convenience.

Figure 4 shows the cost effect of the lens centering tolerances. The centering tolerance is sometimes expressed as light deviation, wedge, or total indicator runout (TIR) at the edge of a lens or window. These different versions of the requirments can be reconciled in the following way. The wedge angle in radians is the same as the TIR divided by the diameter being measured. The deviation depends on the wedge times the refractive index minus one and conversion to arcminutes as needed. The KW-factors in the figure account for this. Plummer's and Fischer's data are in good accord and are well represented by the approximation which we use.

Figure 5 deals indirectly with the cost of radius of curvature tolerances. The work of Thorburn (11) reported in Wiese's collection made it clear that we should refine our approach to radius of curvature. We had previously just used the percentage of the radius as a measure of stringency of the radius tolerance. But it becomes apparent after a little analysis that the delta in sagitta over a surface is much more meaningful. This is what a spherometer or an interferometer can measure. One can derive that the change in sagitta with a change of radius is approximated by Eqn. 8.

$$dZ/dR = .125*(d/R)^2 \qquad (8)$$

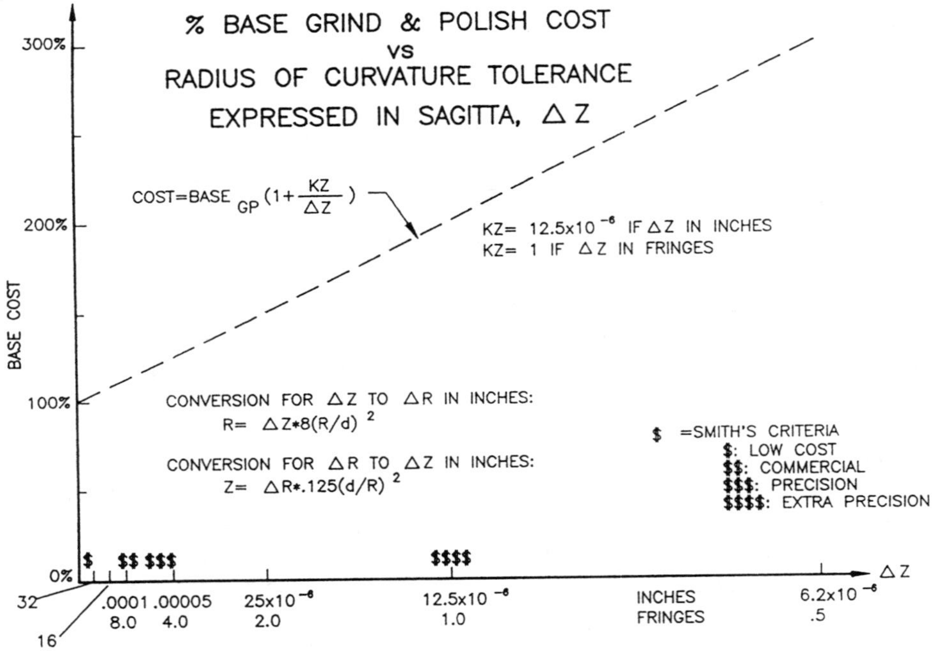

FIGURE 5: RADIUS OF CURVATURE TOLERANCE EXPRESSED IN SAGITTA ΔZ (PLOTTED ON RECIPROCAL SCALE)

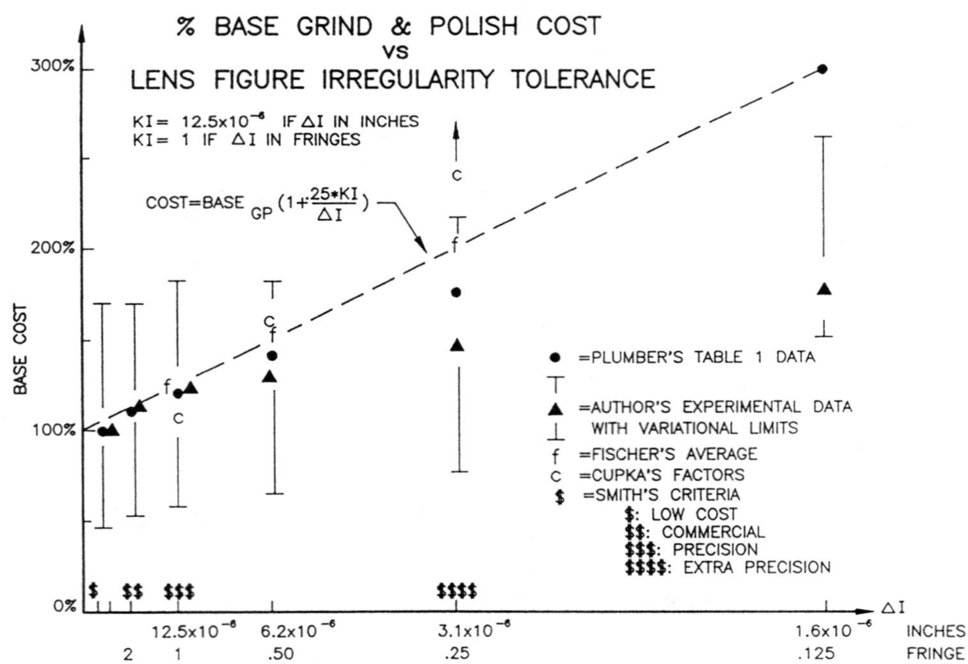

FIGURE 6: IRREGULARITY TOLERANCE ON LENS FIGURE (PLOTTED ON A RECIPROCAL SCALE)

This is what we now use as seen in Fig. 5. The graph is plotted as cost versus the reciprocal of the delta sagitta dZ which is a function of delta radius dR and R and d. The delta R and R alone are not enough to estimate the real cost of the tolerance, the surface diameter must also be taken into account. It might be more appropriate to refer to the sagitta tolerance of a surface than the radius tolerance, but shops are used to seeing the radius tolerance spelled out on a drawing. Equations 9 and 10 can be used to convert either way between dR and dZ (typically in inches).

$$dR = dZ * 8*(R/d)\hat{}2 \tag{9}$$

$$dZ = dR * .125*(d/R)\hat{}2 \tag{10}$$

The cost function chosen is based on Thorburn's (11) statement that a tight tolerance on sagitta is .000005 inch and .0001 inch is loose.

Figure 6 shows the impact of surface figure irregularity tolerances. We described the earlier experimental data in Ref. 2. We have added the points from Fischer and Cupka and a slight reformulation and labeling. The data from the various sources seems sufficently consistent and is reasonably represented by the chosen function.

Figure 7 shows the effects of the diameter to thickness ratio of a lens. This has to do with the flexibility of the part when trying to hold a good surface figure and sometimes the temperature effects when working thin negative lenses. Our simplistic treatments of this to date are probably not as adequate as we would like. This might be worthy of a paper by itself. Fischer (7) reported a much less severe effect than our previous work (3). In our work we estimated that the effect was inversely proportional to the flexibility of a disk which goes as the cube of the diameter to thickness ratio d/T. We will now accept (conditionally) Fischer's data collection as possibly more representative and we have fit our function to a compromise between our old data and that of Fischer where both are seen in Fig. 7. There is little doubt that plano-plano windows can be worked very thin by contacting them to rigid plates which lend the effect of their d/T ratio. In the case of lenses, this is not the practice however. Structural shape and thermal effects, etc. need to be studied in more detail. We will use this function until we or someone else can refine it further.

Figure 8 deals with the cost of center thickness tolerances. A major effect here is that once a thickness goes under the tolerance limit it is lost and must be replaced. We believe that this is what is reflected in the radical change in cost in Plummer's data as a .0005 inch tolerance is approached. At some point, a given tolerance would not allow a scatch or pit to be ground or polished out and the part would be lost due the combination of the thickness and scratch and dig specifications. Plummer's and Fischer's data are in reasonable agreement, and Cupka only

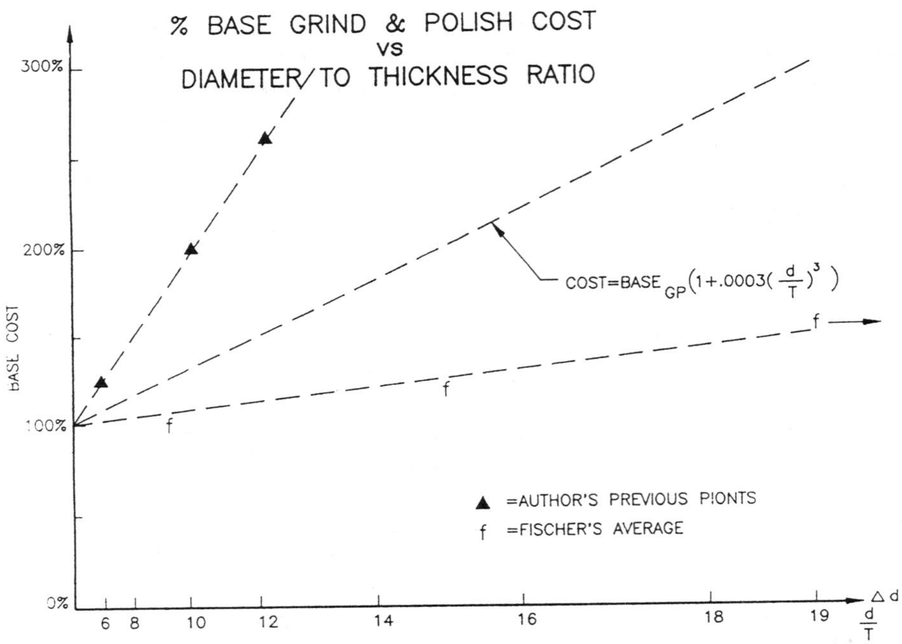

FIGURE 7: COST EFFECT OF DIFFICULTY IN OBTAINING GOOD FIGURE IRREGULARITY DUE TO FLEXURES FROM DIAMETER TO THICKNESS RATIO OF PART. (CUBIC SCALE)

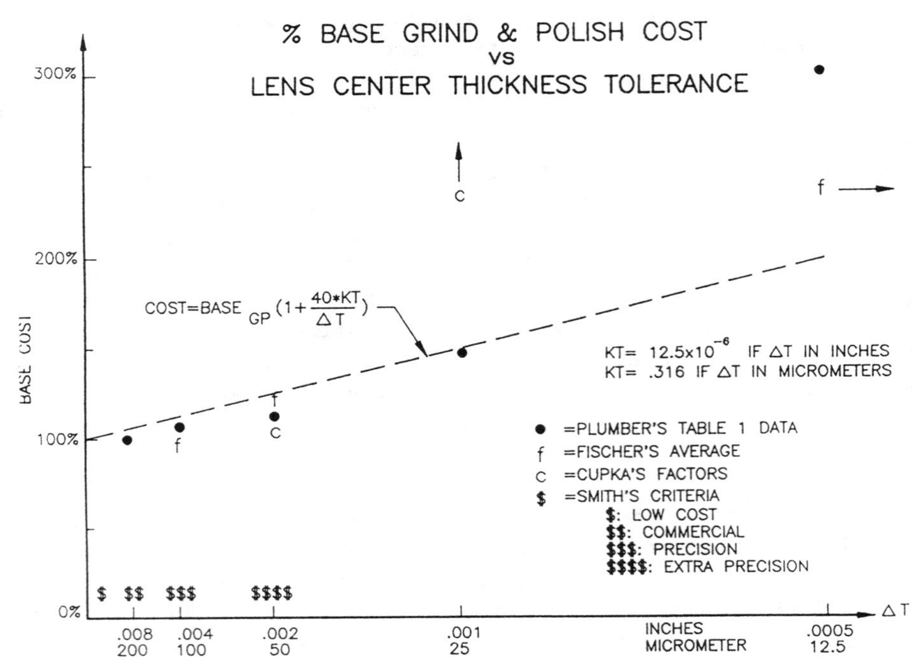

FIGURE 8: LENS CENTER THICKNESS TOLERANCE (PLOTTED ON A RECIPROCAL SCALE)

divided the costs into above and below a .002 inch thickness tolerance.

Figure 9 shows the effects of scratch and dig tolerances. The various data cluster reasonably near the very simple function that we choose to employ.

Figure 10 shows the simple function of cost versus glass stain code that we fit to our previous data and Fischer's. We might comment that we feel that even the 5 code glasses can be worked on a fixed price basis now, but at a significant added cost.

The "polishability" factor should also be applied as a cost impact. Each shop has its own experience that pyrex takes more time to grind and polish than BK7 and fused silica takes longer than pyrex, etc. In an attempt to include this factor, we have collected the estimated time factor for a variety of materials as compared to BK7. Some suggest that germanium has the same polishing time as BK7, and others think it takes significantly longer. This will, of course, vary from shop to shop and the procedures used. The author believes that typically germanium lenses will polish to spec in about the same time as BK7, but that the typical polish spec may be 160/100 for Ge and 80/50 for BK7. Therefore, the Ge will take longer to meet all of the same specs as a piece of BK7. The numbers in Table 1. reflect the author's best estimate based on a variety of inputs, but each shop needs to examine which factors to use in their own case. The grind and polish time will then be multiplied by this material polishability factor P.

TABLE 1.

Polishability Factor of Various Materials

Material	Factor
BK7	100%
SF56	120%
Pyrex	125%
Germanium	130%
Fused Silica	140%
Zerodur	150%
ZnS, ZnSe	160%
FK2, BaF2, Amtir	170%
LaKN9, LaFN21	200%
Electroless Ni	250%
CaF2, LiF	275%
MgF2, Si	300%
Electrolytic Ni	350%
Ruby	700%
Sapphire	800%

Once the lenses are fabricated, they are typically mounted in metal cells. We have to coordinate the tolerancing of the metal and the glass to get the desired results. Similar cost versus tolerance curves can be developed.

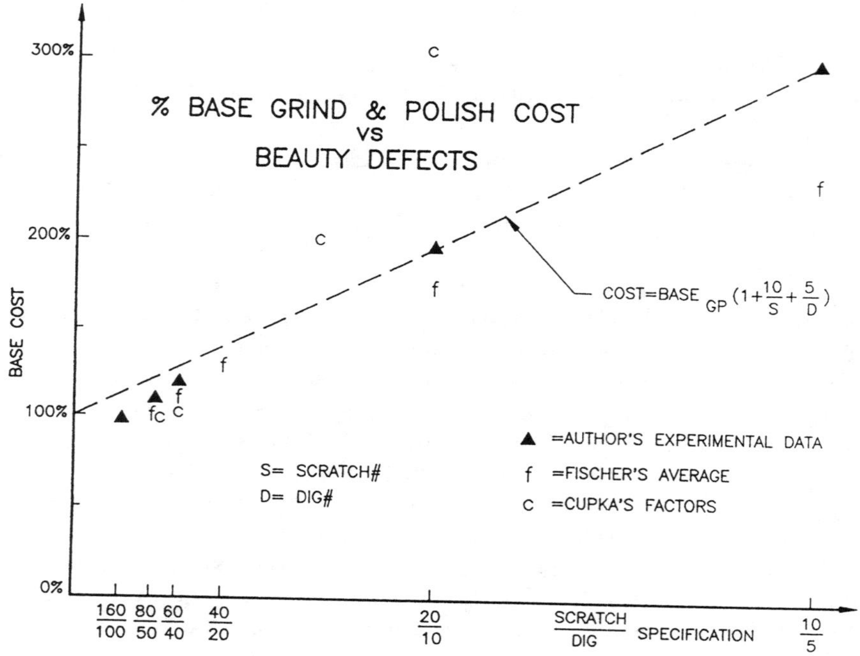

FIGURE 9: EFFECT OF SURFACE FINISH REQUIREMENTS

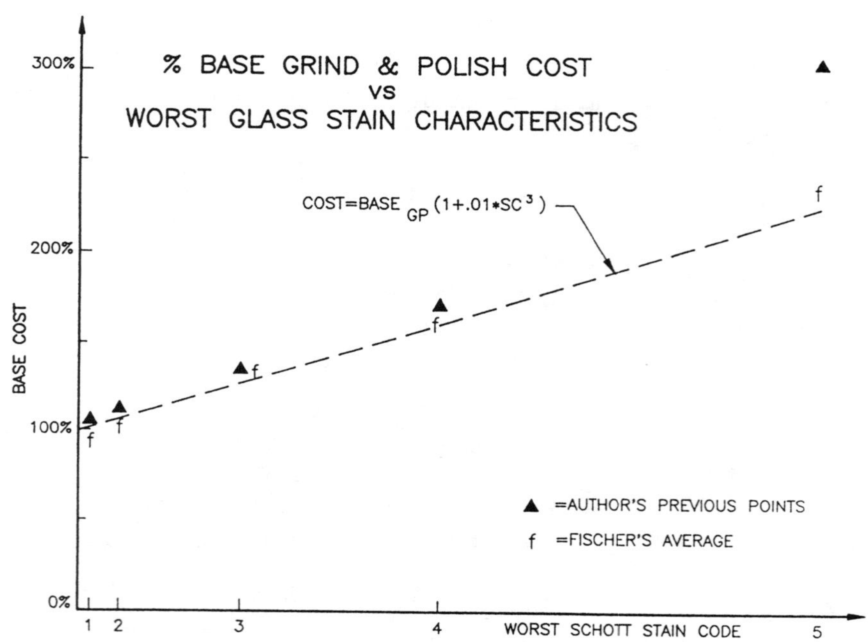

FIGURE 10: EFFECT OF GLASS STAIN CHARACTERISTICS

Figure 11 shows the cost versus lens cell diameter tolerance for both manual and automatic machining. The same curve applies to the length tolerances between bores as in spacers. Once an automatic is setup, the parts repeat within the capability of the machine with little change in the cost versus tolerance. The manual operations are more and more labor intensive as the tolerances increase.

Figure 12 shows the cost versus tolerances of bore concentricity and length run-out such as tilt in a spacer. There is no difference from manual to automatic here. However, the big difference here is whether both bores are cut without removing the part from its holder (chuck). If so, then the concentricity will be limited only by the accuracy of the machine. If the part must be rechucked for the second bore, much more time is consumed to hold a tight tolerance in the rechucking or mandrill type operation. There is a strong motivation to design for single chucking as much as practical.

We will use the results of Figs. 11 and 12 when we come to distributing the tolerances in the instrument design. The base cost for the machining operation will be needed to multiply times the metal tolerance factors in this process. For simplicity we will ignore setup costs and use a base machine fabrication cost of 6 units per bore or spacer cut in the same denomination as those used for the lenses above. For an automatic, we will use 4 units. This aspect could be made much more complex and refined, but this approximation is adequate to allow us to properly distribute tolerances in the later sections, but not estimate total machining costs.

These cost versus tolerance and cost versus other parameter functions are the major ones influencing the optical component fabrication cost. Now that we have them reasonably characterized, we can apply them to the base costs given in Sec. 2 to estimate the total component cost as specified. We can also find (to a sufficient approximation) the change in total component cost with a change in any of the parameters or tolerances. This, coupled with the sensitivity of the system performance characteristics to each of the parameter and tolerance changes, will then allow us to distribute the tolerances in such a way as to minimize the system cost while maintaining the required successful product yield. This is the major goal toward which we now strive.

4. TOTAL LENS COST ESTIMATION

When we combine the base costs given in Eqns. 2, 3, and 7 with the tolerance and other factors on Figs. 3 through 10 and Table 1, we get the total lens cost estimation given in Eqn. 11. This obviously cannot be scientifically rigorous, but is only a practical estimate for engineering or business purposes. The detail factors will vary from shop to shop and time to time. It is the author's belief that this sort of an estimation formula will be more accurate and consistent than most estimators or shop cost records now.

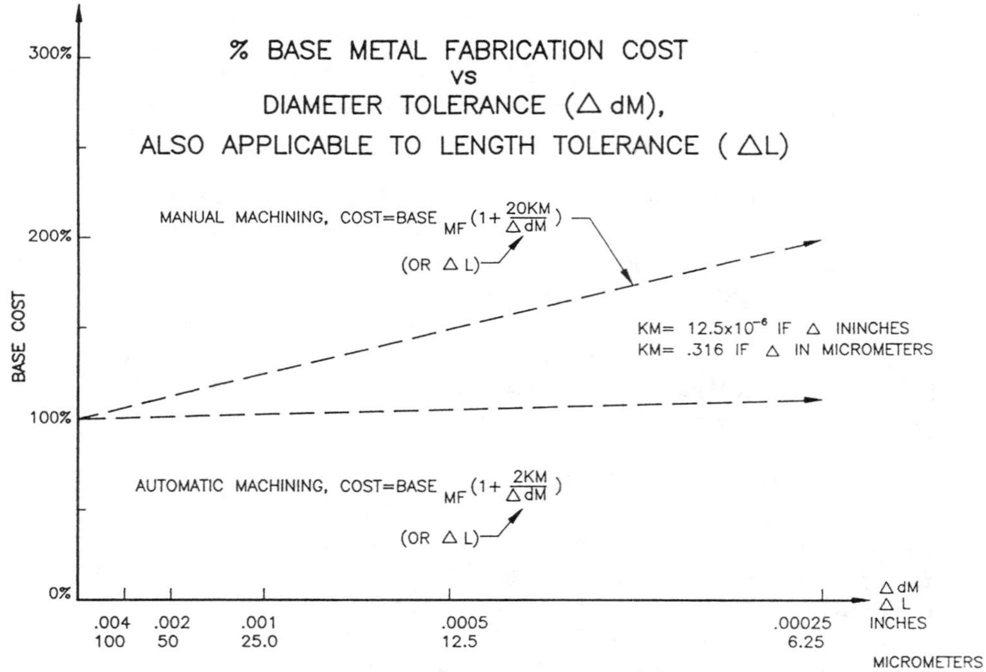

FIGURE 11: LENS BORE DIAMETER OR LENGTH TOLERANCES (PLOTTED ON THE RECIPROCAL SCALE)

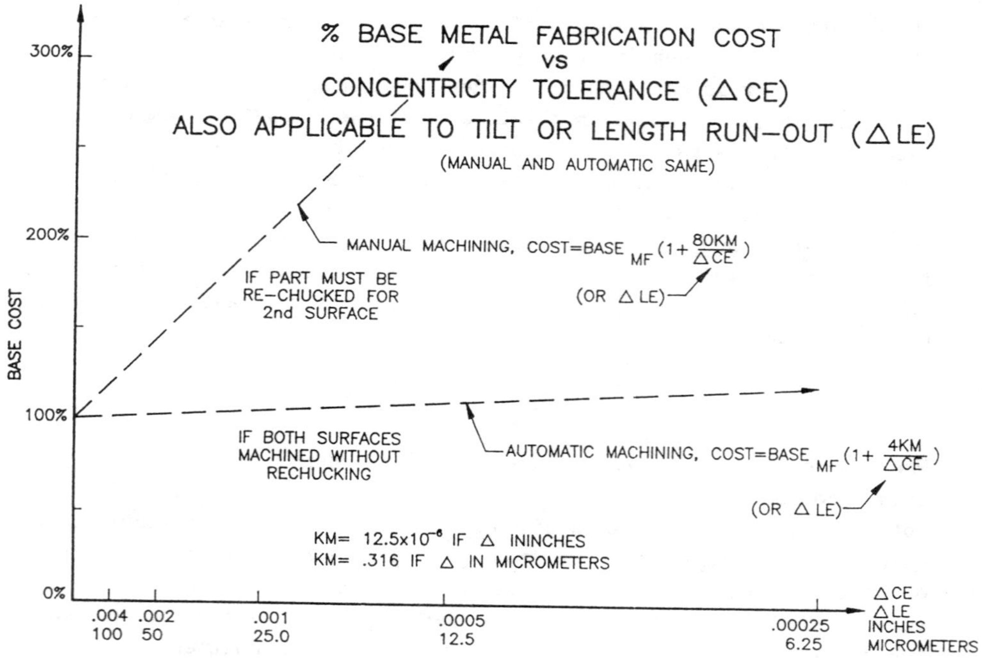

FIGURE 12: LENS CELL BORE CONCENTRICITY OR LENGTH RUN-OUT TOLERANCE (PLOTTED ON RECIPROCAL SCALE)

TOTAL LENS COST= MT=

(GENERATING) + MG

(PART SETUP) + Y*14

(SIDE 1, G&P) + (P*Y*14/N1)*((1+.25*KI/△ I1)*(1+(R1/d)^2 *8*KZ/△R1+.0003*(d/T)^3)
+40*KT/ △T*(1+10/S1+5/D1)*(1+.01*SC^3))

(SIDE 1, G&P) + (P*Y*14/N2)*((1+.25*KI/△ I2)*(1+(R2/d)^2 *8*KZ/△R2+.0003*(d/T)^3)
+40*KT/ △T*(1+10/S2+5/D2)*(1+.01*SC^3)) (11)

(CENTERING) + CE*(1+10*KD/△d+.145*KW/△A)

Although Eqn. 11 is extensive, it is not particularly complex. Table 2 sumarizes the parameters of the equation.

TABLE 2

MT	total lens cost estimate per piece (in relative units)
MG	milling/generating cost from Eqn. 2
LM	number of parts milled in one lot setup
Y	yield factor of parts started/parts acceptable
P	polishabilty factor from Table 1
N1	number of parts/block for R1 side from Eqns. 4 or 5
N2	" " " " " R2 " " " " " "
R1	radius of side 1
R2	" " " 2
G	diameter of largest block to be used for grind and polishing
d	diameter of lens
T	thickness of lens
S1	scratch number spec for side 1
S2	" " " " " 2
D1	dig " " " " 1
D2	" " " " " 2
SC	worst stain class of glass type
K's and delta's as on Figs. 3 to 10	
CE	centering and edging base cost per Eqn. 3
C	number of chamfers on the lens
F	" " flats " " "
LC	number of lenses centered and edged in one lot setup

The generating and certain part setup costs are virtually independent of the tolerances of the part. The centering and edging operation and tolerances are virtually independent of the grinding and polishing operations and tolerances. The operations and tolerances of the two radii of the lens are in most senses independent. This is all reflected in Eqn. 11. There has been some logic applied to whether the tolerance factors are added or multiplied and to what they are applied. The irregularity cost factor multiplies times the base cost of grinding and polishing each side. That irregularity cost is further multiplied by the radius tolerance cost factor plus the flexure (d/T) difficulty factor which make the figure irregularity harder to achieve. The decision to add the d/T influence to the radius tolerance influence was made on the basis that they do not significantly effect each other, but they both effect the cost of meeting the irregularity specifications. The stain class interacts with and effects the scratch and dig requirements and they both increase the difficulty of holding the thickness tolerance, but they do not effect the irregularity degree of difficulty. The edging operation is affected by the diameter tolerance and the wedge or deviation tolerance. We will discuss the unique interaction of the diameter and centering tolerances in the next section.

This formula then reduces the estimation of the fabrication cost of most lenses to a clerical task of entering the parameters from the lens drawing into a simple computer program. This could also, in principle, be worked into a CAD program to allow the designer to see immediately the cost impact of the design and tolerances. The key factor for tolerancing, however, is that we can find the partial derivatives of Eqn. 11 with respect to each of the reciprocal tolerances for use in finding the minumum cost tolerance distribution for a system as discussed in Sec. 6.

5. INTERACTIONS OF LENSES AND MOUNTS

A lens system typically consists of lenses in metal mounts. The mounts are bores that closely fit the lens diameters with "lands" or rings of contact between one or both of the radii and a metal locating surface in the direction of the optical axis. The relationship of one lens to another will be determined by the spacing dimensions and tolerances of the mount and the concentricity and tilts between the locating surfaces. The mounting metal (or other material) must be toleranced to be compatible with the glass tolerances. In this section we will address how we deal with this task.

There are several interesting references (11,12,13) dealing with tilts, decentrations, and rolls (by whatever names) but none seem to have addressed them in an analytic form appropriate to our cost minimization goals.

5.1 Lens Centering

Figure 13 shows a lens in a cell bore where only centering factors are considered, not tilts. It is easy to evaluate the effect or sensitivity of decentering a lens from the intended optical axis in most lens design software. This decentering in a system is the sum of several factors. The decentering of the optical axis of the lens with respect to the outside diameter of the lens is what the optical shop works on. The centering of the mounting bore with respect to the ideal axis is what the machine shop works on. There needs to be some clearance for assembly to insert the lens into the cell, we will call this f. The tolerances of the lens diameter d and the bore diameter dM, delta d, and delta dM give rise to more potential clearance. These clearances will allow the lens to move to extreme

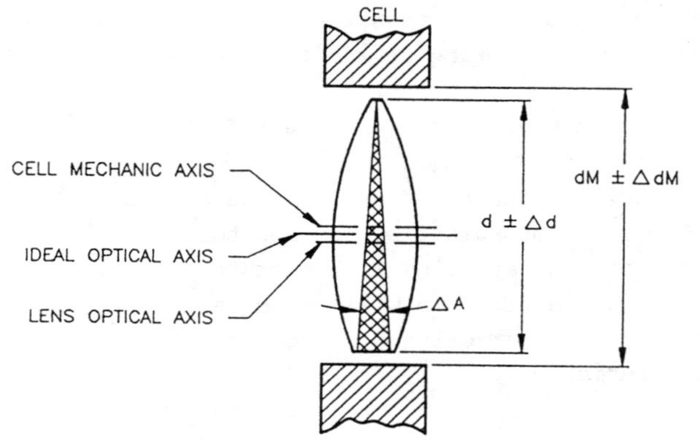

FIGURE 13: DECENTERING FACTORS OF A LENS IN A CELL

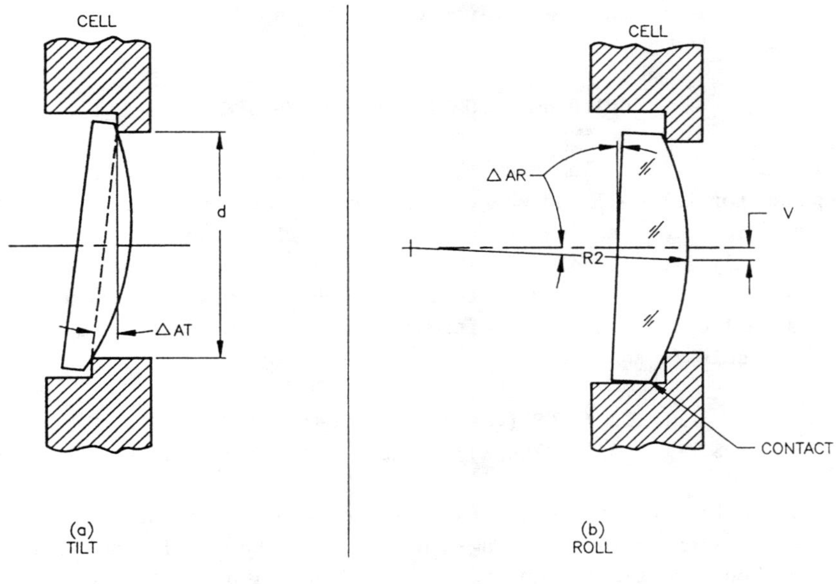

FIGURE 14: TILT AND ROLL FACTORS OF A LENS IN A CELL

positions in the cell, such as fall to the bottom, and cause more decentering. Equation 12 expresses the possible total decentering td as a function of the lens decentering dc and the factors mentioned above.

$$td = dc + (f + delta\ d + delta\ dM)/2 \qquad (12)$$

The costs versus tolerances have been defined and quantified in Sec. 3. We can see that the least cost distribution of tolerances for a lens in a bore with respect to total decentering will dictate a certain ratio between the dc, delta d, and delta dM. We will then reduce the tolerancing of the "set" to a function of only delta d since the dc and delta dM can be taken as the dependent variables.

The dc of the lens is measured by the wedge delta A in Fig. 13 (here we work in milliradians) or in the arcminutes deviation it causes which we relate by the factor KW in Fig. 4. We can show that dc can be expressed in terms of delta A as in Eqn. 13. (Some RSS possibilities will be ignored here.)

$$dc = (R1*R2/(R2-R1))/(1000*KW) * delta\ A \qquad (13)$$

Delta A in terms of delta d is found to be

$$delta\ A = SQR\ (.145*KW/10*KD) * delta\ d. \qquad (14)$$

If we call the base machining cost MF, then delta dM in terms of delta d is given by Eqn. 15.

$$delta\ dM = SQR\ (MF*20*KM/CE*10*KD) * delta\ d \qquad (15)$$

The minimum fit clearance factor f has to be determined by the assembly plans for the cell and whatever allowances are made for differential thermal expansion. At nominal temperature, it will allow a shift in an otherwise perfectly fitting cell of f/2. This decentering will have to come right off the top of the total decentering budget for this lens+bore set leaving the residual budget to be partitioned amoung deltas d, dM, and A. We express the result in Eqn. 16.

$$\begin{aligned} td-f/2 = ((1/2 &+ SQR\ ((MF*20*KM)/(CE*10*KD)) \\ &+ SQR\ (.145/(10*KD*KW)) * ABS(R1*R2/((R2-R1)*1000))\ ) * delta\ d \qquad (16) \end{aligned}$$

This factor times the delta d is to be used in the tolerance allocation process with the decentering sensitivity to determine delta d. The delta d can then be used to assign delta A and delta dM (which are dependent on delta d) in a secondary operation using Eqns. 14 and 15. The diameter of the cell bore dM is also determined by this process as expressed in Eqn. 17.

$$dM = d + f + delta\ d + delta\ dM \qquad (17)$$

Although Eqn. 16 might not appear to simplify anything, it is a relatively straightforward application of the cost and geometric factors which allow us to properly spread the cost and tolerances in the lens and its cell.

5.2 Lens tilt and roll

An otherwise perfect lens might be tilted with respect to the system's ideal optical axis because the metal locating surface of the cell is tilted by an angle delta AT as shown in Fig. 14a. This can be simply dealt with using the cost versus tilt curve in Fig. 12 and the sensitivity of the system to tilt of the whole lens. This assumes that the tilt is not otherwise limited by such factors as a retainer on the other side of the lens or the fit of the cylindrical lens diameter into the cylindrical cell bore which prohibit that much tilt.

The perfect lens might also "roll" in an oversized bore as shown in Fig. 14b. This shows that the left hand surface tilts while the right is correctly located against the "perfect" cell. The lens will roll about the center of curvature of the right hand surface R2. The left hand surface, which we have shown plano for clarity, will tilt through an angle of delta AR which is approximately V/R2 radians. We can show that:

$$V = (f + \text{delta } d + \text{delta } dM) / 2. \qquad (18)$$

We know delta dM in terms of delta d from Eqn. 15. This allows us to express delta AR1 of the left surface R1 as a function of delta d in Eqn. 19.

$$\text{delta AR1} = (f + (1 + SQR((MF*20*KM)/(CE*10*KD)))/ABS(2*R2) * \text{delta } d \qquad (19)$$

This and the system performance sensitivity to a tilt of R1 will allow us to allocate the tolerance budget for a tilt of R1. However, note that controlling this requires the control of delta d which is already determined by the decentering requirements! Generally, one or the other will be the more demanding on delta d. It would appear that we should find which is the more stringent and use it to determine the tolerance allocations. The other would still make some contribution to the error budget, but not be independently determined. As in the tilt case, the ability of the element to roll to the full extent indicated in Eqn. 19 may be otherwise restrained by (but might also be caused by) a retainer ring, etc. Since this can be the case, it may be appropriate for the designer to use some judgement in the application of tilts and rolls after looking at the particulars of the sensitivities and the mounting designs to decide if they should be given the full force of the equations. This unfortunately seems to bring us back a bit from the engineering to the "art" in tolerancing.

Having now dealt with the necessary elements of the cost versus tolerances and other parameters and the applicable geometric simplifications, we will next proceed with how we can apply the foregoing to distribute the tolerances for the maximum performance and cost benefit.

6. ASSIGNING TOLERANCES FOR MINIMUM COSTS, AN EXAMPLE

We described the general principles of distributing the tolerances for minimum cost in our first work (1) on this subject. Adams (6) made some significant additions to our work which we shall draw from below. If there is more than one performance criterion that must enter into the tolerancing process the solution to the equations is somewhat involved, but it can be done. However, many problems, including the one which we will use as an example, have one performance criterion which dominates all of the others as relates to the tolerances. That is, if the tolerances are chosen to meet that performance requirement, then all of the others will also be met. This reduces the computation considerably and makes it easier to visualize. For the balance of this paper, we will use the single requirement case with the understanding that it can be extended to multiple criteria by the methods of the previous papers as needed.

Figure 15 shows the Multi-Focal Length Tracking Telescope (MFLTT) that we will use as an example of the tolerancing process. It has a catadioptric telescope section of 300mm aperture and about 2000mm focal length with a 25% central obstruction due to the secondary mirror. The telescope image is then collimated by a focus lens set. The afocal beam is then imaged by one of three relays to the final focal plane. These relays are alternately positioned in the beam to give the effective focal lengths of 1000, 2000, and 4000mm. Before the final focal plane there is an auto-iris system of variable neutral density filters and a reticle projection unit (AIR). There is also a 500mm system which is partially separated from the others to allow a larger field of view. The 500mm system is folded into the same optical path as the others by a movable prism. There are sealing windows in front of the telescope and the 500mm lens.

In this complex telescope system example, the most stringent requirement of the system is the on-axis MTF at 30 lp/mm. When this is satisfied, the off- axis MTF at 30 lp/mm, the on- and off-axis MTF at 10 lp/mm, and the boresight, etc. requirements will all also be satisfied without additional or adjusted tolerances being required. To be consistent with our previous report (1), we will designate this performance requirement by E which represents the maximum permissible error in MTF from ideal for the system. We will actually convert this E to units of RMS wavefront error (RMSWE) for simplicity. The total E will eventually be partitioned among each of the tolerances which affect it.

To make a tractable example for this paper, we will partition the total E among the various sections of the system. The partial E will then be allocated to the parameter tolerances within one section based on the cost

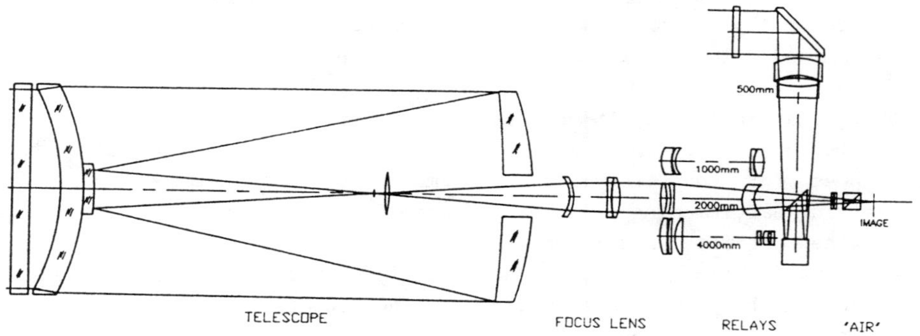

FIGURE 15. EXAMPLE SYSTEM MULTI FOCAL LENGTH TRACKING TELESCOPE

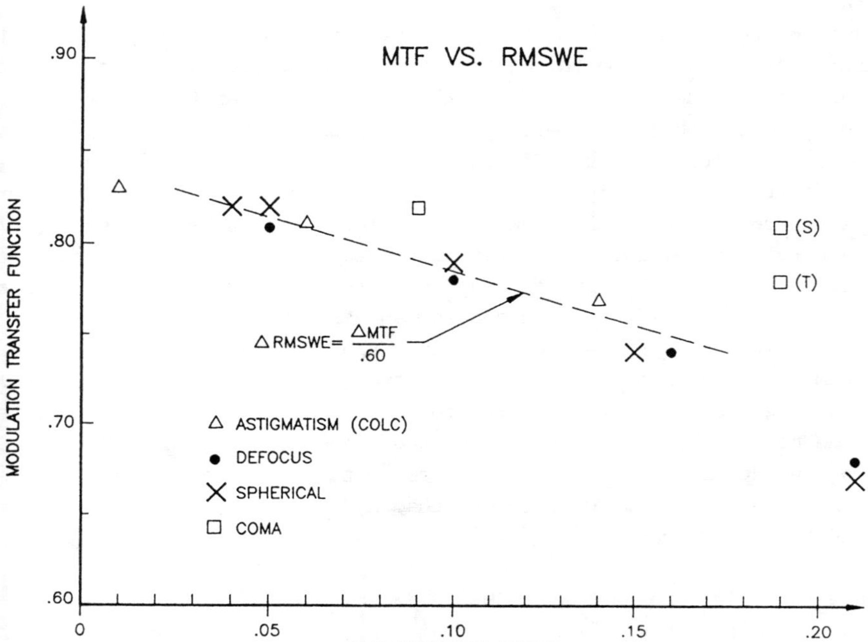

FIGURE 16: DIFFRACTION MTF AT 30 lp/mm VS. WAVEFRONT ERROR FOR 2000mm, f/8 TELESCOPE

minimizing technique. This "divide and conquer" approach is needed here, plus any justifiable simplifications in general, to reduce the overwhelming magnitude of typical problems that have hundreds of component tolerances to be determined. The way to eat an elephant is one bite at a time!

In the final analysis, it is most correct to tolerance the whole optical train from object to image in one operation. This will truly allocate the tolerances for the required performance at the minimum cost. The simplifying partitioning will cause some deviation from the ideal result unless the estimate used in the partitioning was exactly correct. In the example used here, it would be best to tolerance the 4000mm from end-to-end, but the data would be too cumbersome to make a good illustration in this paper.

6.1 Simplifying approximations

The MTF of a system is often the best performance measure to use because it most directly relates in many cases to the performance of an overall system when it is used. It is, however, not generally possible to measure the MTF effect of each component lens of a system in the production process. The characteristics that are readily measured on a lens were discussed in Sec. 3, such as irregularity, radius, centration, etc. We chose to work here with the effects of each tolerance on RMSWE, because we think that it can be reasonably related to the system MTF.

We estimated the reduction in MTF per wave of RMSWE at 30 lp/mm for the 2000mm effective focal length, f/8 system by introducing errors into the system and evaluating it for MTF and RMSWE. With parameter deviations, we produced defocus, spherical aberration, coma, and astigmatism. Defocus was introduced by evaluating the system at different focal planes from the best focus. Spherical aberration was introduced by varying the y^4 aspheric coefficient from the nominal. Coma was evaluated in an "equivalent" (f/8, 2000mm) parabolic mirror system with the stop at the focal plane so that astigmatism was zero. The system was evaluated off-axis to introduce coma. Lastly, astigmatism was introduced in an "equivalent" (f/8, 2000mm) Ritchy-Chretien telescope where coma and spherical aberration were zero. The system was evaluated off-axis to introduce astigmatism. The results appear in Fig. 16. All of the data forms a reasonably consistent pattern except the coma. We do not presently understand this anomaly which may be worthy of a separate study. However, since the effects of coma are less severe than the others, we will ignore them and use the conservative numbers indicated by the others.

Therefore we will use delta RMSWE = delta MTF/.60 as the amount of reduction in MTF that will be accompanied by a corresponding RMSWE. This will allow us to work with the effect of tolerances on the RMSWE which we will assume are quasilinear in the regions where we are applying them. This may be a conservative estimate, but we would like to err on that side.

Another approximation that we will draw upon comes from Smith (10) where

$$\text{RMS} = (\text{ Peak-to-valley }) / 3.5 \qquad (20)$$

approximates the RMSWE expected from most types of error. It would seem that sharp departures over a small portion of the wavefront would violate this rule, but those are not usually encountered. We made a small investigation of our own by comparing the RMS and P-V data on many interferograms from a ZYGO interferometer. This leads us to think that the factor in Eqn. 20 might be more like 7 than 3.5 when small irregularities such as those on surfaces intended to be spheres are examined. For this example, however, we will use Smith's value.

In the example design, the apertures were selected at the first order stage to yield the required MTF when the diffraction effects of the obstruction plus one quarter wave of design and frabication errors were taken into account. This is not much error to spread across the many elements from the object to the focal plane. One help is the fact that certain compensating alignments can be made at assembly since the systems will not be made in very large quantities. We will use the approximation of Eqn. 20 to establish a preliminary total error budget of 0.071 RMSWE (1/4 wave P-V) from all sources in laboratory tests. In the final application, obviously, atmospheric and other effects might influence the results further.

6.2 Error budgets

We need to now decide how to distribute this 0.071 RMSWE among the many facets and tolerances of the system. Smith (10) describes how to work with the root-sum-of-the-squares (RSS) to combine error effects. McLaughlin (14) shows that RSS tends to be too pessimistic and Smith (10) himself concludes that it may err on the conservative side. McLaughlin shows that the total system error will tend to be 0.42 times the RSS prediction if the fabrication errors have a Gaussian distribution which is truncated at the 2 sigma level. Although there is a major move at this time in industry to apply 6 sigma tolerancing, we believe the 2 sigma to be appropriate in this case where individual adjustment and testing is required. We will therefore use McLaughlin's 0.42 factor for the fabrication errors.

To simplify the example, we will partition the 4000mm path of the system. In looking at Fig. 15, we count 32 surfaces through the 4000mm optical path. The authors chose to emphasize the sensitivity effects of mirrors by counting them twice to give 34 as the surface count. Of this 34, 8 are in the telescope, 8 in the focus system, 12 in the 4000mm relay, and 6 in the AIR. The other paths are less complex. This one will be the critical path and set the pace for the telescope, focus system, and AIR tolerances. We will allocate the budget to the four sections of the 4000mm path (telescope, focus, relay, AIR) in proportion to the square root of the number of surfaces in the section divided by the total number of surfaces. This is an engineering estimate of the relative influence of each section.

The division of the system into these sections is also because they can be tested independently by section for RMSWE in production. Figure 17 shows the error budget broken down this way. The top level requirement was determined above to be effectively .071 RMSWE. We know from the design stage that the disign has used up .030 RMSWE. Another analysis indicates that the effects of alignment focus errors and the laboratory environment should be on the order of .009 RMSWE. This leaves .0637 RMSWE to RSS with the other two (three) parts of that level of the budget to give .071 RMSWE. From McLaughlin's information and the assumption of Gaussian errors, we then divide the fabrication budget of .0637 by .42 to give 0.1517 RMSWE which can be distributed over the four sections of the 4000mm system. The bottom four boxes of the budget in Fig. 17 show how these work out when the above argument is applied. We will work through the simplest section, the focus lens with .0736 RMSWE budget, as an example of the procedure for tolerance distribution to give the minimum cost while meeting the performance requirements (to within some statistical uncertainty).

6.3 Derivatives of costs with respect to tolerances

In the assignment of tolerances for minimum cost as we will show below, it is necessary to have the partial derivatives of the total cost with respect to the reciprocal of each tolerance. These are basically derived from Eqn. 11 and the metal tolerance costs in Fig. 12. Since we showed that the metal diameter and lens centering tolerances can be made dependent on the glass diameter tolerance, we only have five types of tolerances to allocate in the framework which we have been discussing. These are irregularity, radius, thickness, diameter, and tilt of a lens due to the errors of the cell. Roll is shown in Fig. 14b and can be derived from delta d. Tilt is just the cell run-out parallel to the optical axis (delta LE of Fig. 12) divided by the diameter d of the lens. We will call the partial derivatives of the total system cost with respect to the reciprocal of these tolerances \$I, \$R, \$T, \$d, and \$L. The \$L applies to both the metal bore diameters and axial run-out. The first three will be functions of the base grind and polishing costs with a common factor that we will call BP which is defined in Eqn. 21.

$$BP = P * Y * 14 / N\# \qquad (21)$$

The N# is the N for the given surface number just as we will use delta I#, delta R#, etc. for those values associated with that surface number. The \$d has as a factor of the base centering cost CE while \$L has as a factor the base machining cost MF. There are additionally three "fudge" factors FI, FR, and FT associated with \$I, \$R, and \$T which can be taken as unity for a simplifying assumption or calculated in each case as we will explain below. Equations 22 through 26 give these partial derivatives which will be needed to allocate tolerances.

$$\$I\# = BP * FI * .25 * KI \qquad (22)$$

$$\$R\# = BP * FR * 8 * KZ \qquad (23)$$

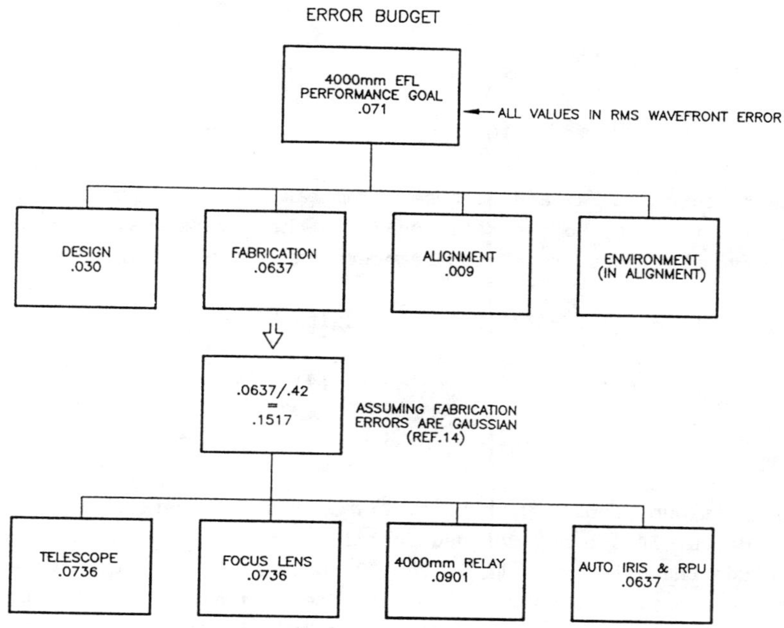

FIGURE 17. ERROR BUDGET ALLOCATION, 4000mm EFL OF MULTI-FOCAL-LENGTH TRACKING TELESCOPE

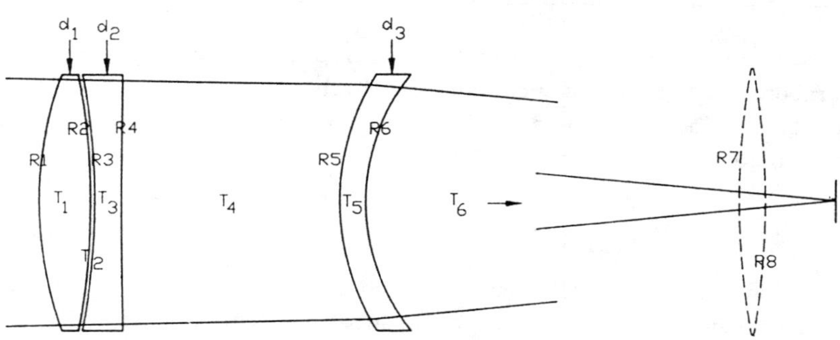

FIGURE 18. FOCUS LENS SET EXAMPLE

```
$T  = BP * FT *  40 * KT        (24)

$d  = CE      *  10 * KD        (25)

$L  = MF      *  80 * KM        (26)
```

The factors FI, FR, and FT come from taking the partial derivatives of Eqn. 11 with respect to the reciprocal of the tolerances. They are due to the modifying effects of other parameters and are given in Eqns. 27,28,and 29.

```
FI = ( 1 + 8*KZ*(R#/d)^2 /(delta R#) + .0003 * (d/T)^2 )     (27)

FR = ( 1 + .25 * KI / (delta I#) )                           (28)

FT = ( 1 + 10/S# + 5/D# ) * ( 1 + .01 * SC^3 )               (29)
```

There is one problem with Eqns. 27 and 28, however. The values of delta R# and delta I# are not defined until after the tolerancing process. We can include the effects of FT in Eqn. 24 because all of its parameters are known at the start. We chose to approximately include the effects of FI and FR after the tolerances are calculated by dividing the resulting tolerances for irregularity and radius by the cube roots of FI and FR respectively. The reasons for this will be more apparent from Eqn. 32 in Sec. 6.4. The other alternative would be substituting FI and FR back in an iterative procedure.

We now only need to address the simple allocation equations and procedures and we can finish the task of finding the least cost tolerances.

6.4 Tolerance allocation process

We showed previously (1) that the total error E was the sum of all of the contributions from each error source which is the product of that tolerance value t(i) times the sensitivity S(i) of the performance to variations of that parameter as seen in Eqn. 30

```
E = sum over all i ( S(i) * t(i) )        (30)
```

We showed that the tolerances could be distributed for minimum cost by applying Eqn. 31 to each tolerance in turn. The A(i)'s are the coefficients of the derivatives of cost given in Eqns. 22 through 26. In Eqn 31, the sum of all the square roots of the products of the cost coefficients A(k) and the sensitivites S(k) is divided into the total error budget E to get a constant which is multiplied by a function of the cost and sensitivity of each tolerance.

$$t_i = \sqrt{\frac{A_i}{|S_i|}} \; \frac{E}{\sum_{K=1}^{n} \sqrt{A_K \, |S_K|}} \tag{31}$$

Adams (6) pointed out that Eqn. 31 gave the solution which represented the tolerances all going to the worst case limit condition, and this is obviously too severe. He showed that the RSS condition would be satisfied by the tolerance distribution if Eqn. 32 was applied instead. We are indebted to Adams for this contribution and will apply it to the problem at hand as the most appropriate solution.

$$t_i = \sqrt[3]{\frac{A_i}{S_i^2}} \; \frac{E}{\sqrt{\sum_{K=1}^{n} \sqrt[3]{(A_K S_K)^2}}} \tag{32}$$

The allocation is then very straightforward in the single E case. It is only necessary to develope a table of A(i) and S(i) for all appropriate i and process the data in accord with Eqn. 32. Table 3 gives these values for the example case and the resulting t(i) which we seek are in Table 4.

The example case has a total of four lenses which we illustrate in Fig. 18. The field lens is close to the focal plane and has negligable sensitivity. We therefore remove it from consideration since we can assign minimum cost tolerances to it without affecting the rest of the task. This then reduces the example to a three element lens where the surfaces and dimensions are numbered as in Fig. 18.

The necessary data processing is very conveniently set up and done using a spreadsheet program for data entry and all of the necessary calculations. Table 3 starts with the parameters determined from the "lens drawing" or design parameters. The number per block for each side is determined from Eqns. 4 and/or 5. The sensitivities of the system performance to each parameter are determined from the lens design program. The base costs and cost versus the reciprocal tolerance derivative coefficients are calculated from Eqns. 22 to 26. The constant multiplier in Eqn. 32 is calculated. The individual factors from the individual sensitivities and cost derivatives are used to compute the tolerances for each of the toleranced parameters. The adjustments to the I and R tolerances are made for FI and FR as mentioned in the previous section.

Table 4 contains the resulting tolerances for which we have been working. It is the set of tolerances for each of the toleranced parameters which will give the least cost solution and meet the performance with some statistical "RSS" certainty. The assumption per Adams (6) is that the errors will be distributed about the norm in a Gaussian manner and the tolerance limits

TABLE 3

TOLERANCE DATA & COMPUTATIONS

E = 0.0736 RMSWE | EXAMPLE FOCUS LENS SET

SURFACE #	1	2	3	4	5	6
RADIUS	3.54	-6.504	-6.346	59.203	2.319	1.941
DIAMETER	2.35	2.35	2.31	2.31	2.04	2.04
THICKNESS	0.472	0.045	0.237	1.969	0.237	10.63
SCRATCH	80	80	80	80	80	80
DIG SPEC.	50	50	50	50	50	50
REFRACTIVE INDEX	1.5		1.6		1.5	
STAIN #	5.2	5.2	5.12	5.12	2	2
POLISHABILITY	3	3	1	1	1	1
CHAMFERS	1	1	1	1	1	1
FLATS	0	0	0	0	0	1
NUMBER/MILLING LOT	7		7		7	
CENTERING LOT SIZE	7		7		7	
YIELD	1.5	1.5	1.3	1.3	1.3	1.3
# PER BLOCK	6.580650	7	7	7	3	1
KI (fringes)	1	1	1	1	1	1
KZ (sag, inches)	0.000012	0.000012	0.000012	0.000012	0.000012	0.000012
KT (thk, inches)	0.000012	0.000012	0.000012	0.000012	0.000012	0.000012
KD (dia, inches)	0.000012	0.000012	0.000012	0.000012	0.000012	0.000012
KM (met dia, in)	0.000012	0.000012	0.000012	0.000012	0.000012	0.000012
KW (mrads wedge)	1	1	1	1	1	1
SI dRMS/FRINGE	0.05	0.05	0.1	0.083	0.05	0.067
SZ dRMS/INCH	78.125	176.471	411.765	250	66.667	220.588
ST dRMS/INCH	1	17	2	1	6	0.001
Std dRMS/INCH	250		150		41.67	
SLE dRMS/radian	375		667		42	
f, CLEARANCE	0	0	0	0	0	0
MF (MFG BASE)	6	6	6	6	6	6
BP (EQN 21)	9.573522	9	2.6	2.6	6.066666	18.2
CE (EQN 3)	7.497619		7.484285		7.394285	
A$I (EQN 22)	2.393380	2.25	0.65	0.65	1.516666	4.55
A$Z (EQN 23)	0.000957	0.0009	0.00026	0.00026	0.000606	0.00182
A$T (EQN 24)	0.004786	0.0045	0.0013	0.0013	0.003033	0.0091
A$d (EQN 25)	0.000937		0.000935		0.000924	
A$L (EQN 26)	0.006		0.006		0.006	
(A(K)*S(K))^2/3 I	0.242823	0.233025	0.161647	0.142763	0.179144	0.452939
(A(K)*S(K))^2/3 Z	0.177502	0.293258	0.225449	0.161647	0.117813	0.544196
(A(K)*S(K))^2/3 T	0.028397	0.180192	0.018904	0.011908	0.069181	0.000435
(A(K)*S(K))^2/3 d	0.380038		0.270026		0.114035	
(A(K)*S(K))^2/3 L	0.971424		1.442507		0.248018	
CONSTANT E/SQR(SUM) =		0.028503				
(A(I)/S^2)^1/3 I	9.853515	9.652704	4.020166	4.551835	8.463641	10.04277
(A(I)/S^2)^1/3 Z	0.005395	0.003070	0.001153	0.001609	0.005151	0.003346
(A(I)/S^2)^1/3 T	0.168561	0.024980	0.068771	0.109163	0.043853	20.87125
(A(I)/S^2)^1/3 d	0.002467		0.003466		0.008108	
(A(I)/S^2)^1/3 L	0.002629		0.001801		0.011863	
delta I (fringes)	0.280863	0.275139	0.114590	0.129744	0.241246	0.286257
delta Z (inches)	0.000153	0.000087	0.000032	0.000045	0.000146	0.000095
delta AT(tilt,rd)	0.000074		0.000051		0.000338	

TABLE 4
RESULTING TOLERANCES AND COSTS

TOLERANCES:

SURFACE #	1	2	3	4	5	6
delta R (inches)	0.003452	0.006652	0.002920	0.344788	0.001924	0.000851
delta I'(fringes)	0.288933	0.344375	0.128449	0.140610	0.249586	0.298252
delta T (inches)	0.004804	0.000712	0.001960	0.003111	0.001250	0.594910
delta AR(roll,rd)	0.000013	0.000007	0.000008	0.000000	0.000051	0.000062
delta td (inches)	0.000070		0.000098		0.000231	
delta LE (inches)	0.000176		0.000118		0.000689	
delta d (inches)	0.000041		0.000050		0.000106	
delta dM (inches)	0.000052		0.000063		0.000135	
delta A (mrad.)	0.001419		0.001715		0.003611	
delta A ('dev)	0.002442		0.003541		0.006212	
Metal Dia.=dM	2.350094		2.310114		2.040241	

BASE COST COMPUTATION BASE=MG+GP+CE

LENS #	1-2	3-4	5-6
MILLING = MG	17.40939	17.39075	17.27330
GR & POL = GP	39.57352	23.4	42.46666
CENT & EDGE=CE	7.497619	7.484285	7.394285
BASE COST TOTAL =	64.48053	48.27503	67.13425

COMPUTATION OF TOTAL LENS COST = MT

LENS #	1-2		3-4		5-6	
MILLING/GENERATE	17.40939		17.39075		17.27330	
PART SETUP	21		18.2		18.2	
GRIND & POLISH	22.62849	699.6288	13.67048	9.800878	18.49274	37.03262
CENTERING	795.6429		658.6013		312.9745	
TOTAL LENS COST	1556.309		717.6634		403.9732	

will be 2 sigma.

However, a significant problem appears in Table 4. Many of the tolerances are well beyond what can be achieved in normal practice; they are off the chart in Figs. 2 to 12! This is a disappointing result for the designer, but hopefully not the end of the road. Finding the problem at the design stage is not nearly as frustrating or expensive as finding it at the production stage.

The lower part of Table 4 shows the application of the above formulas to compute the base cost for each of the lenses and the total costs when the tolerances are included. The latter are about 15 times that of the easy base lens if they could even be made. "Off-the-chart" implies in most cases that it cannot be made or at least it would be much more expensive than the linear chart data would predict. We may have therefore identified here an impractical design. The designer then has the challenge of finding a design and/or an approach which will be less sensitive. The addition of more lens elements is not at all out of the question if they can reduce the sensivity enough. If the added cost of one or more elements reduces the tolerance costs sufficiently and all the lenses can be built, the total cost will be less than the first design. It might be possible to cement a doublet to get rid of a sensitive air-space. It might be practical to make a centering adjustable and/or add other assembly tricks (15). The designer can now evaluate the impact of design changes on cost by using the tools put forth in this paper. The "bottom line" in Table 4 can tell him if he has improved the situation or not.

What we see here is an example of the processes in Fig. 2 where the tolerance sensitivity analysis and distribution feeds into the producibility analysis which sends us back to the detail optical design for further work.

A complex system such as the whole 4000mm example system will have several times as many columns as Table 3, but the process is the same and relatively straightforward to apply. The most difficult aspect can be obtaining the sensitivites of the performance criterion to parameter variations. Existing lens design programs can do this with greater or less facility, but all should be readily modifiable to straightforwardly generate to data needed. This data generation is computer intensive and time consuming, but probably unavoidable. The tolerancing program described above takes only a few seconds on a PC to calculate the tolerances for the six surface case. It should be approximately linear with number of surfaces as long as only one performance criterion is to be considered. Multiple criteria would be more cumbersome to evaluate as we showed in our first work (1). However, Adams (6) shows that it is most likely that one criterion is all that is needed, and more than two is highly unlikely.

7. SUMMARY AND CONCLUSIONS

We had previously (1,2,3) shown the principles of how to assign tolerances to give the minimum production cost and we mentioned the possible application of the results to estimating total lens production cost. In this work, the previous data and principles have been refined and some of the results provided by Adams (6), Smith (10), Parks (8), and Fischer (7) have been incorporated. The lens cost estimating formulas mentioned in the earlier work (3) have been developed into useful tools. A new analysis is presented of the interdependency of the lens and cell diameter tolerances as a result of the cost versus tolerance knowledge. And finally, a minimum cost tolerancing procedure has been reduced to practice in a form which is straightforward and accurate enough for practical engineering application.

It is now practical to estimate the production cost of most lenses by entering the drawing data into a spreadsheet program. This is essentially an "expert system" estimator which can be applied by one with very little training or experience to get as good or better cost estimates for most lenses than an expert. It is also now practical to distribute lens tolerances using a spreadsheet program such that the production costs are minimized and to find the impact of design changes on lens costs. Both of these tools have advanced what was an "art" to an engineering discipline. Both are more accurate than what has been the common practice, and they are more accurate than the cost data as it typically is measured at this time.

The application of the estimating program can reduce the business overhead cost of a production operation, and it can point to the cost drivers of any particular lens such as: setup, milling, polishing, or centering. The application of the tolerancing program can decrease the production cost of systems from what they typically have been in both lens production, assembly and testing. If the tolerances are unnecessarily tight, the lens production cost is wasteful. If the tolerances are not tight enough to give a good yield of deliverable parts, the assembly and testing costs are wasteful. This tool attacks the economic problem at the point of the most potential impact on the life cycle costs as shown in Fig. 1, the detail design and analysis phase.

8. ACKNOWLEDGEMENTS

The author gratefully acknowledges the extensive contributions and discussions of Reinhard Seipp and Steven J. Cupka on the subject of base cost and cost versus tolerences and the assistance of David Poole and Alice Covington in the preparation of the figures.

9. REFERENCES

1. R. R. Willey, R. George, J. Odell, W. Nelson, "Minimized cost through optimized tolerance distribution in optical assemblies," Optical Systems Engineering III, Proc. SPIE Vol. 389, pp. 12-17 (1982).

2. R. R. Willey, "Economics in optical design, analysis, and production," Optical System Design, Analysis, and Production, SPIE Vol. 399, pp. 371-377 (1983).

3. R. R. Willey, "The impact of tight tolerances and other factors on the cost of optical components," Optical Systems Engineering IV, Proc. SPIE Vol. 518, pp. 106-111 (1984).

4. G. E. Wiese (ed.), Optical Tolerancing, SPIE, Bellingham, 1991

5. J. L. Plummer, "Tolerancing for economies in mass production of optics," Contemporary Optical Systems & Component Specifications, Proc. SPIE Vol. 181, pp.90-92 (1979)

6. G. Adams, "Selection of tolerances," Simulation and Modeling of Optical Systems, Proc. SPIE Vol. 892, pp. 173-185 (1988).

7. R. E. Fischer, "Optimization of lens designer to manufacturer communications," International Lens Design Conference, Geo. N. Lawrence (ed.), Proc. SPIE Vol. 1354, pp. 506-522 (1990).

8. R. E. Parks, "Optical component specifications," International Lens Design Conference, Proc. SPIE Vol. 237, pp. 455-463 (1980).

9. R. E. Parks, "Optical specifications and tolerances for large optics," Optical Specifications: Components & Systems, Proc. SPIE Vol. 406, pp. 98-105 (1983).

10. W. J. Smith, "Fundamentals of establishing an optical tolerance budget," Geometrical Optics, Proc. SPIE Vol. 531, pp. 196-204 (1985).

11. E. K. Thorburn, "Concepts and misconcepts in the design and fabrication of optical assemblies," Optomechanical Systems Design, Proc. SPIE Vol. 250, pp.2-7 (1980).

12. B. H. Beyeler und H. J. Tiziani, "Die optische Ubertragungsfunktion von dezentrierten optischen Systemen," Optik Vol. 44(3), pp. 317-328 (1966). (Included in Ref. 4.)

13. T. Kojima, "Yield estimation for mass produced lenses based on computer simulation of the centering tolerance," Optical Systems Engineering, Proc. SPIE Vol. 193, pp. 141-148 (1979).

14. P. O. McLaughlin, "A primer on tolerance analysis," Sinclair Optics Design Notes, Vol. 2, No. 3, (1991) FAX 716/425-4382.

15. R. R, Willey, "Optical design for manufacture," Recent Trends in Optical Systems Design II, SPIE Vol. 1049, pp. 96-102 (1989)

System Stability Considerations

Chair

James B. Breckinridge
Jet Propulsion Laboratory

Key issues in the thermal design of spaceborne cryogenic infrared instruments

Helene Schember and Donald Rapp

Jet Propulsion Laboratory
California Institute of Technology
Pasadena, CA 91109

ABSTRACT

Thermal design and analysis play an integral role in the development of spaceborne cryogenic infrared (IR) instruments. From conceptual sketches to final testing, both direct and derived thermal requirements place significant constraints on the instrument design. Although in practice these thermal requirements are interdependent, the sources of most thermal constraints may be grouped into six distinct categories. These are: (1) Detector temperatures, (2) Optics temperatures, (3) Pointing or alignment stability, (4) Mission lifetime, (5) Orbit, and (6) Test and Integration.

In this paper, we discuss these six sources of thermal requirements with particular regard to development of instrument packages for low background infrared astronomical observatories. In the end, the thermal performance of these instruments must meet a set of thermal requirements. The development of these requirements is typically an ongoing and interactive process, however, and the thermal design must maintain flexibility and robustness throughout the process. The thermal (or cryogenic) engineer must understand the constraints imposed by the science requirements, the specific hardware, the observing environment, the mission design, and the testing program. By balancing these often competing factors, the system-oriented thermal engineer can work together with the experiment team to produce an effective overall design of the instrument.

INTRODUCTION [1-7]

Thermal requirements play an unusually central role in the development process of a spaceborne cryogenic infrared (IR) instrument, and influence the instrument design from conceptual sketches to final testing. The thermal requirements themselves are often poorly defined until well into the instrument design. The design of these instruments, and their interfaces with a telescope and spacecraft, require interaction among many technical disciplines, strongly influenced by a set of driving thermal requirements.

In this paper we will use the descriptor *thermal* to identify the full span of the temperature scale, including the cryogenic region below about 100 K. We have chosen not to conform to the usual distinction between the terms "thermal" (generally applied to spacecraft or observatory bulk temperature control), and "cryogenic" (mainly concerned with supplying very cold temperatures to sensors and optics). Our thesis here is that the engineer designing the instrument must look across the entire temperature scale to understand the system as a whole and produce a balanced overall thermal design. The discussion here is intended to cover sources of thermal requirements for low background infrared astronomical satellites, but may provide helpful commentary for other types of infrared instruments.

The thermal requirements for IR instruments and observatories spring from six main sources:

(1) Detectors. The detectors will have a temperature range for proper operation, as well as a requirement for temperature stability. The associated detector readout electronics (multiplexers or "muxs") will also have required temperature ranges and stabilities, which may well be different from the detector requirements. Both the detectors and the readouts will dissipate heat, and that heat will constitute part of the heat load on the coldest stage of the cooling system. In addition, the wiring and mountings between the cold sensors and the warmer part of the instrument will allow heat to leak back in. The duty cycle of the detectors must be accounted for in providing proper temperatures. Finally, the detectors may require special thermal procedures such as thermal annealing to recover from radiation exposure or exposure to sources which are too bright. Thermal annealing consists of warming the sensor to a specified temperature and holding at that temperature for a length of time.

(2) Optics. The optics of an IR observing instrument usually must be kept cold enough that the telescope sensitivity is limited by external noise rather than by thermal emissions from the telescope optics themselves. Both the telescope noise and the external noise are strongly dependent on wavelength. The choice of optical configuration will be significantly influenced by the predicted temperature distributions of mirrors, baffles, and support structures, and the acceptable telescope noise levels at various wavelengths.

(3) Pointing. The instrument alignment and pointing requirements constrain the allowable thermal gradients as well as long term thermal drifts. These must be maintained within acceptable limits by either passive design or active control. The configuration of the telescope dictates the level of thermal isolation achievable, but must be balanced by structural and observational considerations.

(4) Lifetime. The required mission lifetime will strongly influence the thermal design. A requirement for extremely long lifetimes, say in excess of 3 years, will dictate either severe thermal isolation in an open cycle system, or a highly reliable closed cycle cryocooler. Shorter lifetimes allow use of a standard solid or liquid stored cryogen system, or an off-the-shelf cryogenic refrigerator system.

(5) Orbit. The mission lifetime is closely coupled to the selected orbit. For astronomy missions, longer lifetimes are the direct result of reducing the heat inputs from the sun and the earth. For this reason, high earth orbits (HEO) are often selected for these missions. Heliocentric orbits, especially those beyond 1 AU from the sun, can significantly extend mission life. Other factors must be balanced in the choice of orbit, including launch vehicle lift capability, telecommunications, and power from solar panels.

(6) Testing. The requirements for test, alignment and calibration will also constrain the thermal design. Appropriate testing capability must be incorporated into the design to allow for validation of the hardware. These test capabilities may create new paths for heat to leak in, or may actively dissipate heat while in use. Validation testing, even at the component level, may require separate, dedicated dewars and elaborate cryogenic test setups.

In actuality many of these requirements drivers are related and these relationships may combine or compete in the development of a thermal design strategy. For the purposes of design, thermal requirements may be considered direct or derived. Direct requirements are those which dictate a fundamental temperature or temperature stability. Derived requirements flow from another more elemental requirement, such as a pointing stability requirement, or which spring from a particular instrument configuration or mission assumption.

A more detailed discussion of each of these areas follows, using examples from recent and proposed IR astronomy missions including the Infrared Astronomy Satellite (IRAS) launched in 1983, the Cosmic Background Explorer (COBE) launched in 1989, the European Space Agency's Infrared Space Observatory (ISO) currently in assembly and test and scheduled for launch in 1996, and the proposed Space Infrared Telescope Facility (SIRTF).

1. COOLING OF DETECTORS [8-12]

An IR observatory might contain various instruments which may be classified generically as cameras, spectrometers and photometers. SIRTF, for instance, has three instruments: the Infrared Spectrograph (IRS), the Multiband Imaging Photometer (MIPS) and the Infrared Array Camera (IRAC). Great strides have been made recently in IR detector technology. Progress is also being made in the area of sensor cryogenic readout electronics. These multiplexers ("muxs") may not operate at temperatures as low as their associated detectors. However, new CMOS process technology may allow low noise operation of muxs in the range below 20 K.

It has been common to achieve required detector temperatures and stability by supplying a cold sink below the desired operating temperature, and using feedback thermal control to warm the detectors or readout electronics to the desired temperature. Thermal capacitance is also used to achieve stability, although material heat capacity drops dramatically with decreasing temperature. Because of the variation in thermal conductivity and other properties in the cryogenic regime, thermal analysis must involve the use of properties arrays or curve fits, making simplified design difficult. As another approach to the thermal design, the detectors may be attached to warmer stations on the instrument or observatory. If these warmer thermal stations are used instead, then accurate prediction of the station temperatures and stabilities becomes a major input to the design process, and thermal isolation become a critical problem.

In most IR astronomy applications, certain detectors will require extremely low temperatures, often below 2 K. The bolometers on COBE were operated at 1.55 K, for example. To achieve these very low temperatures, the instruments and telescope are embedded in a cryostat or dewar. This dewar contains a stored cryogen, usually helium, which is vented to space. The heat load into the cryogen causes evaporation and expends the cryogen. Since the cryogen is in a saturated state, its temperature is controlled by the impedance of the ventline and exit valving configuration. Liquid cryogens in space present special venting problems since the liquid and vapor must be separated under micro-gravity conditions. Helium becomes a superfluid below 2.17 K, and in this state the liquid-vapor separation can be achieved using a porous plug.

Detector requirements of 10 K or above may be satisfied by a solid hydrogen stored cryogen system or by the upcoming generation of mechanical and sorption closed-cycle coolers. However, if the required detector temperature is lower than 10 K, fewer options for cooling remain available. In the range from 10 K to about 8 K, a solid hydrogen open cycle (stored cryogen) cryostat may be feasible. The Spatial Infrared Interferometer Telescope III (SPIRIT III) on the Midcourse Space Experiment (MSX) satellite, to be launched in 1993, is an example of an IR instrument using a solid hydrogen cryostat. Also, in the range from 10 K to about 6 K, a supercritical helium stored cryogen system can be used. Below about 6 K, the only cooling options are stored superfluid helium (He II) or a supercritical helium system using a Joule-Thomson expander. It should be noted that, due to the difficulty in positioning the ullage, and accomplishing liquid-vapor separation, only supercritical or superfluid helium systems are used in spaceborne systems. The coldest temperature achieved on orbit has been 1.4 K inside the main COBE superfluid helium tank. Hybrid systems which use a higher temperature, but higher heat capacity, cryogen (such as hydrogen) to intercept the large heat leaks into the cryostat were studied for both SIRTF and ISO. These staged system have generally been found to be too complex, and the thermal designers returned to the single stage superfluid system as the preferred approach.

For very long wavelengths ($> 200\ \mu$m) bolometers are typically used. For highest performance, these detectors must be cooled to very low temperatures (usually < 0.3 K). This is accomplished by thermally staging an ultra-low temperature refrigerator (a sub-kelvin cooler) in the cryostat. The refrigerator uses the stored cryogen as its heat sink. Sub-kelvin cooler concepts have been developed by several NASA Centers, and include ^{3}He evaporation coolers, ^{3}He/^{4}He dilution coolers, and adiabatic demagnetization coolers. The adiabatic demagnetization approach is the most highly developed and has been pursued for the SIRTF MIPS instrument.

In addition to requiring a very low temperature, bolometers also require high thermal stability of the telescope thermal emission. This thermal emission is one component of the total stray light, which is usually required to be less than 10% of the natural background at the wavelengths of interest. (The total stray light is composed of contributions from diffraction, off-axis sources and thermal emission. All vary as a function of wavelength.) The bolometer is typically operated in a chopped mode where the energy received in one chop position is subtracted from the energy received in the other chopped position to cancel the signal from the background (telescope plus sky). However, if the change in chop position causes a change in stray light from telescope emission there will be an added false signal. To cancel this signal, the telescope is nodded between two positions. When the proper subtraction is made between nodded signals, the true signal is obtained.

However, if the temperature of the telescope emission source changes during the course of a nodding operation, the false signal from the emission modulation will not quite cancel when the two nod signals are subtracted. For example, in SIRTF the thermal emissions from the aperture shade (near 60 K), and other elements near 3 K (optical baffles, secondary struts, and mirrors) has been estimated in the 400 to 700 μm bolometer wavelength band. In order to keep the false signal from temperature changes to less than 10% of the true signal, the maximum allowable rate of temperature change ranges from 20 mK/s (for the secondary struts) to 840 mK/s (for the aperture shade). If the temperatures at these locations drops, the allowable temperature variation will increase. Calculations such as these must be performed for each significant change in the instrument/telescope configuration, and usually involve the use of stray light and thermal analysis codes.

Finally, in addition to the focal planes, the thermal designer must also consider focal plane mechanisms such as filter wheels and grating drives which dissipate heat and have operating temperature requirements. Mechanisms in close proximity may be required to be thermally isolated from the detectors. Mechanism problems may have system consequences – a mechanism anomaly on COBE caused excess power to be dissipated and shortened the observing lifetime of that mission.

2. COOLING OF OPTICS [13-16]

A major limit on the sensitivity of an IR telescope is background noise, which limits the ability to detect very weak fluxes from astronomical objects. Background noise can come from thermal emission by the telescope optics or from external sources.

The signal measured by the instrument is a combination of target signal, telescope thermal emission, and signal from external sources other than the target. There is an uncertainty in the background due to the random arrival of photons, which cannot be subtracted out as a constant baseline and is an irreducible noise. Since this noise signal increases with the square root of the integration time, while the signal from an astronomical source increases linearly with time, one can improve the ratio of signal to noise by using longer integration times.

Spatial variations in the background signal can also contribute to the noise. Noise which is not uniformly distributed in space is called confusion noise. Confusion noise arises from true spatial variations in the sky background at spatial scales near those of the objects under observation. In the case of observations of point sources in the IR, the confusion for spaceborne observatories typically comes from a background distribution of unresolved point sources (local stars and extended galaxies) and the filamentary dust structures (IR cirrus) which permeate the interstellar medium in our galaxy. Because the confusion noise is due to real variations in the sky brightness, the noise signal increases linearly with integration time. Thus the signal to confusion noise ratio cannot be improved with longer integration times. These confusion sources put a practical limit on integration times and supply a background noise limit against which to address telescope optics temperatures. Unlike ordinary noise, confusion noise decreases with larger telescope apertures because larger diameters provide higher spatial resolution, pushing the brightness of the average unresolved sources to fainter limits.

In general, the telescope optics must be cooled so that the telescope thermal emission is well below the background noise over the range of included wavelengths. A budget for thermal emission stray light at various wavelengths should be established with the experiment team as part of a program of developing thermal requirements. Both the background noise limits and the stray light budgets will vary depending on the mission design, observing strategy, detector sensitivity and type of telescope. For space telescopes with extremely high sensitivity, the confusion limit must be estimated using various astrophysical models since data are not yet available. Due to their higher sensitivity, optics temperature requirements for the next generation of IR space observatories range between 15 and 3 K.

For an observatory like the Infrared Space Observatory (ISO) or SIRTF, designed for a wide range of wavelengths in both broadband and spectroscopic modes, the simplest way to assure background noise limited operation is to embed the entire telescope within a cryostat. Since a stored cryogen system is already required for the detectors, it is most convenient to simply embed the instruments and the optics within the same cryostat even though the optics may not required to operate as cold.

On the other hand, particularly at long integration times, the confusion noise may be large enough that a moderate sized telescope at 20 to 30 K would be useful across a wide range of wavelengths in the broadband mode. This opens the possibility that a passively cooled telescope might be viable for some IR observations. Passive cooling of the optics would then become a possibility, as well as the use of the developing closed cycle cooler technology. Passive cooling and closed cycle coolers are attractive because they promise lighter weight and longer lifetime than stored cryogen systems. It is important to consider the cooldown times for these systems, since they are launched warm and go through a long approach to their final operating temperature. However, even if these alternate longerlife schemes are used for the optics, the detectors will still require the colder (< 10 K) temperatures.

3. POINTING [17-19]

The usual design strategy of suspending the cryostat, optics and instruments on thermally isolating supports from a surrounding (warmer, passively cooled) superstructure creates special problems for alignment and pointing. Because the supports are designed to minimize thermal conduction, the ratio of their cross-sectional area to length must be minimized. This minimum is determined by the suspended mass, and in the absence of a launch caging system, by the launch loads. Additional lifetime can be gained only by increasing the degree of thermal isolation. Designers have gravitated toward configurations having the lowest possible thermal conduction by suspending the telescope/cryostat on network of thin stretched straps, such as in the design used on COBE and IRAS. The natural frequency of this strap and suspended mass system is typically in the range 20 to 45 Hz. An extremely low frequency could cause undesirable mechanical interactions and affect settling times following either telescope slewing or on-board motions such as secondary mirror chopping or motion internal to an instrument.

The suspension system, furthermore, has tie-down points at the outer supporting structure which are subjected to varying heat loads. The variations in heat load may be due to orbital position, orientation changes (pointing or slewing), or telescope operations which change heat loads onto the outer shell or structure. These heat load variations cause the outer structure to warp and distort, moving the instrument's field of view relative to the spacecraft structure. The heat load variations can be reduced by insulating the outer shell, but there is a penalty for this approach because the outer surfaces act as passive radiators to reduce the outer shell temperature and increase lifetime.

The thermal distortion effect is highly dependent on configuration, and ranges from drastic (worst case boresight shifts of 100 arc-sec were measured on IRAS) to moderate (steady-state models of a SIRTF configuration predict shifts of about 5 arc-sec). To predict these distortions, a design must be iterated across thermal, structural and controls disciplines, often entailing the use of discipline-specific, non-interfacing, modelling routines. The designer must understand the instrument integration time, the slewing time, on-orbit thermal variations, heat dissipation duty cycles, and settling times in order to perform design trade studies to yield improved performance.

4. LIFETIME [20-29]

The series of infrared astronomy telescopes, IRAS, COBE and ISO, used the concept of surrounding most of the telescope and instruments with a superfluid helium cryostat. The cryostat performance was improved by the use of helium vapor cooled shields (VCSs). In this approach, the enthalpy of the helium boiloff gas is routed through plumbing which cools a series of nested shields between the main cryogen tank and the supporting structure. Through the use of the heat capacity of the helium vent gas, the lifetime of the observatory can be extended by reducing the total heat load into the helium liquid. Heat load variations and requirements for temperature stability complicate the use of the higher temperature boiloff gas for instrument and optics heat loads.

Several SIRTF design concepts have been developed to extend the predicted lifetime to 5 years and longer. These designs employed the following techniques to extend lifetime: (1) reducing the heat flux into the stored helium from the outer shell by choosing a high earth orbit and by completely shielding the telescope from the sun, (2) reducing instrument heat loads and leaks (parasitics), and (3) providing improved thermal isolation by either using a high performance suspension system or by allowing a lower natural frequency of the suspension system. The heat input from the outer shell is a steep function of outer shell temperature. In the current SIRTF design, the outer shell has been reduced to about 60 K by using the solar array as a solar shield and by adding an additional shield between the solar array and the outer shell. Also, the earth heat load on the outer shell has been reduced to near zero by adopting a heliocentric orbit.

In the typical cryostat design, a toroidal cryogen tank is used and three VCSs are supported from the dewar outer shell by fiberglass/epoxy straps which are anchored to each of the VCSs. Because helium has small latent heat but relatively large heat capacity, efficient cooling of the shields can reduce the rate of helium evaporated by an order of magnitude when compared to systems which do not use vapor cooled shields.

Insulation is essential to minimize thermal radiation heat leaks. Recent studies show that a single layer insulation (SLI), such as aluminized mylar, used between the helium tank and the inner VCS, as well as between the inner VCS and the middle VCS, provides good thermal isolation. Multilayer insulation (MLI) consisting of layers of metalized mylar with a net spacer between layers, is used in the rest of the dewar system. SLI can be used at the lower temperatures encountered in the inner VCSs where conduction heat transfer is much greater than radiative heat transfer. With SLI, carefully positioned to avoid thermal shorting, the conductive mode of heat flow is eliminated.

5. ORBIT [30]

A low earth orbit (LEO) has the apparent advantages that (1) more mass can be put in orbit, and (2) there is a potential for servicing a telescope. However, LEO has very strong disadvantages for a cryogenic IR telescope. First, there is a large heat load from the earth's thermal emission and reflected sunlight which limits the lifetime and the achievable temperatures. Second, telescope slewing motions will cause varying heat loads on the outer structure which may cause the structure to distort. Third, earth avoidance constraints typically limit integration times to 20 minutes or less, requiring rapid slewing between targets to achieve reasonable observing efficiency. Both IRAS and COBE were placed in a sun-synchronous LEO at an altitude of 900 km. In this altitude range the South Atlantic Anomaly, a region of energetic particles trapped in the magnetic field of the earth, degrades detector performance. Below about 500 km, the presence of atomic oxygen may present a problem.

By contrast, a telescope in a high orbit (HEO), at an altitude of 70,000 km or beyond, would have very small heat inputs from the earth, long periods of good thermal stability, and nearly twice the observing efficiency of LEO. Analyses for SIRTF indicated that the outer shell temperature drops from 200 K in LEO to less than 115 in HEO. The predicted lifetimes for the SIRTF comparison were 2 years for LEO and 5 years for HEO, with the same total cryogen load. A single target could be observed for tens of hours in some cases. Although there is no atomic oxygen problem, there is a moderate radiation environment due to solar protons and galactic cosmic rays. However, such a telescope could not be serviced, and the mass that can be placed into HEO is less than 1/3 the mass that can be placed in LEO. Nevertheless, the combination of much lower heat input from the earth and much higher viewing efficiency allows the designer to use a smaller, less massive cryostat for HEO and still get better observational performance over mission lifetime.

For telescopes where it is impossible to boost the required mass into HEO, a compromise is to use an elongated elliptical orbit (EEO), which permits considerably more mass to be put into orbit than into HEO while allowing high viewing efficiency. For example, in a study of a passively cooled sub-mm telescope it was found that twice as much mass can be put into EEO as can be put into HEO. EEO produces a fairly benign thermal environment for long periods, but it subjects the telescope to a strong radiation environment. A passively cooled telescope can be designed to minimize thermal gradients in EEO. ISO is designed to operate in EEO with orbital parameters of 1,000 km by 70,000 km.

Another orbit possibility currently under consideration for SIRTF is a solar orbit, where the telescope is launched on an earth escape trajectory and slowly moves away from the earth while maintaining approximately 1 AU from the sun. A larger mass can be launched into this orbit than into HEO, and the thermal loads from the earth and moon are negligible. Solar orbits which take the telescope further away from the sun than 1 AU produce even lower heat loads and increase lifetime. These non-earth centered orbits may complicate telecommunications during the mission lifetime.

6. TESTING [31-34]

The development of thermal requirements and design must accommodate a host of validation testing and calibration programs. At the component level, small test dewars must be available to perform acceptance and verification testing. Since the instruments are part of an optical train, the thermal design must also accommodate the need to align the entire system on assembly. Warm up and cool down times may be constrained due to contamination concerns or due to the mechanical stresses induced by large thermal gradients in components with poor thermal conductivity. Transient thermal analyses will be necessary to understand the system response to the test program. Many cryogenic instruments, such as the IR Camera on ISO (ISOCAM) rely on a network of copper straps to provide athermalization of the optics. Testing may be extremely time consuming – spacecraft and dewar integration and test on COBE lasted from July of 1988 through October of 1989. Subsystem level testing on COBE was performed as early as 1983.

In addition to the usual battery of spacecraft thermal tests, the cryogen system will also required verification. This may prove difficult for the next generation of space IR telescopes which will have outer shells running below 150 K, too cold to simulate in an environmental test facility. Even using chambers equipped with large, liquid nitrogen cooled shrouds, the transient cooldown times of these large telescopes makes testing impractical. The approach taken so far has been to design a test program which focusses on careful validation of the thermal analytical models. Several concepts have been suggested to provide additional system level verification testing, such as forcing the outer shell to its predicted operating temperature using liquid nitrogen and then testing the rest of the system for adequate performance.

Vibration testing for validation and qualification must also be performed at cryogenic temperature. For the Short Wavelength Spectrometer on ISO, a special setup was designed and built to expose an instrument prototype to mechanical vibrations comparable to those expected on launch. In this instance, vibration at 80 K rather than 4 K was determined to be acceptable. Some controversy remains as to whether this judgement is satisfactory in all cases.

SUMMARY

The design of spaceborne cryogenic IR instruments is strongly influenced by both direct and derived thermal requirements. The thermal design involves analysis and iteration during all phases, from the preliminary concept phase to the final detailed design. The thermal or cryogenic engineer must understand the constraints imposed by the science requirements, the specific hardware, the observing environment, and the mission design. By balancing these often competing factors, the system-oriented thermal engineer can produce an effective overall design of the instrument.

ACKNOWLEDGEMENTS

The work described in this paper was carried out at the Jet Propulsion Laboratory, California Institute of Technology, Pasadena, California under a contract with the National Aeronautics and Space Administration.

The authors are greatly indebted to T. Nick Gautier, Erick Young, George Rieke and Perry Hacking for their comments and corrections to an early draft of this paper. Dave Elliott and Walt Petrick deserve special thanks for their important contributions.

REFERENCES AND SELECTED BIBLIOGRAPHY

[1] Schember, H., Manhart, P., Guiar, C. and Stevens, J. H., "Space Infrared Telescope Facility Telescope Overview", Technical Conf. 1540, SPIE International Symp. on Optical Applied Science and Engineering, San Diego, CA, 22 July, 1991.

[2] Spehalski, R., Cherniak, A., McLaughlin, Werner, M., Mason, P., and Sohus, A., "Space Infrared Telescope Facility (SIRTF) Implementation Plans," AIAA Space Programs and Technologies Conference, AIAA-90-3545, Huntsville, Alabama, 25-28 September, 1990.

[3] Bothwell, M., "Space Infrared Telescope Facility Science Instruments Overview", Technical Conf. 1540, SPIE International Symp. on Optical Applied Science and Engineering, San Diego, CA, 22 July, 1991.

[4] Gillett, F. and Houck, J., "The decade of infrared astronomy," *Physics Today*, pp. 32-37, April, 1991.

[5] Boggess et al., "The COBE mission: its design and performance two years after launch," *Astrophysical Journal*, to be published, 1 October 1992.

[6] Irace, W., and Rosing, D., "The IRAS telescope," *Journal of the British Interplanetary Society*, vol. 36, pp. 27-33, 1983.

[7] Kessler, M., "The Infrared Space Observatory (ISO) and its instruments," SPIE vol. 1130, *New technologies for astronomy*, 1989.

[8] Lum, N., Asbrock, J. White, R., Augustine, F., Hoffman, A. and Nystrom, S. C., "Low Noise, 256 x 256, 10 K Staring FPA", SPIE Aerospace Sensing Conf., Orlando, FL, April 23, 1992.

[9] Bard, S., Rodriguez, J., Wen, L. and Wu, J.,"10 K Solid Hydrogen Sorption Cooler", JPL Report D-9326, February, 1992.

[10] DeSa, L., and Taride, S., "ISOCAM, the ISO satellite's infrared camera," *Cryogenics*, vol. 30, September Supplement, 1990.

[11] Franco, F., and Sander, M., "Thermal design of ISO payload dual cooling system (LH2 - HeII)," Proceedings of the International Symposium on Environmental and Thermal Systems for Space Vehicles," Toulouse, France, October 1983.

[12] Bernstein, G. et al., "Automated temperature regulation system for adiabatic demagnetization refrigerators," *Cryogenics*, vol. 31, pp. 99-101, February, 1991.

[13] Hacking, P., and Soifer, B.T., "The number counts and infrared backgrounds from infrared-bright galaxies," *Astrophysical Journal*, vol 367, 1 February 1991.

[14] Helou, G. and Beichman, C. A., "The Confusion Limits to the Sensitivity of Sub-Millimeter Telescopes", Proc. 29th Liege International Astrophysics Symposium, Institut d'Astrophyique, Liege, Belgium, July 3-5, 1990.

[15] Rapp, D. "Potential for Active Structures Technology to Enable Lightweight Passively Cooled IR Telescopes", JPL Report D-9449, March, 1992.

[16] Dinger, A., "STRAY An interactive program for the computation of stray radiation in infrared telescopes," SPIE vol. 675 Stray Radiation V 1986.

[17] Ledeboer, W. C., Elliott, D. and McNeal, P., "Space Infrared Telescope Facility Thermal Distortion Analysis", JPL Report D-9264, December 13, 1991.

[18] Chang, L. and Mansfield, J., "SIRTF on-orbit thermal bending analysis," NASA/ARC system engineering report, 10 August 1988.

[19] Karsten, L. and Teule, F., "IRAS thermal misalignment – a lesson for the future," *ESA Journal,* vol. 8, pp. 245-259, 1984.

[20] Murakami, M. et al., "Design of cryogenic system for IRTS," *Cryogenics*, vol. 29, May, 1989.

[21] Bhandari, P., Petrick, S. W. and Schember, H., "Thermal Systems Analysis for the Space Infrared Telescope Facility Dewar", Technical Conf. 1540, SPIE International Symp. on Optical Applied Science and Engineering, San Diego, CA, 22 July, 1991.

[22] Lee, J.H. and Brooks, W.F., "Cryogenic and thermal design for the Space Infrared Telescope Facility (SIRTF)," *Journal of Spacecraft and Rockets*, vol. 23, no. 1, January-February, 1986.

[23] Urbach, A., Hopkins, R., and Mason, P., "One year of stars, provided by the IRAS cryogenic system," Proceedings of the International Symposium on Environmental and Thermal Systems for Space Vehicles," Toulouse, France, October 1983.

[24] van Leeuwen, P., "Thermal design and performance of the IRAS S/C system," Proceedings of the International Symposium on Environmental and Thermal Systems for Space Vehicles," Toulouse, France, October 1983.

[25] Hopkins, R. and Payne, D., "Optimized support systems for spaceborne dewars," *Cryogenics*, vol 27, pp. 209-216, April 1987.

[26] Birur, G. and Tsuyuki, G., "A simplified generic cryostat thermal model for predicting cryogen mass and lifetime," presented at the Space Cryogenics Workshop, Cleveland, OH, 18-20 June 1991.

[27] Hopkins, R., Nieczkoski, S. and Volz, S., "Performance predictions for spaceborne, long-lifetime helium dewars containing large aperture telescopes," SPIE Conference, San Diego, July 1990.

[28] Satter, C., "Large Deployable Reflector (LDR) Structural Design Studies", JPL Report D-7686, July, 1990.

[29] Tsuyuki, G. and Cohen, E., "Thermal Design Optimization of a Segmented GFRP Primary Reflector for a Submillimeter Telescope", SPIE Aerospace Sensing Conf., Orlando, FL, April 23, 1992.

[30] Lee, J.H., Ng, Y., and Maa, S., "Comparison of SIRTF dewar performance in the 900 km and 100,000 km orbits," Advances in Cryogenic Engineering, vol. 35, 1990.

[31] Volz, S. et al., "Cryogenic on-orbit performance of the NASA Cosmic Background Explorer (COBE)," Proceedings of the International Symposium on Optical and Opto-Electronic Applied Science and Engineering, SPIE, San Diego, 1990.

[32] Mosier, C., "Thermal design of the Cosmic Background Explorer Cryogenic Optical Assembly," AIAA 29th Aerosciences Meeting, Reno, NV, January 1991.

[33] Taride, S., "Shaking in cryogenic conditions," Proceedings of the 3rd European Symposium on Space Thermal Control & Life Support Systems, Noordwijk, The Netherlands, 3-6 October 1988.

[34] Wildeman and Ploeger, G., "Cryogenic vibration set-up for space qualification," *Cryogenics*, vol. 29, pp. 559-562, May 1989.

Thermo-elastic analysis of large optical systems

Earl Pearson

National Optical Astronomy Observatories††
Tucson, Arizona 85726

ABSTRACT

Once the temperature distribution is known for a large optical system, there are various methods to predict its effect on the optical performance. The thermal distribution is assumed known by measurement, heat-transfer analysis, or supposition. A system consisting of reflective and refractive elements, their supporting structure, and the surrounding medium will all be affected. The reflective optical elements and structure are usually analyzed for their thermo-elastic response, while the refractive elements are subject to both elastic distortions and refractive index changes.

While it might appear almost hopeless to look for theoretical (closed form) solutions, there are some available that are both powerful and practical. Most finite difference and finite element elastic solutions can incorporate temperature effects and are used for a wide range of opto-mechanical structures. In addition to the temperature, one must also know the corresponding material parameter (for example, the coefficient of thermal expansion) These parameters are ofter themselves temperature dependent and are not constant either throughout the structure or even within a single (non-homogeneous) component.

Since temperature distributions can be irregular, variable and difficult to predict exactly, orthogonal functions can often be analyzed. Then the thermal distribution can be approximated by a sum of these functions thereby predicting the whole response.

1. INTRODUCTION

This discussion centers on large optical telescope systems. Most of the comments are just as applicable to other optical systems but the examples are expressed in terms of telescope technology. One way to start is to describe a system that would not require a thermo-elastic analysis . Such a system would be obtained if all components were made, assembled, tested and used at one single, constant temperature. Some systems are built and used in an optical laboratory where the environment is controlled. If the excursions of the actual temperature are small enough not to cause any noticeable effects, then this is a reasonable approximation to this idealization.

The special temperature in the idealization above can be called the reference temperature, Tref. This could be 22 degrees Celsius. Tref, a constant, could have been chosen as a different value BUT it is implied that you know all the necessary conditions pertaining to the system at the Tref temperature you choose. The difference of the actual temperature and Tref is the temperature distribution, T, used to analyze

††Operated by the Association of Universities for Research in Astronomy, Inc. under cooperative agreement with the National Science Foundation.

the system. Then,

$$T\ (x,y,z,t) = \text{actual temperature}(x,y,z,t) - \text{Tref} \tag{1}$$

where x,y,z are rectangular Cartesian coordinates (or any coordinate system you prefer) and t is time. As long as the actual temperature is equal to Tref, T will be zero, i.e., there are no thermal effects to be added to those we already know about at Tref.

Temperature, a scalar, has magnitude but no direction associated with it and its distribution is always in three dimensional space even though it can be useful at times to describe its value in only one or two directions. The units of temperature can be confusing since only differences are described relative to Tref. Only if Tref were the zero point of a known scale (e.g., Celsius) would T have the same scale values. Instead, the differences of temperature above (positive) or below (negative) the chosen reference value are used. The units are C for Centigrade degree (not degree Celsius), or F, for Fahrenheit degree, etc.

While this paper does not treat heat transfer in itself, certain ideas are assumed common knowledge; heat will transfer from a higher temperature to a lower, the heat transfer can be described as conduction, convection or radiation or some combination thereof, etc. It is assumed that the thermal distribution is known or taken by hypothesis. It is also assumed that the governing thermo-elastic equations, finite element approximations, or physical laws are known.

2. CONSTANT TEMPERATURE IN A LINEAR, HOMOGENEOUS, ISOTROPIC MATERIAL (A CLASSIC)

The body in question has a constant but non-reference value such that,

$$T(x,y,z,t) = C0 \tag{2}$$

The displacements and stresses that result from T can only be determined if the boundary conditions are known. There are two idealized conditions that can be described easily: the completely unconstrained body and the completely constrained body. The former is probably better known than the latter but we can consider them as the two extremes; intermediate cases of partial constraints would fall between them.

As long as the body is free to expand and contract (unconstrained) there is no stress induced in the body due to the temperature change. All dimensions change in a manner that is proportional to their previous values. U, v, and w are the x,y,z displacement components resulting from T:

$$u(x,y,z,t) = CTE * C0 * x$$

$$v(x,y,z,t) = CTE * C0 * y$$

$$w(x,y,z,t) = CTE * C0 * z \tag{3}$$

The linear coefficient of thermal expansion (CTE), describes the change in length per unit length per unit temperature change, or "thermal strain", for an unconstrained body. For example, take a circular concave mirror of diameter 100 cm., edge height of 10 cm. and focal length of 500 cm. Then if C0=10 C and CTE = 1.0 E-5/C, the diameter increases by 0.1 mm, the edge height increases by 0.01 mm. and the focal length increases by 0.5 mm.

For an unconstrained body, rigid body displacements (RBD's) can be added to (3) depending on the choice of the origin and any other arbitrary defining conditions. A body can be defined in space and move about as a rigid body and be unconstrained thermally as long as the temperature change does not introduce "thermal stresses". Assuming the mirror above is supported and defined in the gravity environment there are elastic stresses in the body but no change in stress due to temperature changes.

For the other idealization of the body totally constrained there are no strains and the thermal displacements are all zero but the normal stresses are

$$\text{stress} = -\,\text{CTE} * E * \text{C0} \,/\, (\,1. - 2.*\,\text{PR}\,) \qquad (4)$$

where E is the modulus of elasticity and PR is Poisson's ratio. The negative sign is conventional for compressive stresses. Again this case is never found in practice, but indicates the upper bound of thermally induced stresses which can be quite high. In the example above, if E = 10.0E+6 psi and PR is 0.25, the normal stresses are 2000 psi.

Most real systems fall somewhere between the two idealized cases and can require considerable analysis for individual situations. We also assume that when thermal stresses are induced, they are added algebraically to the other stresses. The unconstrained case is more easily achieved in practice and is usually not a problem when allowances are made for it in the system.

An example of a theoretical solution for partial constraints that might be of interest to optical engineers is the following. Assume a circular, flat optical window, constrained only at its outer edge (normal to the surface, no shear stress) with the T = C0 value. Then the normal in-plane stresses are;

$$\text{stress} = -\,\text{CTE} * E * \text{C0} \,/\, (\,1. -\ \text{PR}\,) \qquad (4a)$$

while the axial normal stress is zero. The change in the window thickness is

$$\text{change in thickness} = \text{CTE} * \text{C0} * \text{z0} *(\,1. + \text{PR})/(1. - \text{PR})$$

where z0 is the original thickness. The stresses are reduced while the displacement is increased.

Of even more practical interest would be the case of the window above being constrained by the structural material (of a different CTE) in which it is mounted when the window stress has a negligible effect on the structure. Now the window stresses are

$$\text{stress} = (\text{struct.CTE} - \text{CTE}) * E * \text{C0} \,/\, (\,1. -\ \text{PR}\,) \qquad (4b)$$

and as a further note, the window would not change thickness if

$$\text{struct.CTE} = \text{CTE} * (1. + \text{PR})/ (2.* \text{PR}) \qquad (4c)$$

even though there is still a stress in the window. If PR = .25, the struct.CTE = 2.5 * CTE.

2.1. Constant temperature changes (a contradiction in terms)

If the temperature of a solid body were truly constant (isothermal), there would be no heat transfer within the body nor any heat transfer to or from its surroundings. It is difficult to maintain constant temperature in a body (within a small error) and it is impossible to do even that while the temperature is required to change in a finite amount of time. While this seems too obvious to state, it is a request often made for thermally sensitive optical elements in a changing temperature environment. Given a value for specific heat and conductivity, an estimate of temperature gradient can be made to accomplish a temperature change in a given time frame.

3. LINEAR TEMPERATURE GRADIENTS IN A LINEAR, HOMOGENEOUS, ISOTROPIC MATERIAL

Since gradients will normally also be present in the body, it would be advantageous to understand the effect of linear gradients. We can start with the premise that

$$T(x,y,z,t) = C1 * x \qquad (5)$$

For an unconstrained body, the thermal displacements (sans RBD's) are:

$$u(x,y,z,t) = .5 * \text{CTE}* C1 * (x*x - y*y - z*z)$$

$$v(x,y,z,t) = \text{CTE} * C1 * x*y$$

$$w(x,y,z,t) = \text{CTE} * C1 * x*z \qquad (6)$$

Again all the thermal stresses are zero. If the gradient were in the y direction, then a similar set of displacements are obtained with the role of x above replaced by y and similarly for a gradient in the z direction. A gradient of a scalar is a vector so that a gradient in an arbitrary direction can be composed of components in the three coordinate directions. Since the equations are linear, the solutions can be superimposed. Thus,

$$T(x,y,z,t) = C0 + C1 * x + C2 * y + C3 * z \qquad (7)$$

results in displacements of

$$u(x,y,z,t) = CTE *[C0*x + .5*C1*(x*x -y*y -z*z) + C2*x*y + C3*x*z]$$

$$v(x,y,z,t) = CTE *[C0*y + C1*x*y + .5*C2*(y*y -x*x -z*z) + C3*y*z]$$

$$w(x,y,z,t) = CTE *[C0*z + C1*x*z + C2*z*y + .5*C3*(z*z -y*y -x*x)] \quad (8)$$

The stresses are all zero.

Because all the stresses are zero, this result applies equally well to any homogeneous body regardless of its geometry. The same solution applies to a steel frame, a cube, or a structured, cast honey-comb mirror. While just the constant term is normally a poor approximation of T, the addition of the gradients allow one to come much closer to real situations.

We can also extend the above solution to geometries expressed in other coordinate systems. One case of very practical importance to telescope engineers is a primary mirror with a parabolic upper surface. The upper surface can be expressed most easily in cylindrical coordinates, r,θ,z, as

$$z(r) = z0 \ + r*r/(2. * R) \quad (9)$$

where R is the radius of curvature (twice the focal length), and z0 is the z coordinate of the mirror vertex. The relation between the two systems of coordinates is given by,

$$x = r * \cos(\theta)$$

$$y = r * \sin(\theta) \quad (10)$$

Then w(r,θ,z,t) is found by substituting the coordinate transformation relations (9,10) into the expression for w of (8) and after collecting "optically significant" terms together, one can find that the axial surface displacements for a parabolic mirror due to a constant and linear temperature gradients is (reference 1):

$$\begin{aligned} W(r,\theta) = \ & (\alpha C_3/8R^2)r^4 && \text{(spherical)} \\ & +(\alpha C_1/2R)r^3\cos(\theta) + (\alpha C_2/2R)r^3\sin(\theta) + && \text{(coma)} \\ & +(\alpha C_3 Z_0/2R - (\alpha C_3/2 + \alpha C_0/2R)r^2 && \text{(focus)} \\ & +(\alpha C_1 Z_0 - C_4)\, r\cos(\theta) + (\alpha C_2 Z_0 - Cr_5)\, r\sin(\theta) && \text{(tilt)} \\ & +(\alpha C_3 Z_0^2/2 + \alpha C_0 Z_0 - C_6) && \text{(piston)} \quad (11) \end{aligned}$$

where

$$W = \text{mirror surface deformation caused by temperature}$$

$$r = \text{radius position on surface } (r = \sqrt{x^2+y^2})$$

$$\theta = \text{angular position on surface } (\theta = \arctan (y/x))$$

Here, C4, C5, C6 include arbitrary tilts and piston that depend on the mirror defining system for completeness. As noted in the reference, we used this theoretical solution to verify the finite element solution of a structured mirror with this same temperature distribution. The same procedure can be used to rewrite (8) for other special geometries as well.

It must again be stressed that the homogeneous, unconstrained conditions must obtain. In addition, a mirror of non-uniform thickness with a linear gradient implies that while a flat back could have a uniform T, the curved front could not. The temperature difference from front to back is not the same everywhere. Steady-state heat conduction requires linear temperature gradients. With the assurance that no thermal stresses are introduced and that the displacements are known, it may be possible to change the temperature of elements with fewer optical "side effects".

3.1. Some comments about CTE

For most materials, CTE is a positive, non-zero number; for positive T there is expansion. The term linear can be misleading. For a small range of T, the CTE can be approximated as a constant but for large T (e.g., cryogenic temperature-room temperature) it is not normally a good approximation. If CTE(T) is known, the effect can be integrated, but only if the body remains unconstrained through the entire range of T. It can happen that boundary conditions change and an additional non-linearity is suddenly introduced in mid-range. This has been used to "lock" optical elements at one temperature while allowing them freedom at another. When it is an unplanned event, it can change the elastic properties of the system.

The term homogeneous is also an idealization. A material can have variations of CTE from one point to another. Then the variations of CTE at different temperatures exhibit strain/stress patterns similar to those due to T alone. As an example, we have determined that a pyrex mirror with CTE = 3.0 E-6/C can have only .02 C temperature variation to meet our requirements. If we then change its temperature 20 C the variation of CTE tolerable is then 3.0E-9/C (i.e., one part in 1000) to satisfy the same requirement. As another example, the Hubble space telescope primary mirror is kept at the temperature of the polishing shop where it was made to avoid unknown effects at the lower space temperature.

One can also speculate on a possible solution to at another problem facing telescope design. Boules of glass can be fused together to make large mirrors but the individual pieces often have variations of CTE (reference 2). Since the role of CTE variation is essentially the same as T, it might be worthwhile to arrange the pieces so that approximately linear variations of CTE are present along the coordinate directions. To a first approximation, little thermal stress would be introduced and the thermal distortion effect could be calculated as an equivalent temperature gradient to see its effect on the error budget.

3.2. Extension to systems of multiple components

The solution above also applies to an assembly of components as long as the individual members are all of the same, homogeneous material. If a telescope structure is made entirely of steel components (sans optical elements) then any of its dimensions also change according to (8) as long as this does not introduce constraints.

4. ORTHOGONAL T FUNCTIONS

One can devise additional T functions and a common approach is to use power series. It can also be shown that past the linear terms there are always thermally induced stresses even for unconstrained bodies. If theoretical solutions could be found, they would be geometry dependent and would lack the universality of those above. However, modern FEA programs have temperature inputs that give good approximations to thermo-elastic problems. However, many runs would be required to solve all the possible temperature distributions possible in a practical situation.

Orthogonal functions have the advantage that individual terms are independent of each other and thus can be included, or not, in a sum of component terms. If any arbitrary T can be decomposed into a finite series of orthogonal functions, and each of the individual terms is run as a unit T case, then the stress-displacement result is the sum of unit cases weighted by the component terms of the thermal distribution of T. An example of this technique is presented in reference 1. The functions used were two dimensional Zernike polynomials. The two dimensional (x,y) functions could be used in this case since the nodes of the structured mirror were located in z=constant planes. Zernike terms are also used for optical surfaces but there is no direct relation between the thermo-elastic results and the optical surface terms.

4.1. Non-homogeneous and anisotropic materials.

At some small error level, all materials exhibit both characteristics. This can cause errors in optical elements and should be accounted for. Some processes, such as powdered, sintered beryllium, improve the isotropy of the material. Composite materials are designed to be anisotropic to enhance certain directional properties. This increases the analysis difficulty, but can result in light-weight, stiff, thermally stable structures. When we can afford them, an added advantage is the internal damping afforded for higher frequency regimes now required of telescopes.

4.2. Finite element models.

Most problems will require the use of finite element analysis (FEA). In fact, there are FEA for heat transfer problems and the thermal results can be used as the input for the thermo-elastic solution. There are even software programs that will do both solutions, sequentially, for the same model. There is then a tendency to use the same model for both. This saves time but few, if any, heat transfer problems are the same as elastic problems from the point of view of modeling. The choice of nodes, type of elements, input conditions, etc., have to be good representations of the problem to be solved. The only single-model-solution is to combine the complexities of both such that you have "over-modeled" both regimes. This requires a sensitivity to both models, time-consuming model building, and longer solution runs.

4.3. The time factor, t

Any idealization that T is not changing with time is suspect for the same reasons that "isothermal" and "steady state" are. It makes the analyst's life a little easier but real systems never like to cooperate. FEA heat transfer analysis will solve for time dependent conditions, T(x,y,z,t), and elastic FEAs also find dynamic and time response solutions. HOWEVER, not only is there the modeling problem, but a mathematical and physical distinction between the two time dependent solutions. For the time intervals, ti, i=1,n, each resulting T(x,y,z,ti) can be considered a "snapshot" of thermal input for thermo-elastic FEA. This is a laborious process for most software.

5. SUMMARY

1. Theoretical solutions can be easily misapplied to the wrong conditions but they also can produce good results quickly in many cases.
2. FEAs can be used for both the estimation of T as well as the thermo- elastic stress/displacement relations but dissimilar models probably are required and time dependent cases are truly "time consuming".

6. REFERENCES

1. E. Pearson and L. Stepp, "Response of large optical mirrors to thermal distributions," *SPIE Proceedings*, Vol. 748, pp. 215-228, January 1987.
2. M. H. Krim, G. P. Ruthven, and B. F. Skrobacz, "Accommodating CTE discontinuities in a ULE mosaic mirror," *SPIE Proceedings*, Vol. 1236 (Part 2), pp. 605-614, February 1990.

Athermalization of optical instruments from the optomechanical viewpoint.

Thomas H. Jamieson
Lockheed Missiles and Space Co., Palo Alto, California.

ABSTRACT

The refractive properties and physical dimensions of optical components change with temperature, and hence also do the characteristics of optical systems. Athermalization is the principle of stabilizing the optical performance with respect to temperature, either by designing the optical elements and mounts to be mutually compensating, or by including movable corrective mechanisms. For refractive materials, two coefficients can be defined which characterize the thermo-optical sensitivity, one applicable to uniform temperature changes, and the other to spatial temperature gradients. For normal optical glasses, the effects are small, but for plastics, infrared materials, and liquids, the thermal effects can be so great as to limit their usefulness. Passive athermalization is analogous to achromatism and optical systems can be designed simultaneously achromatic and athermal. The use of composite, and high expansion mounts employing plastics or fluids, gives greater control of thermo-optical effects. Active athermalization uses auxiliary power to drive compensating elements to maintain optical performance. Commercial optical design programs can be used to model and analyze thermally perturbed systems accurately.

1. INTRODUCTION

The theme of this session is stability, and in fact thermal stability is only one more requirement imposed upon the opto-mechanical design of imaging systems. The aim of the optical designer is to ensure that the geometrical aberrations are small and stable over the entrance aperture, over the required field of view, over the waveband of interest, over the range of magnifications in the case of zoom systems and, in addition, over the required range of environmental temperature.

The temperatures that optical instruments (even those produced for the consumer market) experience can be extreme. A furnace camera must endure several hundred degrees, while others operate near the absolute zero of outer space. At such temperature extremes, mere survival of the optics may be as important as maintenance of image quality. All optical instruments are designed to operate satisfactorily over some specified range of temperature, which in the case of military systems is typically from -40°C to +70°C, and even commercial cameras and binoculars can be used in Arizona in summer and in Alaska in winter.

The temperature changes experienced by optics may be with respect to time, in which case the temperature is uniform throughout the optical system, or with respect to space, if there are temperature gradients over the optical elements. In general there is a combination of both. Temperature variations cause changes in the dimensions of the elements, mount and support structures, and in the refractive properties of the optical materials. The actual changes which can occur with increasing temperature, include:-

1) surface radii increase,
2) spherical surfaces become aspheric,
3) inter surface separations normally increase,
4) refractive indices of optical materials increase or decrease,
5) refractive index of air decreases,
6) strain on the optical elements causes warping and bi-refringence.

1.1 Physical Constants

The basic physical coefficients of the materials which control the magnitude of temperature effects are the **linear coefficient of thermal expansion** α defined as:-

$$\alpha = 1/L\,(dL/dT)$$

where L has the dimensions of length and T is temperature, and the **thermal coefficient of refractive index** β defined as:-

$$\beta = dN/dT$$

where N is the refractive index of the transmitting material.

The values of α for virtually all optical materials are positive, vary considerably in magnitude, and for a given material are fairly constant over a wide temperature range. The values of β vary in sign and magnitude for different materials and vary considerably with temperature and wavelength. Care should be taken to distinguish between the value of β relative to air and relative to vacuum (absolute), which can differ significantly due to the large change in the index of air. For any refracting material the index relative to vacuum N(abs) is related to that relative to air N(rel) by:-

$$N(abs) = N(rel) \times N(air)$$

where N(air) is the index of air relative to vacuum. Differentiating gives:-

$$dN/dT(abs) = dN/dT(rel)\,N(air) + N(rel)\,dN(air)/dT$$

Since N(rel) is very close to unity we have:-

$$\beta(abs) = \beta(rel) + dN(air)/dT$$

The value of N(air,T) varies with temperature T according to Penndorf (1957) as follows:-

$$N(air,T) = 1.0549\,(N_{15} - 1)/(1 + 0.00366T).$$

Differentiating with respect T gives dN(air,T)/dT as:-

$$dN(air,T)/dT = -0.003861(N_{15} - 1)/(1 + 0.00366T)^2$$

where N_{15} is the index of air at 15°C, and this according to Edlen (1957) varies with wavelength as:-

$$(N_{15} - 1)\times10^8 = 6432.8 + 2949810/(146 - \nu^2) + 25540/(41 - \nu^2)$$

where $\nu = 1/\lambda$, and λ the wavelength is in μ and T in centigrade. The value of dN(air,T)/dT at 15°C and pressure 760 mm is -0.964×10^{-6} for the wavelength 0.55μ.

2. SPATIAL TEMPERATURE DISTRIBUTION

All portions of an optical system may not be at identical temperatures. Change of ambient conditions, varying solar load, removal of the optics to different environments etc., lead to quasi-stationary temperature differences between elements, and over the element surfaces themselves. An example is that of the optical mount at a different temperature from the optical element and the environment, which occurs frequently in both military and commercial instruments. It has been shown that, for the ideal case of a plate held in a circular mount of infinite extent and conductivity at elevated temperature, a radial quadratic thermal gradient exists over the surface of the plate. This is approximately true for finite mounts, finite conductivities and for lenses as well plates.

Most optical systems only have to endure changing environments but in a few, the elements are deliberately cooled. In high sensitivity longwave infrared sensors some optical elements as well as the detectors are cooled to cryogenic temperatures to increase the signal to noise, and in some cases this cooling must be effected within an few minutes. Mirror surfaces are cooled from the rear, whereas, lenses can only be cooled from the periphery. Athermalization in these cases corresponds to minimizing the magnitude, duration, and effects of thermal gradients.

2.1 Refractors

Consider the effects of temperature change on an optical element of thickness S, small compared to its diameter. This implies that longitudinal temperature gradients are being neglected. The change in optical path difference dW along a reference ray through the center of the element at temperature T, and that along an arbitrary ray through a part of the element of thickness S+dS at temperature T+dT, is:-

$$dW = (N - 1)\,S - (N + dN - 1)\,(S + dS)$$

Now since $dS = \alpha\, S\, dT$ we may rewrite this as:-

$$dW = S\,(N - 1)\,\gamma$$

where:-
$$\gamma = \beta/(N - 1) + \alpha$$

is a thermo-optical coefficient, which depends only upon the physical properties, and describes the sensitivity to spatial temperature distributions. Thus the magnitude of wave aberration dW for a plate or thin lens due to a temperature gradient depends only upon the maximum temperature difference, the element thickness and the optical material, and to this approximation is independent lens diameter and shape.

The majority of optical glasses have values of γ in the range 5 to 25×10^{-6}, except the special fluor crown (FK) and phosphate crown (PK) glasses from Schott and Ohara and the ATF glasses from Hoya. For the common optical glass BK7 ($N = 1.518$ and $\gamma = 9.9 \times 10^{-6}$), of thickness S=10mm, and dT = 10°C, the wave aberration induced is 1.25λ ($\lambda = .546\ \mu$). This is a significant aberration! Figure 1A shows the wave aberration calculated by finite raytrace of a simulated quadratic thermal gradient of 10°C between center and edge of a 10mm thick lens of BK7. This result is typical of standard glasses. A few special glasses, however, have been developed with small and even negative values of γ to desensitize them to gradients, such as FK51, PK51, and FK03 from Ohara. Figure 1B shows the corresponding wave aberration for a lens of FK03 of the same thickness, with the same thermal gradient. Those glasses with small values of γ are those whose increase in physical path length is balanced by the decrease in refractive index.

For optical plastics the γ values are much greater, in the range -100 to -200×10^{-6}, and since the heat capacities and thermal conductivities of plastics are low, the probability and magnitude of potential temperature gradients are high. Figure 1C shows the corresponding wave aberration for the same gradient applied to a polycarbonate lens ($\gamma = -117 \times 10^{-6}$) of the same dimensions. Note the scale on the wave aberration curve in this case is greater by a factor of 10.

For optical materials in wavebands other than the visible, the effects are also large. Section 4.1.2 gives a fuller description of infrared materials, but for typical materials such as germanium and chalcogenide glasses (AMTIR1, ZnS etc.) in the infrared, the γ values range from 50 to 150×10^{-6}, and for the alkali metal halides such as NaCl, KBr etc., the values are mostly negative from -10 to -120×10^{-6}. The wave aberration (for $\lambda = 10\ \mu$) of a 10mm thick germanium singlet subject to the above gradient, is illustrated in Figure 1D. The very large value of

γ for germanium creates a significant aberration, but the combination of thermal capacity, conductivity, and long infrared wavelength, means that thermal gradients are generally not so serious in germanium systems.

The negative values of γ give rise to the possibility of balancing positive and negative γ materials together, to produce a resultant zero effect. As stated above, only a few catalog optical glasses have negative γ values, one such being FK03. To illustrate the technique of desensitizing designs, we compare the effects of the above thermal gradient on a common BK7/SF2 doublet, to that of a doublet of FK03/SF2, in Figures 2A and 2B. In the latter example the relative thickness of either element is varied to achieve the balance, and the thermal changes to the wavefronts are greatly reduced, but some higher order effects are becoming evident.

We have noted the deleterious effect of thermal gradients on the optical performance, but an equally important aspect of athermalization is the prevention of gradients in the first place. Gradients are minimized in magnitude and duration by selecting materials which can absorb large amounts of heat with small resulting temperature rise, and conduct this heat away rapidly. A measure of this ability is the quantity, the thermal diffusivity, defined as:-

$$k/(\rho\ c)$$

where k is the thermal conductivity, ρ is the density, and c is the specific heat.

The values of the diffusivity for a material give a comparative indication of its tendency to create and sustain temperature gradients, and this value can be easily calculated from glass catalog data. Optical glasses have quite high values, optical plastics are low, and germanium is high compared to other infrared materials.

2.2 Reflectors

The thermal properties of pure mirror systems are determined by their expansion coefficients, since their dN/dT values are zero. Modern mirrors, especially those for large aperture, astronomical, or space applications, are complex structures, and their thermal design and analysis is beyond the scope of this paper, except for a few general comments. Solid monolithic construction for mirrors becomes increasingly impossible as their diameter increases. The weight and rigidity of mirror materials cause them to deform under their own weight. Lightweighting is achieved by, say, molding or machining recesses in the mirror rear surface, or by employing a composite structure, of a thin mirror face and rear sheet, fused to a ribbed or "egg-crate" type support structure. This technique simultaneously reduces the mirror weight and by reducing the heat capacity, improves the temperature uniformity across the surface. Thermal analysis is further complicated by the fact that the cross section of the support structure may vary in thickness between the mirror center and edge, in the shape of a single or double "arch", to increase the mirror rigidity.

As stated in Section 2, in sensitive infrared systems, the optical elements both refractive and reflective, are often cooled to prevent the thermal background radiation from swamping the sensor signal. The elimination of surface temperature gradients depends upon material choice, and additionally, upon the design of the cooling system, which may simply be radiation cooling to vacuum in space applications, or mechanical refrigerators or fluid transfer cooling for rapid cooldown applications. Uneven mirror cooling results in surface deformations, and athermalization may be a very involved procedure. The choice of the face sheet and support structure materials is particular important, with beryllium being particularly advantageous. The method and area of attachment of face sheet to support structure can greatly affect thermal uniformity of the mirror surface and the consequent wave aberrations.

Modern mirror optical designs often employ a high degree of decentering to avoid central obscuration problems, and in this case thermal gradients or even uniform temperature change, cause deviations of the line of sight, which have to be taken into account at the design satage. For fast cooldown mirrors, note that the mirror surfaces only are cooled, and the mount and spacer remain at ambient temperature.

3. TEMPERATURE CHANGE WITH RESPECT TO TIME

Different considerations apply in the case of thermal "soak", where the complete optical system is at a uniform temperature, but this varies with time. The primary effect is to produce a change in focus position, and to a lesser extent, to change the magnification and upset the balance between geometrical aberrations, all of which cause serious degradations in image quality.

It is instructive to calculate the individual contributions to focus shift for a typical single lens, due to radius change, thickness change, and index change. We shall consider three materials, namely, BK7, polycarbonate, and germanium, since they span the range of available thermo-optical constants. The results, as a percentage of focal length, calculated from simple first order ray tracing, are given below:-

	radius	thickness	index	total
BK7	.0143	-.00045	-.0114	.00238
polycarbonate	.1399	-.00983	.4746	.6056
germanium	.0118	-.00059	-.2644	-.2532

We see that the effects are additive; the radius effect is always positive; the index effect can be of either sign and is usually the greatest in magnitude; and the thickness effect is an order of magnitude smaller. Consequently, we can consider, without large error, only "thin" lenses.

Each lens has two radii R_1 and R_2, refractive index N, spectral dispersion ν, linear expansion coefficient α, and temperature coefficient of index β. As a general rule, the optical designer chooses the values of R_1, R_2, and N to permit the lens to operate over a finite aperture and field, ν for operation over a finite waveband, and α and β for operation over a finite temperature range. The converging power K of a thin lens is given by the well known expression:-

$$K = (N - 1)\,(1/R_1 - 1/R_2)$$

Differentiating with respect to temperature we have:-

$$dK/dT = (1/R_1 - 1/R_2)\,dN/dT - (N - 1)\,(1/R_1{}^2 dR_1/dT - 1/R_2{}^2 dR_2/dT)$$

Since $\alpha = 1/R\ dR/dT$ we can rewrite this as:-

$$dK/dT = (N - 1)\,(1/R_1 - 1/R_2)\,(\beta/(N - 1) - \alpha)$$

Thus:-

$$dK/dT = K\,\delta$$

where:-

$$\delta = \beta/(N - 1) - \alpha$$

is a second thermo-optical coefficient which characterizes the change of converging power of the lens with respect to temperature. We can write this relationship in terms of the lens focal length f, which is equal to 1/K, as:-

$$df = -\,\delta\, f\, dT$$

and thus $-\delta$ is analogous to a "focal length expansion coefficient", namely the fractional change in focal length per degree of temperature change.

The value of α is positive for all refracting materials, and the value of δ for optical glasses range from -32×10^{-6} to 22×10^{-6}. Those optical glasses with small δ values, are those for which the increase in focal length, due to expansion of surface radii, is balanced by a corresponding decrease, due to higher index. The δ values of optical plastics and infrared materials are (similar to γ), more extreme than optical glasses.

We see that there are two coefficients, γ and δ, both functions of α and dN/dT, which are useful in describing optical temperature effects.

Note the relationship between them, namely, that $\delta = 2\,\alpha + \gamma$. Thus glasses with small values of γ, have large values of δ, and we have the somewhat paradoxical situation that those glasses which are insensitive to temporal thermal gradients are highly sensitive to spatial thermal gradients. For example the Schott glass FK51, which has the one of the smallest numerical values of γ in the catalog, has one of the largest values for δ. For plastic and infrared materials the values of dN/dT dominate, and there is little difference between γ and δ.

4. ATHERMALIZATION OF FOCUS - TECHNIQUES

Uniform temperature change causes optical performance degradation, primarily through image defocus. The means employed to counteract the effect come under three general principles.

Passive Athermalization - Optical
No moving parts are employed, and the onus is on the optical designer to choose appropriate optical materials to follow the mechanical expansion.

Passive Athermalization - Mechanical
Moving parts may be employed, but the energy required for motion comes from the thermal expansion itself, and the onus is on the mechanical designer to choose the structure to match the focus shift.

Active Athermalization - Electromechanical
No optomechanical compensation is attempted and supplied energy provides motion by means of an electromechanical servo system, to track the focus change.

The necessity and/or choice of athermalization method depends upon the application of the optical instrument, its size, weight, cost, required performance, and environmental specification.

4.1 Passive Athermalization - Optical

The concept of designing optics to be insensitive to temperature change is not new. A paper of Perry dates from 1943, and Grey applied the ideas to military optical plastic designs as early as 1948. Volosov dealt with thermal soak in 1958, and Kohler and Strahle(1968) describe various lens designs, some athermal with respect to soak and others with respect to spatial gradients. The results of these investigations, indeed, were probably behind the development of some special glasses types at Schott Glaswerk. Straw and Estelle (1980) in separate papers investigated the plastic/glass Cooke triplet system used in inexpensive commercial cameras. Several review papers have appeared, Jamieson (1981), and, devoted to the infrared Roberts (1989) and Povey (1986), who emphasizes active techniques.

The goal of athermalization is that the focus position should remain fixed with respect to the mount structure, not necessarily fixed with respect to the lens, i.e. we match the focus change of the optical system to the thermal expansion of the mount. To achieve this in single lens systems, we choose a glass with the same value of δ as the value of $-\alpha$ for the support structure. The expansion coefficients of metal optical mount materials range from almost zero for invar steel to $26\text{x}10^{-6}$ for magnesium. Modern construction plastics have much larger coefficients, and although they have undesirable properties such as hysteresis, and moisture absorption, they are increasingly employed in medium quality optics because of low cost, and ease of fabrication.

Aluminum is probably the most common prototype mount material, and the value of α ($23.6\text{x}10^{-6}$) is large compared to the δ value of common optical glasses except for the special FK and PK glasses. Note that these glass types (say FK51) have two useful thermal properties, namely values of γ close to zero making them insensitive to spatial thermal gradients, and values of δ , which closely match the properties of aluminum. The use of optics of a single optical material, however, is normally permissible only where chromatic aberration is unimportant, such as laser optics.

4.1.1 Doublets

Achromatic operation (i.e. over a finite range of wavelengths) demands that refractive systems employ multiple materials of differing dispersive powers. The powers K_1 and K_2 of the individual elements of an achromatic doublet of total power K , are given by the well known relations:- -

$$K_1 = \nu_1 K/(\nu_1 - \nu_2) \quad \text{and} \quad K_2 = -\nu_2 K/(\nu_1 - \nu_2)$$

where ν_1, ν_2 are the Abbe dispersion coefficients of the doublet materials. We can define δ_D , as the thermo-optical coefficient for the achromatic doublet by:-

$$K\,\delta_D = K_1\,\delta_2 + K_2\,\delta_2$$

where δ_1, δ_2 are the individual thermal- optical coefficients, which gives:-

$$\delta_D = (\nu_1\,\delta_1 - \nu_2\,\delta_2)/(\nu_1 - \nu_2)$$

The doublet will be athermal as well as achromatic if the value of δ_D is the same as $-\alpha$ for the mount. A search of the glass catalogs enables the value of δ_D for achromatic material pairs to be calculated and matched to the mount to some tolerance. Approximately 150 doublet pairs from the Schott catalog, were found with a $-\delta_D$ value within $0.5x10^{-6}$ of aluminum. A requirement, however, on the glass choice is that the individual lens powers be small, otherwise the surface curvatures become impracticably steep, and the high order geometrical aberrations unacceptably large. This secondary requirement in the search reduces the number of solutions. Table 2 shows the individual lens powers for K equal to unity, of glass pairs matched to aluminum to the above tolerance, with the sum of the absolute values of K_1 and K_2 less than 3.5. Figure 3A gives the wave aberrations, at 20°C and 60°C, for one solution, namely the athermal pair FK5/SF14, of 250mm focal length at F/5, compared to a common BK7/SF2 doublet of the same aperture and focal length.

A similar search for solutions can be carried out for any specified mount material. Since there is normally a greater choice of optical glasses than mount materials, it is more reasonable that the optical designer should conform to the mechanical specifications. If, however, the glass choice is restricted by, say, special transmission requirements, the mechanical designer may have to employ unusual materials or even a composite mount of multiple materials (see section 4.2) to achieve the deired expansion.

4.1.2 Infrared Imaging Systems

Probably more effort has been devoted to athermalizing infrared optical systems than to all others, because dN/dT of most materials suitable for use in this waveband, is extremely high, and this type of system frequently must conform to full military environmental temperature range. The materials used fall into three groups, namely semiconductor types (germanium and silicon), chalcogenide glasses (ZnSe,AMTIR1), and crystalline alkali metal halides (NaCl, KBr etc.). Germanium is a most desirable infrared optical material because of its high refractive index and low spectral dispersion, but it has the great disadvantage that its refractive index and even transmission are highly temperature sensitive. To quantify the effect, we compute the temperature range dT over which a germanium lens maintains diffraction limited performance, which by convention implies a defocus aberration of $\lambda/4$. For a lens of focal length f, the focal shift due to a temperature increment dT is -d f dT, and this gives rise to a wave aberration:-

$$dW = -\delta f\, dT/(2\, F\#)^2$$

where F# is the relative aperture (ratio of focal length to entrance aperture) and if $dW = \lambda/4$, then solving for dT, we have:-

$$dT = 2\,\lambda\,(F\#)^2/(f\,\delta)$$

For a typical infrared lens of germanium of 500mm focal length at F/5, the allowable temperature change is only 8°C. If we compare this to a similar doublet lens of BK7/F2, the corresponding temperature tolerance in wavfront terms is comparable, and athermalization is important in both types of system. Note that for germanium, the value of $-\delta$ is $-136\text{x}10^{-6}$, which implies that as the temperature increases, the focus moves in towards the lens, and thus it will match no ordinary spacer. Table 3 quotes a collection of optical materials and their coefficients, suitable for the 8 μ to 12 μ thermal infrared waveband.

We may add an element of a different material to form a doublet to athermalize the germanium. A suitable material is KRS5, which, although a double halide (thallium bromo-iodide), is insoluble in water and can be optically polished. Figure 3C shows the wave aberrations for a germanium single lens of 400 mm focal length at F/4 at 20°C and 60°C contrasted with a similar Ge/KRS5 doublet. We see that the temperature change has added 2λ to the germanium lens, whereas it produces no measurable aberration change in the doublet. The lens powers appropriate for athermalization mean, however, that both elements are of the same sign, so that the chromatic aberration of the doublet is uncorrected and greater than germanium alone.

Reliable data on the thermal coefficients of optical materials is often difficult to obtain. The optical glass manufacturers quote dN/dT values for their products, although not all glasses are included. Schott and Ohara give values for up to six temperature ranges from -40°C to +80°C, and at five wavelengths in the visible waveband, but Hoya quotes the value at 0.6328μ only. The situation is more unsatisfactory with other materials and wavebands especially at low temperatures. The values for optical plastics are somewhat uncertain due to their variable composition. Although some chalcogenide glasses may be of varying composition, other infrared materials are either simple elements (Ge and Si), or stoichiometric compounds (KBr, ZnSe) with well defined indices. Probably the main reason for the lack of reliable thermo-optical data of infrared materials, is insufficient measurements. This lack of reliable infrared thermal data is especially significant, since cooling is often employed, and the optical properties change considerably on cooling to cryogenic temperatures, and even the expansion coefficients of mount materials decrease significantly, approaching zero at 0°K.

Because of the importance of germanium as an infrared optical material, we repeat its thermal (from 100 - 293°K) and spectral variation (from 1.9 to 18μ) of the index $N(\lambda,T)$, which according to Li (1980), is as follows:-

$$N^2(\lambda,T) = e(T) + L(T)\,(A_0 + A_1T + A_2T^2)/\lambda^2$$

where $L(T) = \exp(-3\,M(T))$,

$$A_0 = 2.5381,\ A_1 = .001826,\ \text{and}\ A_2 = 2.8888\times10^{-6},$$

$$e(T) = 15.2892 + 1.4549\times10^{-3}T + 3.5078\times10^{-6}T^2 - 1.2071\times10^{-9}T^3,$$

$$M(T) = -.00089 + 2.626\times10^{-6}(T-100) + 1.463\times10^{-8}(T-100)^2 - 2.221\times10^{-11}(T-100)^3,$$

with λ in μ and T in degrees K . Although this very complicated expression is claimed to be valid only down to 100°K, we have used it at 80°K and 40°K without noticeable error.

4.1.3 Triplets

By employing three materials, it is always possible to achromatize and match exactly the thermal defocus to the expansion of the mount. For three thin elements close together of total power K, the system will be achromatic and athermal if:-

$$K_1 + K_2 + K_3 = K$$

$$K_1/v_1 + K_2/v_2 + K_3/v_3 = 0$$

$$K_1\delta_1 + K_2\delta_2 + K_3\delta_3 = K\alpha$$

For visible waveband systems employing conventional glasses, this degree of athermalization is rarely necessary, but it may be used with advantage in the infrared waveband, where the materials have large values of α and dN/dT. Applying the above equations to the materials in Table 3, generates 364 solutions, from which we select those with the smallest absolute values of lens powers (Σ K), in order that the systems be reasonably practical. A series of achromatic and athermal triplets with small absolute powers, is given in Table 4. The smallest power solution is the triplet Ge/NaCl/CsI with a value for Σ K = 1.11, for a total power of unity. For durable practical solutions, we should probably not use crystalline halides, which are brittle and water soluble, except perhaps KRS5. With this restriction the best solution is Ge/ZnS/KRS5 (Σ K of 1.15), and Figure 4A shows its wave aberrations at 20°C and 60°C. The best non-halide combination is ZnS/AMTIR1/Ge with Σ k of 3.09.

Infrared materials other than germanium and silicon, are expensive and difficult to fabricate, so that if we must use elements of those materials, there is some advantage in keeping their diameters small. We can achieve this if we use a single germanium lens and a small sub-aperture doublet of the other materials, separated by a distance d from it. The above equations now become:-

$$h_1 K_1 + h_2 K_2 + h_2 K_3 = h_1 K$$

$$h_1^2 K_1/\nu_1 + h_2^2 K_2/\nu_2 + h_2^2 K_3/\nu_3 = 0$$

$$h_1^2 K_1 \delta_1 + h_2^2 K_2 \delta_2 + h_2^2 K_3 \delta_3 = K\alpha$$

where h_1, and h_2 are the ray heights at the single lens and doublet respectively. Taking $h_2/h_1=0.5$ we cut in half the size of the non germanium elements and search for solutions using the above equations. Note that the separation d is given by:-

$$d = (1 - h_2/h_1)/K_1$$

and we choose only those solutions which give rise to positive d values.

Using the 14 materials, there are 91 such systems, 47 of which have a positive d and the solution with the smallest value of Σ k (2.02) is Ge-ZnS/KRS5. The sub-aperture doublet provides three benefits; it balances the chromatic, and the thermal aberrations of the germanium single lens, and corrects the spherical aberration, eliminating the need for aspherizing the germanium. The design is shown in Figure 4B, and although rather long (1.7 times the focal length), shorter systems of this type do exist with higher lens powers.

4.1.4 Multiple Lens Systems

A general expression for the thermal defocus of a system of separated lenses is very complex, but a good approximation can be derived by consideration of the wave aberration dW_i due to a change df_i in the focal length f_i of each lens. The value of dW_i is given by:-

$$dW_i = 0.5\ (h_i/f_i)^2\ df_i$$

where h_i is the paraxial ray height at each lens, and summing over all lenses, gives the total wave aberration as:-

$$\Sigma\ dW_i = 0.5\ \Sigma\ (h_i^2 K_i \delta_i)\ dT$$

Compound lens systems consist of units of single lenses, doublets or triplets and the total thermal defocus can be calculated, by applying this relationship, with the values of δ_i corresponding to each unit. Note that this method takes no account of the changing mechanical spaces between the lenses. In addition, for a general set of lenses the length of the mount is not even approximately equal to the focal length. An improved procedure is to convert the wavefront $\Sigma\ dW_i$ to an effective focus shift, and subtract from it, the expansion of the actual lens mount. The mount length is approximately equal to the overall length of the optics from first element to final focus, and varies with the type of lens design.

As an example, the telephoto lens, commonly used because of its compact nature, has an overall length less than its focal length, and consists basically of two lens groups, with powers K_1 and K_2, which are of opposite sign. For a total power of K, typical values of K_1 and K_2 are K_1=2K and K_2=-K with a separation of 0.25/K, which gives $dW = 0.5(1.75h_1{}^2K\ \delta)$, where h_1 is the semi-aperture and δ is some approximate average for the glasses in the system. Thus the thermal defocus for a telephoto lens is greater than for the equivalent single lens, while the length of the mount is smaller, and for the above case is 0.75/K. On the other hand, in a Petzval type lens, used for its wide aperture capability, the converging power is shared equally between two groups of the same sign. Typically K_1 = K/2 and K_2 = K and in this case the thermal defocus is smaller than the focal length, while the overall length is longer, namely $0.75h_1{}^2K\delta$ and 1.5/K, respectively.

Zoom lenses are extreme cases of multi-lens systems, consisting of several groups of elements, some of which, move longitudinally to alter the gross focal length and others to re-focus the system for varying conjugate distances. The overall length bears little relationship to focal length and completely passive athermalization techniques are generally not possible in this type of system. Zoom lenses achieve their variable focal length effect by balancing the power of one group of elements against others, and even small thermal changes to the focal lengths or image positions of individual groups can greatly upset the overall zoom effect and the residual balance of chromatic and geometrical aberrations.

4.1.5 Plastic and Liquid lenses

Plastics are becoming increasing popular as optical materials because of the low cost of material and manufacture. The four most common types are acrylic, polystyrene, polycarbonate, and the co-polymer SAN. As was noted in section 2.1, their major disadvantage, is their temperature sensitivity, since both α and dN/dT values are all very large. To demonstrate this sensitivity, an acrylic lens of 100mm focal length has 0.55mm of thermal defocus for 20°C temperature change, and in an aluminum mount at F/5 this corresponds to 2.4 λ aberration. Although only a few materials exist, it is possible to athermalize all-plastic designs, if we employ three different plastics to satisfy the equations of section 4.1.3, and a highly expansive plastic mount. Figure 5A shows a close triplet design of acrylic, polystyrene, and SAN designed for a polycarbonate (α = 67.5×10^{-6}) plastic mount. The design is achromatic and athermal and although individual lens powers are high, it can operate at F/5 over a large temperature range with respectable performance. The wave aberrations for this design are shown in Figure 5B.

The all-plastic separated (Cooke) triplet has been investigated in considerable detail, because of its importance as a lens for small format cameras. At low relative apertures, say F/8, the thermal defocus can just be tolerated, but wider apertures demand athermalization. According to Estelle, it is not possible to athermalize the all-plastic design. In order to correct the aberrations over a substantial field with three elements, the choice of powers for the elements is severely restricted, and the actual values required are approximately +2, -3, +2 , for a total triplet power of unity. When all elements are plastic, the resulting thermal defocus is of the same magnitude as a single plastic lens of unit focal length. Partial athermalization occurs if we choose plastic for one positive and one negative element, their defocus balances one another, and use ordinary optical glass, typically up front, for the third unbalanced defocus.

Recently, liquids have been proposed as optical elements, sandwiched between the elements of doublet or triplet lenses. The dispersive properties of optical liquids are even more extreme than optical plastics. Liquids of adequate consistency and specification for use in optical design, are manufactured by Cargille for use in refractometry, chemical analysis, engineering and forensic applications. The perceived advantage of optical liquids is that their abnormal optical properties permit their use as substitutes for expensive rare earth glasses, which are necessary in wide spectral waveband apochromatic lens designs. In effect, since these liquids are relatively inexpensive and only small amounts are required, we gain extra optical elements at little cost by filling the air spaces, in doublet and triplet designs, with them. This benefit is partially off-set, when we take into account the extra complexity of mount, required to retain the liquids in the appropriate space. An additional advantage is that coatings are not required on glass to liquid interfaces, which may reduce scatter fron the system.

We propose an additional use for such liquids, namely to athermalize plastic optical designs. An achromatic doublet of acrylic and polycarbonate gives medium quality performance, which degrades rapidly with temperature. If we add a liquid to the air space, we can achromatize and athermalize using the equations of Section 4.1.3. The effective expansion coefficient for the liquid is zero, and the radii and thickness of the liquid element are controlled by the plastics containing it. The doublet becomes a "cemented" triplet, and since we must maintain the individual optical powers, the triplet has only one degree of freedom, and can only be bent to minimize, but not correct, geometrical aberrations. The liquid we have used, has the extreme value of dN/dT -900×10^{-6} , and the wave aberrations of this design are shown in Figure

6A. The optimum bend has larger residual spherical aberration than the equivalent "dry" doublet, but has an order of magnitude higher thermal stability. The spherical aberration can be removed by aspherizing one surface, which may add little to the cost, since plastic is molded rather than ground and polished. In laser diode optics achromatization is not required, so we can maintain the athermalization yet reduce the spherical aberration, by adjusting the power distribution between acrylic and the polycarbonate elements. For a doublet of this type with 10mm focal length at F/3, the r.m.s. aberration is as low as 0.025 λ over a range of plus or minus 20°C. The system and corresponding wave aberrations is illustrated in Figure 6B.

Several hundred of these liquids are available, together their refractive index, dispersions and dN/dT, from their manufacturer, but increased measurement accuracy is necessary, if they are used for crtical designs.

4.1.6 Reflective systems

For pure mirrors, there is no value of dN/dT to consider, and the value of δ for an individual mirror is simply $-\alpha$ of the substrate. In general reflective systems consist of multiple mirrors separated by large separations, and the change of these spaces has a significant effect on the total thermal defocus. Consequently the relationships of Section 4.1.4 are not generally applicable, since they ignore varying separations. If we consider even the simplest multiple mirror system, namely a two surface cassegrain or gregorian separated by a distance d , the expression for the thermal defocus is a complicated function of d , the two radii, and the two expansion coefficients. This expression is of limited value in the choice of mirror substrate materials for athermalization, but we can gain some insight into the possibilities if we consider the particular case of optical surface, spacer or mount material of being identical. All points on the optics and mount expand at the same rate and the system is self-athermalized since the focus remains precisely constant with respect to the mount. A typical example is diamond machined aluminum mirrors supported in an aluminum housing. A more general system is a two element cassegrain, with dissimilar mirror surface materials, held in a mount of a third material. The primary mirror has positive converging power and the secondary is diverging, so that it is somewhat analogous to a telephoto type construction. The mirror substrates might be of lower expansion materials (optical glass, ULE) than the mount (aluminum), or conversely, higher expansive substrates (aluminum) in a mount of lower expansion material (steel). Simple paraxial ray tracing shows that in the former case, the focus moves in towards the mirrors, while in the latter, the movement is away from the mirrors, and the actual movement can be quite large due to the telephoto effect.

If we calculate an actual case, namely, F/4 300mm focal length cassegrain with a separation and back focus both equal to 75mm, i.e. the image is located at the vertex of the primary mirror. With a steel mount ($\alpha = 17.3\text{x}10^{-6}$) a 40°C temperature change produces 0.0519mm change in the length of the mount. With both mirrors of aluminum ($\alpha = 23.6\text{x}10^{-6}$) the defocus is 0.248mm, whereas with beryllium mirrors ($\alpha = 11.6\text{x}10^{-6}$) the defocus is -0.125mm. The corresponding aberrations are -2.68λ and +2.44λ respectively. In principle, an appropriate choice of substrate materials, (the two mirrors need not be identical) and spacer material, allows any cassegrain design to be athermalized. Alternately the focal position can be moved back or forth to a specified position in either direction with respect to the mirrors. In practice the choice of substrate, and spacer materials, is driven by other considerations, and a perfect match may be impossible.

From the above, we see that an effective method of athermalizing an infrared telescope system is to convert a refractive design to reflective! The inverting telescope is a common component in thermal imagery, and is used to magnify (or minify) the system field of view to that of the scanner, typically 30° to 60°. The telescope consists of an objective and "eyepiece", the latter being invariably refractive and often of germanium, because its high refractive index allows the eyepiece to operate over this large field of view. If the objective is also germanium, the total system is very temperature sensitive. From section 4.1.2, the temperature rise causes the focus of germanium lenses to move inwards, while the mount moves outwards, we see that the thermal defocus of objective and eyepiece add to produce a very large effect in the direction opposite to the mount. If the objective is reflective, then as described above, by proper choice of surface and mount materials, the defocus of the objective outwards from the mirrors, can be matched to the eyepiece defocus in towards the germanium lenses.

4.2. Passive Athermalization - Mechanical

The optical materials are fixed and the positions of the optical elements are controlled by choosing the mechanical structure such that its thermal expansion matches the thermal defocus of the optics. A large number of materials can be employed as spacers or structural elements, with a correspondingly wide range of expansion coefficients, from invar steel with α of zero or slightly negative, to plastics such as CR-39 with $\alpha = 138\text{x}10^{-6}$. A list of potential structural and spacer materials is given in Table 5A.

A suitable spacer material, with exactly the correct value to match the thermal defocus in any particular case, is not always available. The effective expansion of the spacer can be modified by various techniques.

For example it may be set at an angle to the optical axis, say, by incorporating the expanding member into a strut type arrangement supporting the secondary mirror of a cassegrain telescope. The expansion of the strut is scaled by the cosine of the angle to produce a smaller displacement along the axis.

The expanding element may be in the form of a circumferential ring or series of rings in tandem. Their expansion can be made to drive a screw thread connected to a focusing element. In this arrangement, the effective thermal expansion can be negative, since the distance between element and focus can be decreased with increasing temperature by employing a screw thread of the appropriate direction. If the expansion rings are high expansion material, then considerable focussing movement can be effected.

More than one material may be used to form composite or multiple element structures. By combining spacers of length L_1 and L_2 of materials of expansion coefficients α_1 and α_2, we produce an effective expansion coefficient for the composite spacer of :-

$$(\alpha_1 L_1 + \alpha_2 L_2)/(L_1 + L_2)$$

Note that if L_2 is negative and greater numerically than L_1 then the effective expansion is negative, and corresponds to a re-entrant mount. The lengths L_1 and L_2 are of low and high expansion materials respectively, and if, say $L_2 = -.9\,L_1$, the value of α is $10\,(\alpha_1 - .9\alpha_2)$, based on the net spacing of $.1L_1$, which is of course much smaller than the overall mount length. For L_1 and L_2 of SS304 and Al1100 respectively, using Table 5A gives an effective α value of $-39.4\text{x}10^{-6}$. The lengths L_1 and L_2 of expanding material may be in the form of cylinders, one inside the other, or simple stacked rods. This technique requires an athermalizing mount length much greater than the system focal length. Consequently the method is most suitable for optical designs in which the overall length is considerably longer than the focal length, such as the inverted telephoto.

Another use of the expansion of composite material mounts is to control the position of a compensator element by means of a bi-metal element. Equal lengths of metals of differing expansion coefficients are fused together along their length, and when the temperature changes, the differential expansion of each material causes the device to distort into a circular arc. This distortion can be made to control the optical element position. A suitable bi-metallic arrangement to control a lens position is an annulus, or washer, which distorts from flat to spherical or conical. The outer periphery is fixed to the lens mount and the inner rim of the annulus drives the focusing element back and forth.

Another effect which has been proposed for mechanical athermalization, is the expansion of fluids such as waxes, liquids or gases contained in a cylinder and piston arrangement or thin metal bellows. The high expansion coefficients of the fluids in this type of arrangement, can provide a great magnification of the focus compensator travel. The reservoir containing the fluid should have a large surface area to ensure fast heat transfer and prompt response.

All mechanical athermalization is passive in the sense that it is automatically controlled by the temperature and requires no power to be supplied. The methods are relatively simple and have high accuracy, but on the other hand, they have disadvantages of extra bulk, weight, cost, and suffer from high hysteresis and often low reliability.

4.3 Active Athermalization - Electromechanical

No attempt is made to thermally desensitize the optomechanical design and relies on compensator element(s) driven in a temperature controlled fashion using information from separate temperature sensors. This is properly a subject for a session on electronic or servo mechanism design, and is a brute force type solution to athermalization, which should be used only when other techniques fail. It is most suitable in complex optical systems such as multi configuration optics or zoom lenses, in which an electromechanical focus mechanism already exists. An electric motor drives the focusing compensator, usually through a servo loop, which receives control signals from temperature sensors, placed in several locations. Positional control can be simple analog direct drive or via digital processor, the latter being a much more flexible approach. In complex systems, the focus position is usually a non-linear function depending on many factors, among which is conjugate distance, and atmospheric pressure, so that the addition of temperature as a variable simply represents a perturbation to the control signals. The increased availability of electronic micro-circuitry renders the use of microprocessors and sophisticated software control, increasingly attractive. The penalty to be paid for this added flexibility, is considerable extra cost and weight, and lower system reliability.

5. THERMAL MODELING IN OPTICAL DESIGN CODES

Whatever method of athermalization is chosen, an accurate assessment of its effect should be performed by geometrical ray tracing. A number of comprehensive optical design programs are available commercially to analyze in detail the performance of optical systems. All such programs are capable of calculating the effects of temperature change (at least temperature soak), but some require considerably more user input than others. The capabilities of the programs vary considerably, and the procedures described here are based on those available in Optima, the Lockheed in house software, and CodeV from Optical Research Associates. Both programs include comprehensive thermal modeling features, including uniform temperature change and radial gradient modeling. Of equal importance to the user, is the availability of a large number of stored physical constants for optical glasses, mirror substrate, and mount materials.

The detailed changes required to made to the optical system to accurately simulate the effects of temperature, are the following. The refractive index of air is converted to a nominal design temperature (usually 20°C), using equations such as those in Section 1.1. The indices of all refractive materials are converted to absolute values by multiplying them by the air index. The air indices at the new temperature, follow from the above equations and the indices of the refracting materials are calculated using the values of the stored expansion and dN/dT coefficients. The value of dN/dT used, is interpolated according to the actual wavelength and temperature used. Optima stores the coefficients for 5 wavelengths and 6 temperature ranges, while CodeV has 5 wavelengths and a single temperature value.

The radius of each optical surface is increased according to its stored expansion coefficient. The effect of temperature on inter surface distances is less well defined. The elements are usually held in a single tube mount, and kept separated by means of spacers, which may be of a different material from the mount. Thus it may be difficult to determine whether the total mount controls the expansion of the optics and the spacers become loose and/or compressed by this expansion, or whether each lens and spacer expands according to its own coefficient and the mount "stretches" to accommodate their expansion. The mechanical designer may decide which is the more accurate assumption for his particular layout, and either option can be modeled by specifying the surface separations to the program in the appropriate manner.

Radial thermal gradients are modeled in an approximate fashion, by assuming that the refractive index and thickness of a lens element changes as a function of the radial distance from the optical axis. The temperature profile is converted to a thickness increment and fitted to an effective aspheric figuring function, added equally to the sag of both surfaces of the lens. Similarly, the temperature variation is converted into refractive index profile and fitted to a multi-term polynomial, which is used to compute the actual index at any ray height. For reasonably thin lenses, the index is constant along any ray through the lens, and conventional ray trace procedure is approximately valid, but for thick lenses a true gradient index ray trace should be used. Similarly axial thermal gradients may be approximated by dividing the lenses into thin sections by introducing extra dummy surfaces at differing temperatures.

The design examples of thermal soak and thermal gradients, given in this paper were computed according to the above methods.

6. SUMMARY

We have described the effects of spatial and temporal thermal variations on the performance of a broad number of types of optical systems, and the computational procedures required to simulate accurately the performance reduction.. We have described some optical and mechanical design techniques, which we can use to stabilize or desensitize the image quality with respect to temperature. Whether and what type of athermalization is used, depends on a host of factors, including:- relative aperture, waveband, materials employed, single or multi focal length, image quality, environmental specification, and considerations of size, weight, cost, power, and reliability. As a word of warning, it should be remarked that systems can behave in an unpredictable thermal fashion, and some experimental verification may be advisable.

7. REFERENCES

R. Penndorf, J.O.S.A. **47**, 176-182 (1957)
B. Edlen, J.O.S.A. **43**, 339-344 (1953)
J. W. Perry, Proc. Pays. Soc. (London) **55**, 257-285 (1943)
D. S. Grey, J. O. S. A. **38**, 542-546 (1948)
D. S. Volosov, Opt. Spectrosc. **4**, (1958)
H. Kohler and F. Strahle, Proc. I. C. O. **9**, 116-153 (1972)
K. Straw, Proc. S.P.I.E. **237**, 386-391 (1980)
L. R. Estelle, Proc. S.P.I.E. **237**, 392-401 (1980)
T. H. Jamieson, Opt. Eng. **20**, 156-160 (1981)
V. Povey, Proc. S.P.I.E. **655**, 142-153 (1986)
M. Roberts, Proc. S.P.I.E. **1049**, 72-75 (1989)
H. H. Li, J. Pays. and Chem. Ref. Data **9**, 561-658 (1980)

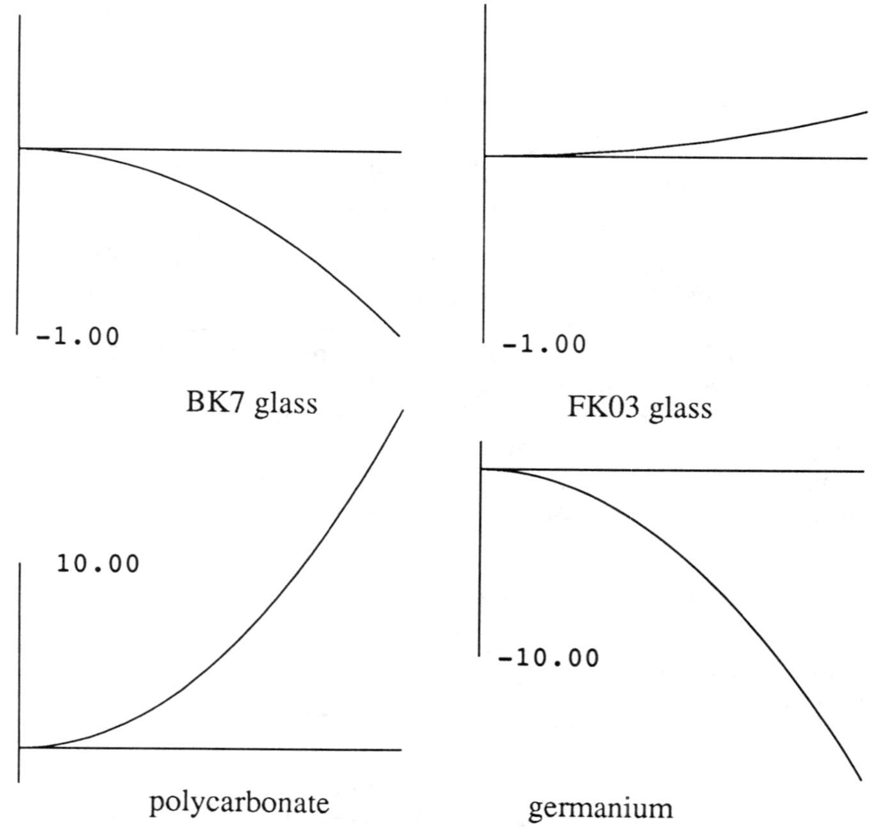

Figures 1A to1D. Wave aberration due to quadratic thermal gradient of 10 °C in 10mm thick BK7, FK03, polycarbonate and germanium.

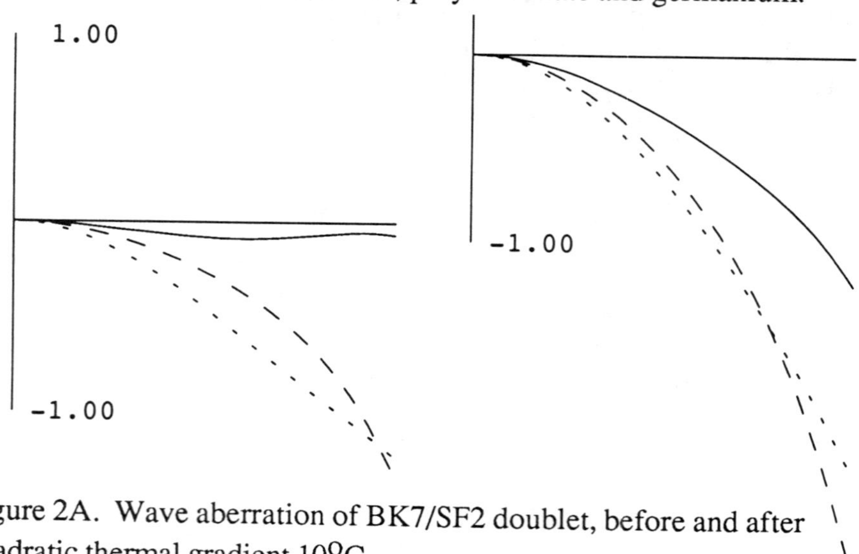

Figure 2A. Wave aberration of BK7/SF2 doublet, before and after quadratic thermal gradient 10°C

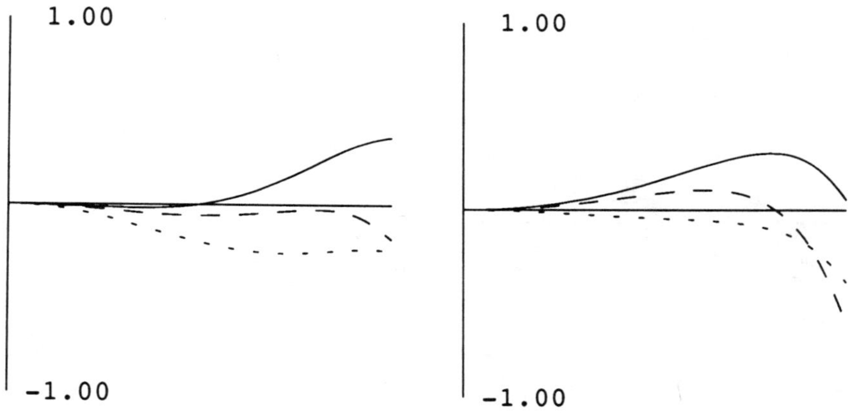

Figure 2B. Wave aberration of FK03/SF2 doublet, at uniform temperature and with quadratic thermal gradient 10°C.

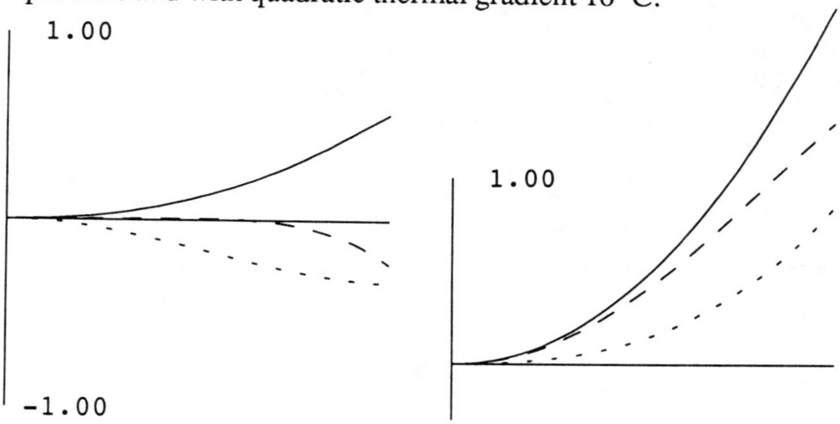

Figure 3A. Wave aberration of BK7/SF2 doublet lens(F/5, 250mm focal length) at 20°C and 60°C.

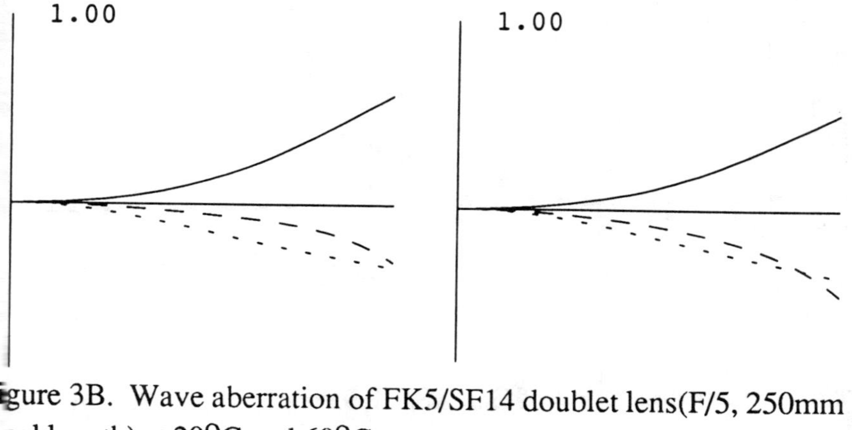

Figure 3B. Wave aberration of FK5/SF14 doublet lens(F/5, 250mm focal length) at 20°C and 60°C.

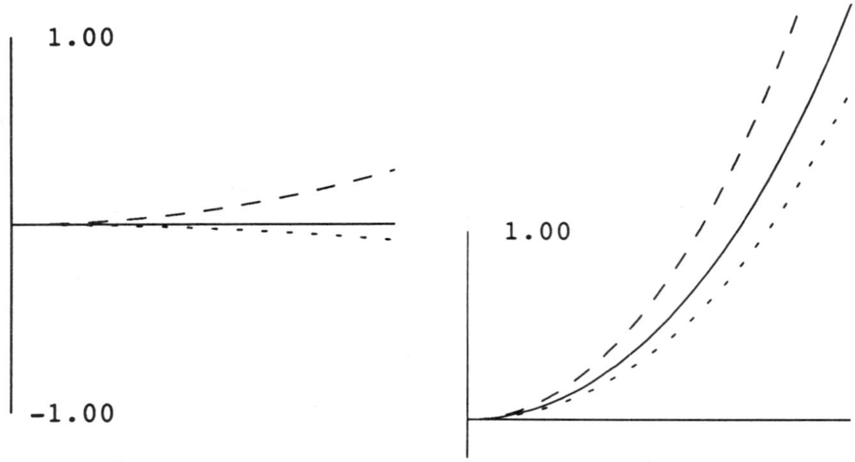

Figure 3C. Wave aberration of germanium single lens, (400mm focal length at F/4) at 20°C and 60°C.

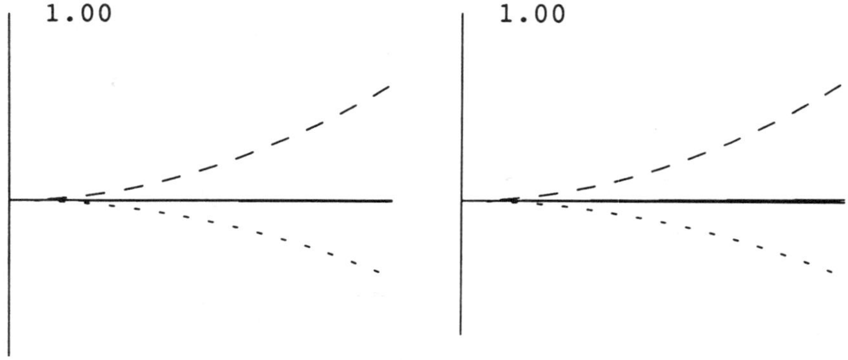

Figure 3D. Wave aberration of athermal, non-achromatic Ge/KRS5 doublet lens, (F/4, 400mm focal length) at 20°C and 60°C.

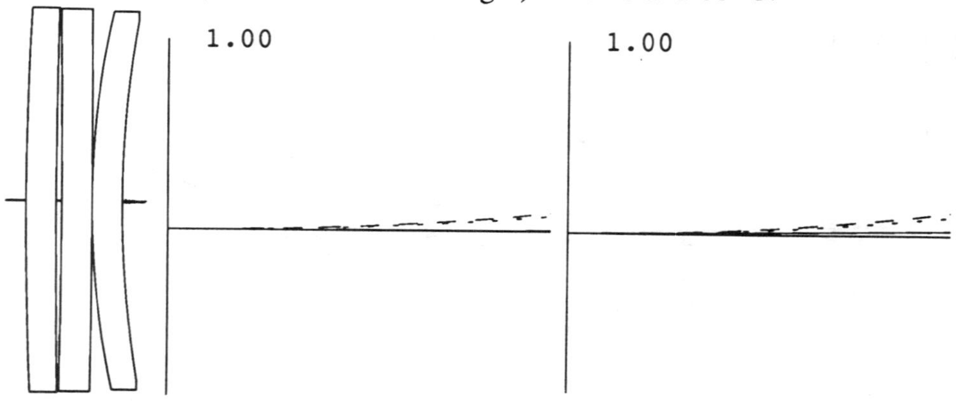

Figure 4A. Wave aberration of Ge/ZnS/KRS5 triplet (F/4, 400mm focal length) at 20°C and 60°C.

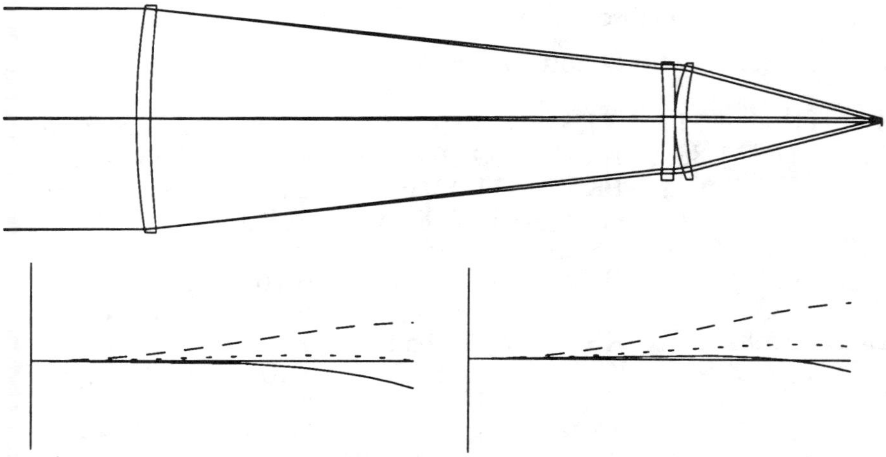

Figure 4B. Ge single lens plus ZnS/KRS5 doublet (F/4, 400mm focal length) at 20°C and 60°C.

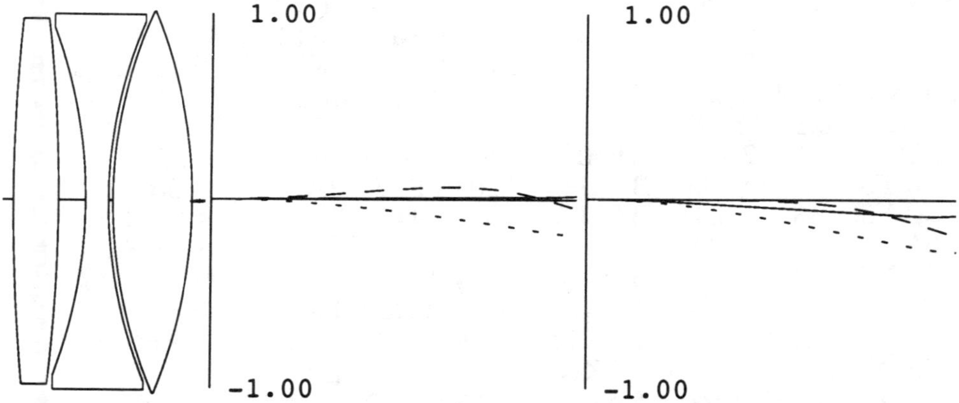

Figure 5A. 100mm F/6.3 all plastic triplet, at 20°C and 40°C

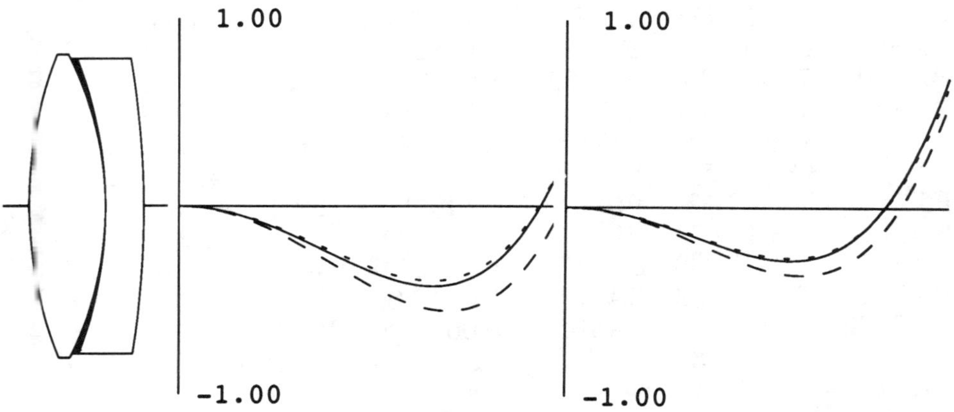

Figure 5B. 100mm focal length F/5 acrylic/liquid/polycarbonate triplet lens at 20°C and40°C

Table 1. Values for absolute thermo-optical coefficients δ and γ ($x10^{-6}$) respectively for Schott glasses.

FK3	-11.66	4.74	FK5	-14.89	3.51	FK51	-29.35	-2.15
FK52	-30.68	-1.88	FK54	-30.66	-1.46	PK2	-3.95	9.85
PK3	-4.58	9.62	PK50	-11.42	6.18	PK51	-31.82	-5.62
PK51A	-28.12	-2.72	PSK2	-3.97	8.83	PSK3	-3.58	8.82
PSK50	-11.64	5.56	PSK52	-11.55	5.45	PSK53	-15.37	3.43
BK1	-6.54	8.86	BK3	1.81	12.41	BK6	-5.88	9.72
BK7	-4.33	9.87	BK10	-1.49	10.11	BALK1	-10.94	7.26
BALKN3	-5.82	9.98	K3	-7.08	9.52	K4	-3.01	11.59
K5	-7.76	8.64	K7	-8.03	8.77	K10	-1.04	11.96
K11	-2.31	10.49	K50	-2.18	11.82	K51	5.75	14.35
ZK5	-6.80	10.60	ZKN7	6.07	15.07	BAK1	-5.72	9.48
BAK2	-6.96	9.04	BAK4	-3.00	11.00	BAK5	-6.73	8.87
BAK6	-4.91	9.69	BAK50	8.99	16.39	SK1	-1.62	10.58
SK2	-0.67	11.33	SK4	-4.22	8.58	SK5	-1.67	9.33
SK6	-0.94	11.46	SK9	-0.26	11.74	SK10	-5.03	8.97
SK11	-3.51	9.49	SK12	-3.54	9.26	SK14	-3.60	8.40
SK15	-5.89	7.91	SK16	-5.68	6.92	SKN18	0.18	12.98
SK19	-4.14	8.66	SK20	-3.38	9.42	SK51	-12.69	5.11
SK52	0.34	12.34	SK55	-3.05	8.95	KF3	-3.96	12.24
KF9	-1.42	12.18	KF50	-1.63	12.97	BALF4	0.26	13.06
BALF5	-7.27	8.93	BALF6	-2.03	11.37	BALF8	-6.50	10.10
BALF50	-8.19	8.41	BALF51	-6.75	9.45	SSK1	-1.40	11.20
SSK2	-1.51	10.89	K3	-1.84	11.36	SSK4	-3.46	8.74
SSKN8	-4.94	9.26	SSK51	-6.86	8.34	SSK52	-3.53	9.87
LAKN7	-7.86	6.34	LAK8	-1.07	10.13	LAK9	-2.89	9.71
LAK10	-1.15	10.25	LAK11	-7.96	6.44	LAKN12	-9.69	5.51
LAKN13	-11.03	5.77	LAKN14	-2.27	8.73	LAKN16	1.25	11.85
LAK21	-7.63	5.97	LAKL21	-3.28	8.92	LAK23	-11.19	4.61
LAK28	-0.73	10.67	LLF1	-5.01	11.19	LLF4	-5.73	10.67
LLF6	-3.53	11.47	BAF3	-4.09	11.51	BAF4	-4.04	11.76
BAF5	-2.74	11.26	BAF8	-1.69	12.31	BAF9	-0.60	12.40
BAF50	-7.17	9.43	BAF52	-8.00	8.80	BAF53	-1.92	11.08
BAF54	-1.96	10.44	LF1	-6.45	10.55	LF3	-5.07	11.13
LF5	-6.75	11.45	LF8	-6.76	10.24	F1	-5.32	12.08
F2	-3.82	12.58	F3	-2.58	13.42	F4	-4.37	12.23
F5	-2.48	13.52	F6	-3.79	13.21	F9	-2.20	13.20
FN11	-3.93	11.07	F14	-2.36	13.44	F15	-3.26	12.94
BASF1	-5.11	11.89	BASF5	6.36	22.16	BASF6	-3.69	11.11
BASF10	-5.38	11.82	BASF13	-2.17	12.03	BASF51	6.32	17.12
BASF52	4.51	14.91	BASF54	-0.06	14.54	BASF55	5.63	15.83
BASF56	-2.36	13.84	BASF57	-2.21	11.99	BASF64	-3.68	10.92
LAF2	-9.02	7.18	LAF3	-7.33	7.87	LAFN7	3.46	14.06
LAF9	3.11	17.51	LAFN10	-0.15	11.25	LAF21	-1.90	9.90
LAF22	-4.24	9.76	LAFN23	-9.08	7.12	LAFN24	0.85	11.65

LAF25	2.79	14.39	LAF26	3.35	14.55	LAFN28	-1.18	10.42
LASF3	0.66	11.66	LASFN9	-3.18	11.62	LASF11	-0.14	11.46
LASF13	3.10	15.50	LASFN15	-2.33	10.67	LASFN30	-2.04	10.36
LASFN31	-2.64	10.96	LASF32	-4.37	11.43	SF1	0.11	16.31
SF2	-3.33	13.47	SF3	0.33	17.13	SF4	2.51	18.51
SF5	-2.18	14.22	SF6	3.72	19.92	SFL6	-9.61	8.39
SF7	-1.82	13.98	SF9	0.97	17.37	SF10	1.05	16.05
SF11	8.01	20.21	SF12	-1.89	13.71	SF13	2.96	17.16
SF14	5.75	18.95	SF15	-1.15	14.65	SF16	-3.01	13.79
SF18	1.02	17.22	SF19	-0.27	15.13	SF50	-8.73	11.47
SF53	0.42	16.82	SF54	1.69	17.09	SF55	0.64	17.04
SF56	-0.10	15.70	SFL56	-8.36	9.04	SF57	4.81	21.41
SF58	4.09	22.09	SF61	1.62	17.42	SF62	-1.17	15.23
SF63	0.82	17.22	SFN64	-7.30	9.70	TIK1	-17.85	2.75
TIF2	-9.87	7.33	KZFN1	-3.39	10.81	KZFN2	-1.54	10.66
KZFS1	-2.01	7.99	KZFSN2	1.82	10.82	KZFSN4	1.89	10.89
KZFSN5	2.04	11.04	KZFS6	-0.96	9.24	KZFSN7	2.61	12.21
KZFS8	3.30	13.90	LGSK2	-20.31	3.89			

Table 2. Achromatic glass pairs of unit power, athermal to aluminum ($\alpha = 23.6 \times 10^{-6}$) to within 0.5×10^{-6}.

FK5	F2	2.059	-1.059	FK5	BASF1	2.226	-1.226
FK5	BASF10	2.240	-1.240	FK5	BASF54	1.832	-0.832
FK5	LAF9	1.670	-0.670	FK5	SF7	1.958	-0.958
FK5	SF9	1.788	-0.788	FK5	SF12	1.917	-0.917
FK5	SF14	1.599	-0.599	PK50	LASF13	2.096	-1.096
PK50	KZFSN7	2.135	-1.135	PSK53	SF56	1.692	-0.692
SK51	FN11	2.489	-1.489	SK51	LAF9	1.885	-0.885
SK51	SF12	2.271	-1.271	LAKN13	SF13	2.067	-1.067
LAKN13	SF18	2.209	-1.209	LAKN13	SF62	2.485	-1.485
LAK23	SF9	2.183	-1.183	LAK23	SF14	1.854	-0.854
LAK23	SF19	2.345	-1.345	BAF50	SF14	2.468	-1.468
BAF52	SF14	2.327	-1.327	LAF2	SF6	2.315	-1.315
LAFN23	SF14	2.138	-1.138	SF6	TIF2	-1.235	2.235
SF61	TIF2	-1.486	2.486				

Table 3. Refractive index N, Abbe ν value, expansion coefficient α, absolute thermo-optical coefficients β, δ, and γ ($\times 10^{-6}$) for infrared, plastic and some liquid optical materials.

Infrared:-	N	ν	α	β	δ	γ
GERM	4.00	861.03	5.70	392.14	124.87	136.27
SILICON	3.43	241.00	2.50	156.69	61.93	66.93
ZNSE	2.41	57.84	7.60	58.68	34.12	49.32

ZNS	2.20	22.76	7.90	40.88	26.17	41.97
CDTE	2.68	148.79	4.50	95.42	52.46	61.46
AMTIR1	2.50	119.97	13.00	69.59	33.47	59.47
GAAS	3.28	107.00	5.90	146.84	58.56	70.36
KRS5	2.37	164.73	61.00	-236.29	-233.22	-111.22
KCL	1.46	30.27	36.00	-28.90	-99.25	-27.25
KBR	1.53	61.45	27.60	-41.47	-106.44	-51.24
KI	1.62	89.26	42.60	-51.56	-125.76	-40.56
NACL	1.49	18.65	44.00	-26.44	-97.44	-9.44
CSBR	1.66	132.30	47.90	77.40	68.84	164.64
CSI	1.74	236.20	50.00	-91.68	-174.05	-74.05

Plastic:-	N	ν	α	β	δ	γ
ACRYLIC	1.4918	57.46	62	-105	-278.4	-154.4
POLYCARB	1.5855	29.91	68	-107	-253.4	-117.4
POLYSTYR	1.5905	30.87	50	-140	-289.7	-189.7
SAN	1.5674	34.81	50	-110	-246.5	-146.5

Liquids:-	N	ν	α	β	δ	γ
CG305974	1.30	97.44	0.00	-328.26	-1076.26	-1076.26
CG505257	1.50	25.80	0.00	-466.45	-923.66	-923.66
CG710209	1.71	20.93	0.00	-657.65	-926.27	-926.27
CG810184	1.81	18.41	0.00	-901.74	-1113.27	-1113.27

Table 4. Achromatic athermal triplets and powers K_1, K_2, K_3 for total power of unity.

			K_1	K_2	K_3
GERM	ZNSE	CSI	0.63	-0.078	0.440
GERM	ZNS	KI	0.497	-0.161	0.664
GERM	ZNS	CSI	0.541	-0.065	0.524
GERM	KRS5	KCL	0.624	0.488	-0.112
GERM	KRS5	NACL	0.606	0.459	-0.065
GERM	KCL	CSI	0.521	-0.091	0.571
GERM	KI	NACL	0.416	0.726	-0.142
GERM	NACL	CSI	0.511	-0.054	0.543
GERM	CSI	AGCL	0.545	0.633	-0.179
ZNS	CDTE	KRS5	-0.176	0.888	0.288
ZNS	CDTE	CSI	-0.156	0.795	0.362
ZNS	KRS5	CSBR	-0.194	0.341	0.853
ZNS	CSBR	CSI	-0.169	0.344	-0.139
CDTE	KI	NACL	0.592	0.590	-0.182
CDTE	NACL	CSI	0.698	-0.121	0.423
AMTIR1	KRS5	NACL	0.867	0.302	-0.169
AMTIR1	NACL	CSI	0.772	-0.150	0.378
GAAS	KRS5	NACL	0.799	0.383	-0.183

GAAS	NACL	CSI	0.692	-0.157	0.465
KRS5	NACL	CSBR	0.396	-0.151	0.755
KI	NACL	CSBR	0.655	-0.196	0.541
NACL	CSBR	CSI	-0.130	0.651	0.479

Table 5A. Coefficients of thermal expansion of common structural materials ($x10^{-6}$), values approximate. From CodeV manual.

ABS (medium impact	86.4	Al 356 (cast)	21.4
Al 713 (Tenzalloy)	24.1	Al 771 (Precedent 71A)	24.7
Al 1100 (wrought)	23.6	Al 2024	23.2
Al 6061	23.4	Al 7075	23.6
Be-Cu 17200	17.8	BeO	6.7
Be S-200F impact ground	11.2	Brass, Cartridge 26000	20.0
Brass, free cutting	20.5	Cordierite (CG4, Ge)	-0.07
CR-39 (Allyl diglycol)	138	Cu 10200	17.6
DUCTILE cast iron nodule	11.9	Granite	8.3
Graphite Epoxy GY70/X30	0.2	Invar 35	0.56
Invar 36	1.5	Kovar	5.5
Polycarbonate 141 grade	67.5	Mg AZ31B-F or H24	25.2
Metal Matrix Gr/Al (40%)	0.4	Moly (Low Carbon Grade)	5.0
Moly TZM w/Ti & Zr	5.2	Nylon (transparent)	77.6
Plexiglass G	83.0	SS303/303Se (Austenitic)	17.3
SS304 (Austenitic)	17.3	SS316 (Austenitic)	16.0
SS416/416Se Martensitic	9.9	SS430 (Ferritic)	10.4
SS440S/B/C Martensitic	10.1	Polystrene, 30% glass fiber	86.4
SiCAl Metal Matrix Bars	14.8	SiCAl Metal Matrix Co	12.4/10.8
Table/Bench, 400 Series	10.0	Table/Bench - (6061T6)	23.0
Table/Bench, Super Invar	-0.18	Table/Bench - Steel	11.0
Titanium	10.8	Ti - 6Al - 4V	9.5

Table 5B. Coefficients of thermal expansion of mirror substrate materials ($x10^{-6}$), values approximate. From CodeV manual.

Cer-Vit (O/I TM) C-101	0.3	Fused Silica CGW 7940	0.52
Fused Silica Amersil Types	0.51	Zerodur	0.05
Be I-70A	11.6	Be I-220	11.6
Be I-400	11.2		

Dimensional instability of materials: how critical is it in the design of optical instruments?

Roger A. Paquin
Advanced Materials Consultant
Tucson, Arizona

Abstract

Dimensional instability exists to some extent in all components no matter what the materials may be. So the question is not "how can we eliminate instability?" but rather, "how can we reduce it to a tolerable level?" The maximum allowable dimensional instability will vary with application and depends on the particular component and its role in the optical instrument. The purpose of this paper is to provide the basis for deciding how much can be tolerated and for making intelligent choices in the selection of materials and processes for components that will achieve stability design goals with which to meet optical instrument performance specifications. This basis is a better understanding of the causes of instability and methods for minimizing instability.

After a discussion of tolerable levels of instability, four types of dimensional instability are defined: thermal, temporal, cycling and hysteresis, with examples given for each. The principal causes of these instabilities: external stress, changes in internal stress, microstructural changes and inhomogeneity/anisotropy of properties, are explained in some detail along with a discussion of material types and properties. Most importantly, methods for minimizing the instabilities are shown. This discussion includes specific recommendations for commonly used materials including: processing techniques to minimize instability, specific problems observed in some materials and how to avoid the problems, and some general guidelines regarding the effects of fabrication methods on stability.

It is most important to realize that increasingly tighter specifications for optical instruments mean that the optomechanical designer must work concurrently with other engineering disciplines, particularly materials and processes engineers, to insure the desired thermal and temporal stability of the product.

1. INTRODUCTION

All materials are dimensionally unstable to some degree. In preparing to design and fabricate dimensionally stable components, it is important to realize that this implies controlling the sources of dimensional instabilities to a level such that any dimensional changes that occur are kept within specified tolerances. To be able to accomplish this requires an understanding of the sources of these instabilities. The primary purpose of this paper is to impart such an understanding.

The key to building stable instruments is knowing the requirements. Stable materials can then be chosen from which components can be fabricated utilizing methods that minimize introduction of dimensional instabilities. But what exactly is dimensional instability? My definition is simply the dimensional change that occurs in response to internal or external influences.

The problem is to control this dimensional change, a distortion or strain in the component of interest, to levels that will not compromise instrument requirements. If we require stability on the order of machining tolerances, strain of approximately 10^{-3}, there is no serious problem. If parts must maintain dimensional tolerances to parts per million, e.g., micrometers/meter, then care and consideration must be given to materials selection and processing steps. This is the realm of precision optical instruments. If nanotolerences are required, e.g., nanometers/meter, there is little help available, although this has recently become the subject of increasing study. For nanostability, the principles of controlling the sources of dimensional instability must be diligently applied – and then hope for the best.

Once a tolerance has been specified to bound the amount of allowable instability, the next step is determining the potential sources of instability and controlling them to meet the specification. The balance of this paper explains the common types of instabilities and their sources, and gives suggestions as to how they can be controlled. While this paper is, of necessity an overview, I recommend for further study Marschall and Maringer's excellent book[1] on the subject, although it is unfortunately now out of print, and the two volumes of SPIE Proceedings dealing specifically with dimensional stability, volumes 1335[2] and 1533[3].

2. TYPES OF INSTABILITY

The evidence of dimensional instability is all around us and takes many forms, from the warping of a 2x4, which is readily apparent, to the figure change in a mirror subjected to a change in environment. I have chosen to classify instabilities into four categories which I call:

- temporal instability,
- thermal/mechanical cycle instability,
- thermal instability and
- hysteresis.

Each of these can have magnitudes ranging from sub-nanostrain to very large numbers, as the following examples show.

Temporal instability is the change that takes place in a component as a function of time in a fixed environment. It is a permanent change. For example, Figure 1 shows the behavior of two sets of nominally similar one inch gage blocks tested at NBS over a period of roughly thirty years[4]. The best sample, No. 2 in set B, changed only one microinch in 22 years, while the worst changed 24 μin in 25 years. Note that the worst sample was changing at the rate of 10^{-6}in./in./yr. That is a very small amount and yet it is totally unacceptable for the application. This kind of behavior is usually associated with relaxation of residual stress.

Thermal/mechanical cycling instability is the change measured in a fixed environment after exposure to a variable environment. It too is a permanent dimensional change. A common example is the change that takes place in fiber reinforced composites as shown in Figure 2 for boron reinforced aluminum[5]. The figure shows the behavior of three samples which changed length by over 0.02% on the first cycle, but the amount of change decreased with each succeeding cycle, approaching an asymptote. Later I show this kind of behavior in other materials. For composites, the cause for the dimensional changes is usually internal micro-cracking and in single phase materials it is usually some other form of internal stress relief. Similar behavior has been observed with mechanical cycling and vibration[6].

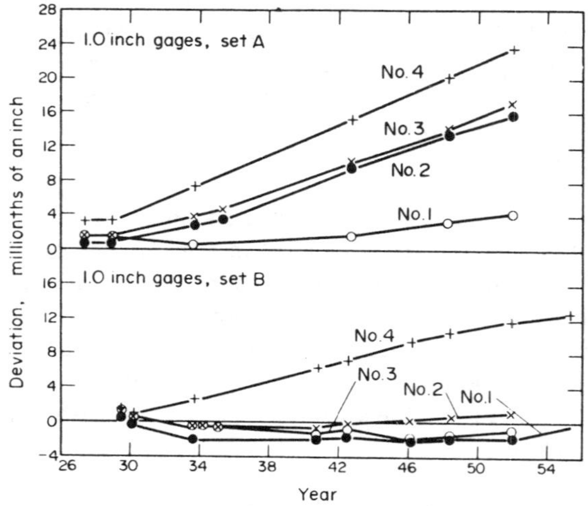

Figure 1. Example of temporal instability of two sets of nominally similar one inch steel gage blocks[4].

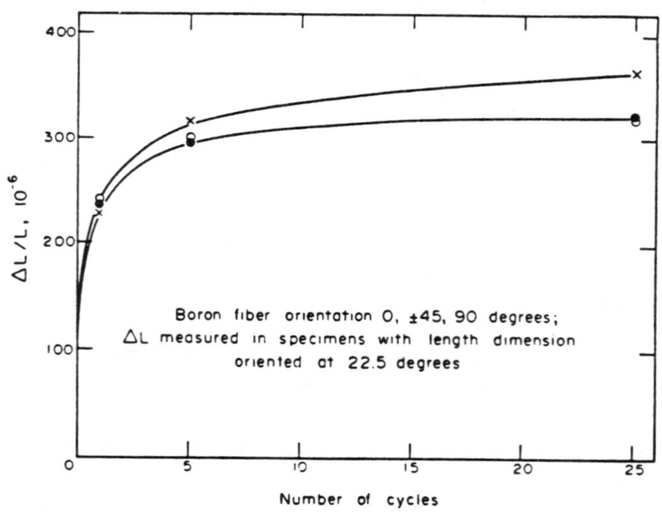

Figure 2. Example of thermal cycling instability of boron/aluminum composite samples thermally cycled from -55°C to +55°C[5].

Thermal instability, as I choose to define it, is the dimensional change measured in a fixed environment after a change from another fixed environment, independent of the environmental path. This dimensional change is reversible upon returning to the original conditions. Figure 3 shows evidence of just such a change. This beryllium (Be) mirror was made from an experimental billet produced in the late 60's which had a substantial amount of thermal expansion inhomogeneity[7]. It was interferometrically tested many times over a period of ten years and always exhibited the same distortion when heated as is shown in the figure, and always returned to the same optical figure at room temperature, within the 0.02 wave accuracy of the instrument. This behavior has been virtually eliminated in modern Be materials.

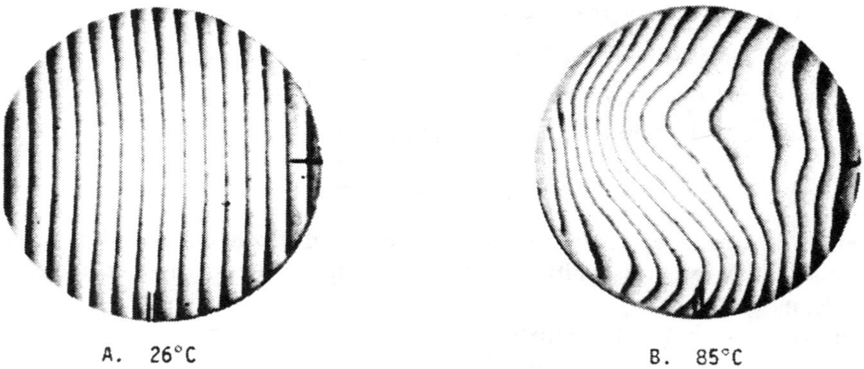

Figure 3. Optical interferograms of an electroless nickel coated experimental beryllium alloy mirror (ca. 1968) showing a reversible thermal instability of approximately two waves[7].

What I call hysteresis is the change measured in a fixed environment and dependent on the environmental path to another fixed environment. This type of distortion can be permanent or reversible. For example, in Figure 4, the length of Zerodur on cooling from 300°C to 20°C depends on the cooling rate[8]. This is typical behavior for glasses containing MgO. But note that the curves are parallel below 150°C indicating that the temperature range of sensitivity is between 150°C and 300°C. This behavior has been eliminated in a new version of this material called Zerodur M.

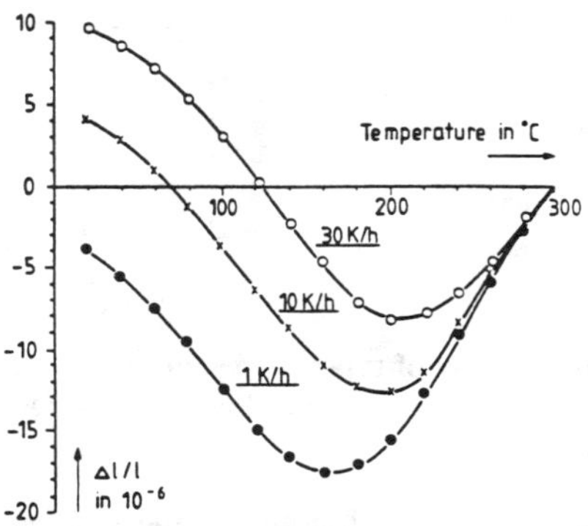

Figure 4. Thermal length contraction of Zerodur for three cooling rates from 300°C to 20°C illustrating a hysteresis type of dimensional instability[8].

These are the major types of dimensional instabilities that can be encountered in components for optical instruments. I will show later that almost all observed instabilities can be placed into one of these four categories and that the sources of these instabilities are readily understood so that their effects can be minimized.

3. SOURCES OF DIMENSIONAL CHANGE

The sources of dimensional changes such as those described above can be attributed to one or more of the following factors:

- externally applied stress,
- changes in internal stresses,
- microstructural changes and/or
- inhomogeneity/anisotropy of properties.

I will describe each of these in some detail in the following sections.

3.1 External Stress

When an external stress is applied to a component, if it behaves according to Hooke's law, it should deform elastically no matter how long the stress is applied and return exactly to its original shape when the stress is removed. But this being an imperfect world, where all practical materials are imperfect, there are other responses to externally applied stress[9] as shown in Figure 5.

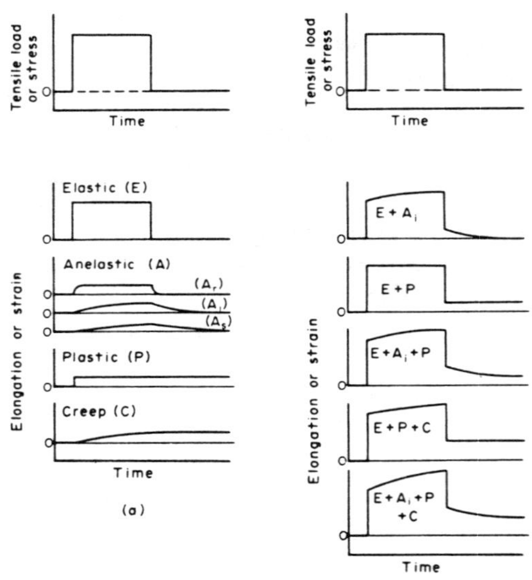

Figure 5. Schematic illustration of types of response to an instantaneously applied and removed tensile stress: (a) individual types of response, and (b) combined responses[9].

If a load is applied suddenly, held for a length of time and then released as shown at the top of the figure, the elastic response is as shown for curve (E), a typical response for glasses, ceramics, and brittle metals. A time dependent type of elastic response is shown in (A) and is called anelastic strain. The three curves show rapid, intermediate and slow response times. For this type of behavior there is no strain when the load is first applied, but it increases toward an asymptote with time; when the load is removed, the strain asymptotically returns to zero. Anelastic behavior has been observed in some glass ceramics at low temperature, but is more commonly observed in polymers, and particularly in polymer matrix composites where both the anelastic effect and mechanical properties are effected by moisture content and temperature. This behavior has been summarized by Wolff[10]. A combination of elastic and anelastic behavior is shown in the E+A curve. Plastic strain is permanent and does not decrease as the load is removed, as shown in P. The most common behavior for metals, a combination of elastic and plastic response to stress is shown in the E+P curve. Time dependent plastic strain is called creep, C in the figure. Other possible combinations are also shown.

There are a number of material properties that are important to dimensional stability. A listing for a number of optical and structural materials is given in Table 1. Among the important thermal properties are: coefficient of thermal expansion (CTE), more precisely called instantaneous linear coefficient of thermal expansion (α'), specific heat (c_p), thermal conductivity (K) and thermal diffusivity (D). Thermal figures of merit are the steady state (CTE/K) and transient (CTE/D) distortion coefficients. The important mechanical properties are elastic modulus (Young's Modulus, E) or stiffness represented by the slope of the stress vs. strain curve as in Figure 6; Poisson's ratio, the relationship between tensile (or compressive) and shear strain; yield strength (at 0.2% offset), the stress to cause 2 x 10^{-3} permanent or plastic strain; and ultimate or fracture strength. Fracture toughness (K_{Ic}) is related to the energy to propagate a crack. Specific stiffness (E/ρ) is a measure of structural efficiency. Particularly important for precision applications are microyield strength, the stress to cause 1 x 10^{-6} plastic strain (one microstrain); and microcreep strength, which has no acceptable definition other than that it is less than the microyield strength and is a constant stress that produces microstrain after some period of time.

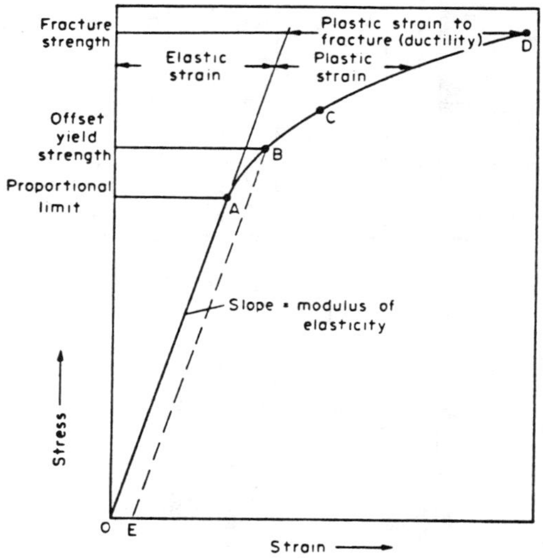

Figure 6. Schematic stress-strain diagram to illustrate typical material behavior[11].

Microyield behavior can not be inferred from the macro-mechanical properties of either yield strength or modulus. For example, in Figure 7 the behavior of I-400 Be and 2024-T4 aluminum (Al), metals with approximately the same yield strength, are compared. The Be sample exhibits a microyield strength of only 6.8 ksi, but with increased

Table 1. Properties of selected Materials

MATERIALS	ρ DENSITY	E YOUNG'S MODULUS	E/ρ SPECIFIC STIFFNESS	MYS* MICROYIELD STRENGTH	Klc FRACTURE TOUGHNESS	CTE THERMAL EXPANSION	K THERMAL CONDUCTIVITY	Cp SPECIFIC HEAT	D THERMAL DIFFUSIVITY	DISTORTION COEFF CTE/K STEADY STATE	DISTORTION COEFF CTE/D TRANSIENT
	lb/cuin *g/cc*	Msi *GPa*	10E8 in	Ksi	Ksi in	ppm/F *ppm/K*	Btu/hr-ft-F *W/cm-K*	Btu/lb-F *W-sec/g-K*	sqin/hr *sqcm/sec*	*um/W*	*usec/sqcm-K*
PREFERRED	Small	Large	Large	Large	Large	Small	Large	Large	Large	Small	Small
Fused Silica	0.079 *2.19*	10.4 *72*	1.32	[1.2]	<1	0.28 *0.5*	0.8 *0.014*	0.178 *0.745*	4.72 *0.0085*	*3620*	*59*
ULE	0.8 *2.21*	9.7 *67*	1.21	[1.2]	<1	0.008 *0.015*	0.76 *0.013*	0.183 *0.77*	4.32 *0.0077*	*115*	*19*
Zerodur	0.019 *2.53*	13.3 *92*	1.46	[1.5]	<2	-0.05 *-0.09*	0.91 *0.016*	0.194 *0.812*	4.27 *0.0076*	*-570*	*-12*
Beryllium I-70A	0.067 *1.85*	42 *287*	6.3	2	>10	6.3 *11.4*	112 *1.93*	0.436 *1.825*	319 *0.572*	*590*	*20*
Aluminum 6061-T6	0.097 *2.7*	9.9 *68*	1.02	10	>20	13.1 *23.6*	96.5 *1.67*	0.214 *0.896*	385 *0.69*	*1410*	*34*
Copper	0.323 *8.95*	17 *117*	0.53	0.5	>20	9.2 *16.6*	230 *3.98*	0.092 *0.385*	644 *1.155*	*410*	*14*
Stainless Steel 304	0.29 *8.02*	28 *193*	0.97	7	>20	9.6 *17.3*	9.4 *0.162*	0.12 *0.5*	22.5 *0.04*	*10600*	*428*
Invar 36	0.291 *8.05*	20.5 *141*	0.7	3	>20	0.7 *1.26*	6 *0.104*	0.12 *0.5*	14.3 *0.026*	*1190*	*48*
Super Invar	0.294 *8.13*	21.5 *148*	0.73	5	>20	0.17 *0.3*	6 *0.105*	0.12 *0.51*	13.8 *0.025*	*290*	*12*
Ti6Al4V	0.16 *4.43*	16.5 *114*	1.03	10	>15	4.9 *8.8*	4.2 *0.073*	0.135 *0.56*	16.2 *0.029*	*12100*	*303*
Silicon	0.084 *2.33*	19 *131*	2.26	[>10]	<2	1.39 *2.5*	85.7 *1.48*	0.17 *0.71*	497 *0.892*	*170*	*3*
SiC RB-30% Si	0.104 *2.89*	47.9 *330*	4.61	[>40]	<5	1.47 *2.64*	94.8 *1.64*	0.17 *0.7*	455 *0.81*	*161*	*3*
SiC RB-12% Si	0.112 *3.11*	54.1 *373*	4.83	[>40]	<5	1.49 *2.68*	85 *1.47*	0.16 *0.68*	393 *0.7*	*180*	*4*
SiC CVD	0.116 *3.21*	67.4 *465*	5.81	[>40]	3	1.33 *2.4*	110 *1.9*	0.17 *0.72*	462 *0.82*	*126*	*3*

* for brittle materials, the values in parentheses are fracture strengths

12/9/91

stress yields little more, whereas the Al alloy resists yielding for a high microyield strength of 38.8 ksi but then continues yielding readily.

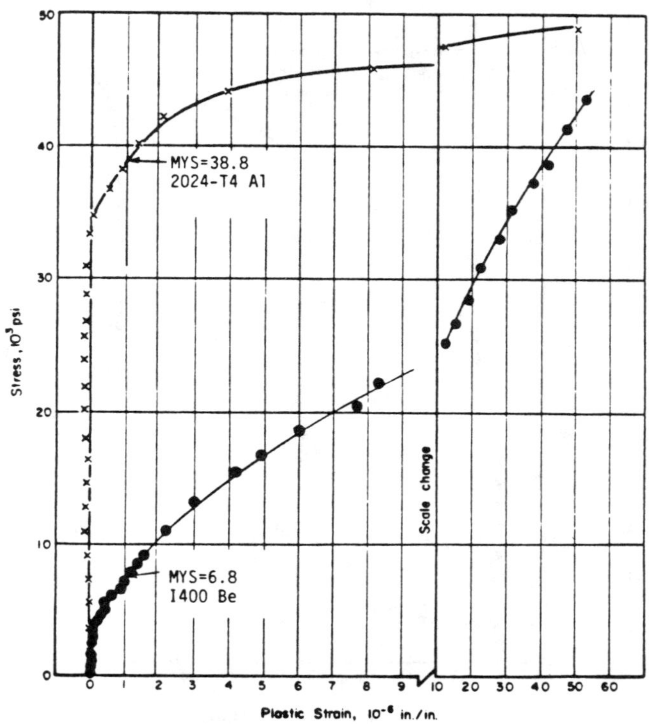

Figure 7. Comparison of a beryllium and an aluminum alloy that have similar yield strengths but very different microstrain behavior[12].

Microyield strength is strongly dependent on the prior history of the material. If it has been annealed, microyield will be lower than in almost any other condition. Conversely, if there has been prior straining, either through intentional or inadvertent application of stress, the microyield will be raised as demonstrated for a type of stainless steel in Figure 8. While prestraining produces a stronger material, it also leaves a level of residual stress which may be detrimental. Residual stress is discussed further in the next section.

Since high microyield strength (MYS) is a desirable property, and since many materials have relatively low MYS, it is important to know that there are methods for increasing it. Prestrain, as mentioned above, is one method, but it has its disadvantages. Many aluminum alloys, after rolling to plate form, are stretched a few percent to both straighten and level the stress through the thickness of the plate, which also increases MYS; but this process also seems to lower microcreep strength. Since the process of microyielding occurs, at least in the early stages, by movement of dislocations, anything that pins or prevents dislocation movement will increase MYS. By reducing the grain size of a material, dislocations are more readily pinned, as they are when particle or fiber reinforcement is added to a single phase material. Multi-phase materials almost always have higher MYS than similar single phase alloys. Thermal treatments that precipitate a second phase or produce a metastable phase tend to increase strength, and alloying a pure material usually produces dislocations and lattice strains that likewise increase MYS.

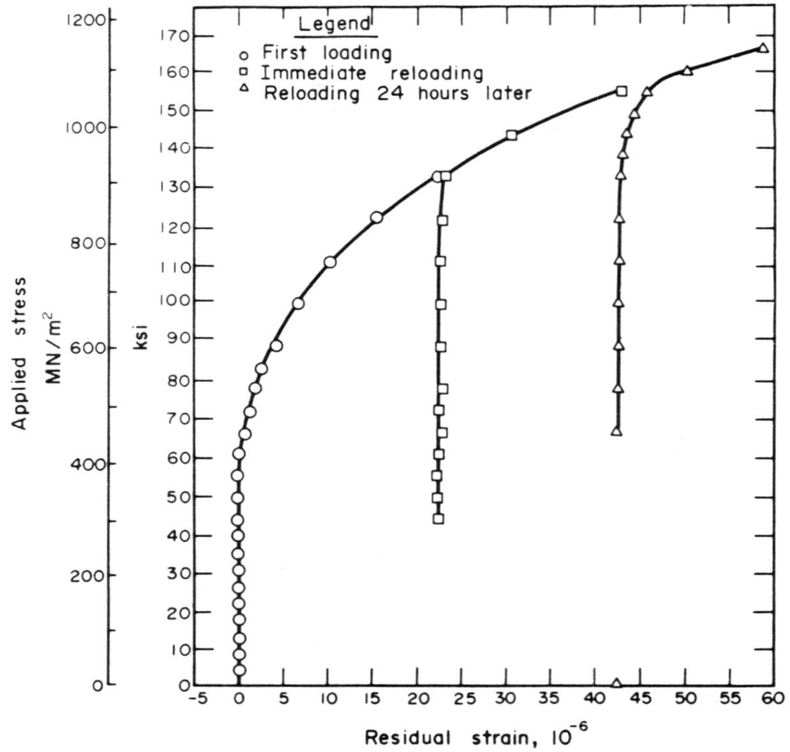

Figure 8. Effect of prestrain on microyielding behavior of 440 C stainless steel[13].

3.2 Changes in Internal Stress

While external stress is applied and removed from a component and is readily observed and measured, internal stress is not obvious. A component can be free of external attachments, even floating in a zero-g environment, and have internal stresses. They are in equilibrium and consist of balanced tensile and compressive stresses.

There are two types of internal stress called short range and long range. The spatial extent of these is, as the name implies, microscopic and macroscopic respectively. The long range internal stress is better known as residual stress. To illustrate both types, consider a component machined from an inherently anisotropic metal like Be. If the component was annealed prior to machining, the bulk of the material will consist of Be grains that vary in crystallographic orientation, and therefore in CTE at the grain boundaries. This produces short range internal stress at the grain boundaries as further described below. At the surface of the part, the machining will have plastically deformed a surface layer within which the residual or long range stress will be approximately the yield strength, approximately 40ksi for Be. Below the deformed layer there would be a partially deformed transition layer where the residual stress level would decrease rapidly from 40ksi (probably tensile), through zero to a low level compressive stress.

Short range internal stress can also result from unequal amounts of distortion between neighboring crystals in plastically deformed material. It primarily arises from inhomogeneous CTE: in a two phase material, between adjacent crystals with anisotropic CTE as described above for Be, or in a matrix with a dispersed phase or reinforcing particles, whiskers or fibers. Table 2 lists the theoretical maximum values of thermally induced microstrain due to CTE mismatch between adjacent grains of a few non-cubic materials[13,14]. In practice, the average values are approximately one third of these calculated maxima. You can see that for Be, one third of 63.8 psi/°C over a 100°C temperature change results in a short range stress of over 2000 psi, a value that exceeds the microyield strength of some Be alloys.

Table 2. Theoretical Maximum Values of Short Range Internal Stresses Due to Thermal Expansion Anisotropy[14,15]

Material	Lattice	psi/°C
Zinc	Hexagonal	177.
Calcite	Rhombohedral	165.
Cadmium	Hexagonal	91.4
Beryllium	Hexagonal	63.8
Quartz	Rhombohedral	43.1
Indium	Tetragonal	32.6
Magnesium	Hexagonal	2.74
Graphite	Hexagonal	0.94

Long range internal stress, residual stress, is usually the result of processing operations such as forming, heat treating, welding, machining or plating. As you would expect, the level of the stress is dependent on the severity of the operation, as shown in the following examples. Figure 9 shows how the temperature of the quench water effects the residual stress in Al alloys. The yield strength also drops when the water quench is less severe. Polymer quenchants are available that can provide the low residual stress of a boiling water quench with the strength of the cold water quench.

The effects of grinding parameters on residual stress in steel are shown in Figure 10. Here we see that abusive conditions, heavy feed, high speed and no coolant, produce high compressive stress with a deep layer, while a gentle cut produces a shallow layer with a small compressive stress[17]. Similar effects take place in glasses and ceramics. Sharp milling cutters generally produce compressive stresses with conventional feeds and speeds, but a surface layer with high tensile stress occurs when the tools are dull[18].

The introduction of residual stress, or any change in the balance of the stress will cause changes in dimensions of the component. This means that removal of a stressed layer will cause dimensional changes as demonstrated in Figure 11. When two specimens with surface residual stress are acid etched, they both shrink, the one with the higher stress (deeper cut) shrinking more. This principle is used in the manufacture of dimensionally stable components where after heavy material removal operations the surfaces are acid etched to restore unstressed dimensions to the part[20].

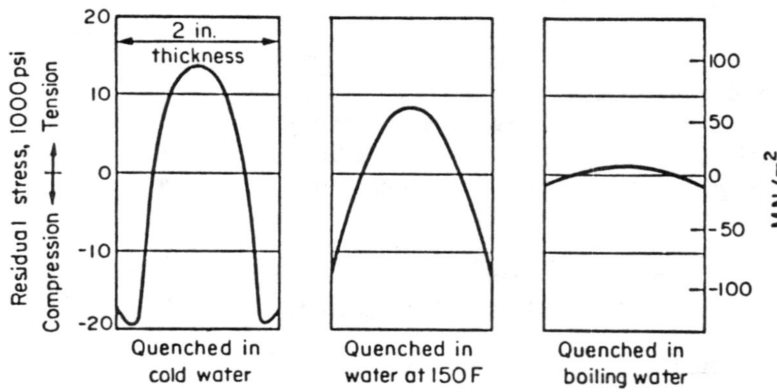

Figure 9. Residual stresses in specimens of 7075-T6 aluminum plate quenched in water at different temperatures[16].

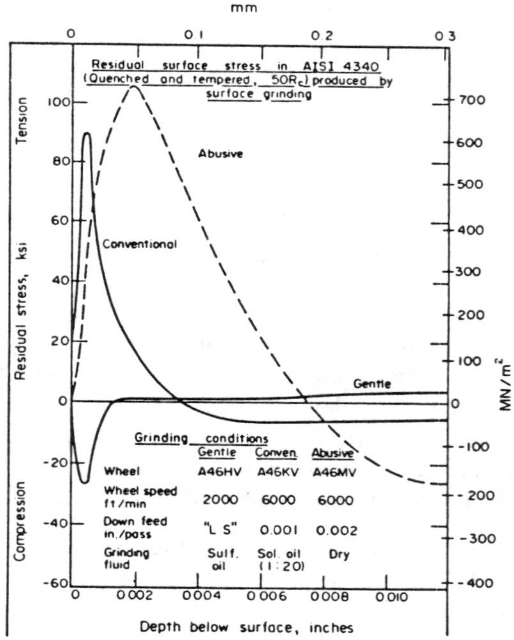

Figure 10. Residual surface stress in AISI 4340 steel produced by surface grinding[17].

Residual stress can also decrease spontaneously with time with a related change in dimensions. This effect is called stress relaxation and the decrease in stress is proportional to the stress level as shown in equation (1), where s is stress, t is time and τ is the relaxation time.

$$(-ds/dt)\tau = s \qquad (1)$$

But stress also decreases exponentially with time as shown by equation (2).

$$s = s_o e^{-t/\tau} \tag{2}$$

Note that when time, t, is equal to the relaxation time, τ, the ratio of stress to original stress, s/s_o, is equal to $1/e = 0.37$. This behavior is shown in Figure 12, where a Be mirror was fabricated with no treatment for stress relief after annealing the rough blank. Note the exponential shape to the optical figure change curve.

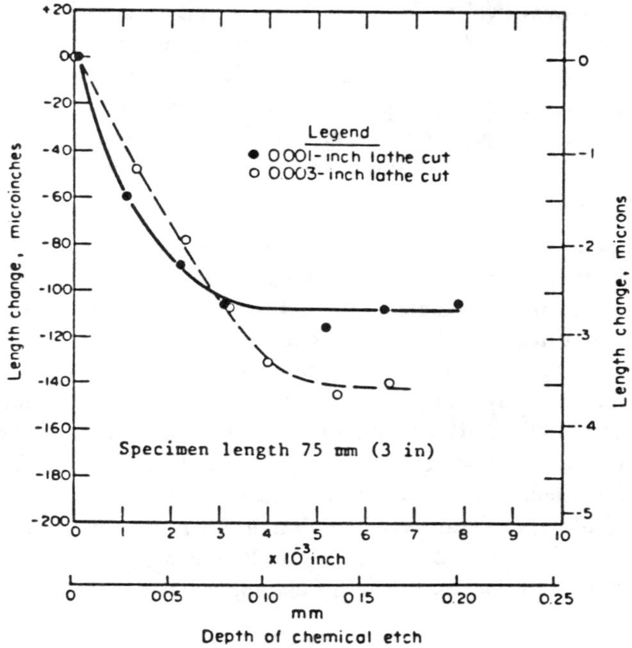

Figure 11. Dimensional changes in 3.0" long specimens of Ni-Span-C on chemical removal of residual stresses due to machining[19].

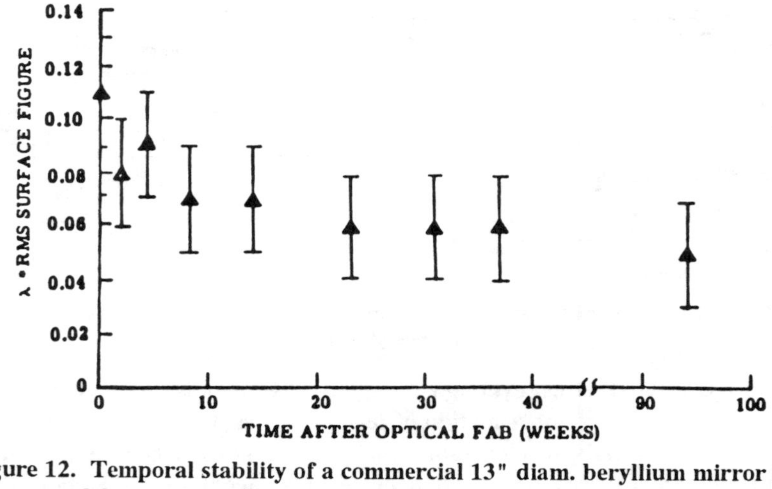

Figure 12. Temporal stability of a commercial 13" diam. beryllium mirror fabricated from vacuum hot pressed block.

Stress relaxation is also a thermally sensitive process, behaving according to the Arrhenius relationship of equation (3), where E is the activation energy, k is Boltzmann's constant and T is absolute temperature.

$$1/\tau \propto e^{-E/kT} \qquad (3)$$

This phenomenon can be used to reduce the stress level with thermal treatment - stress relief. However the question is often raised whether an isothermal treatment for some reasonable time or thermal cycling between elevated and reduced temperatures is the more effective stress relief treatment. Much has been written on this subject as summarized in Chapter 6 of Marschall and Maringer[1], but the best answer is, "it depends". It depends on the crystal structure and purity of the material; it depends on the prior thermo-mechanical history of the component; it depends on the temperature, time and rate of change of temperature; and of course it depends on the level, type and distribution of the internal stresses.

One example of what can happen is shown in Figure 13 where both isothermal and thermal cycling treatments were given to Be specimens previously stressed to 11.2 ksi[21]. In this case, the low annealing temperatures of 100° and 190°C do very little but relieve peak stresses. The best treatment is a 600°C stress relief treatment, but this is a higher temperature than most designers would want to subject a semifinished component. A temperature of 400°C still only removes 40% of the stress, but note that thermal cycling from 400°C to either -70°C or -196°C provides a 55% reduction. The cycling is more effective than the isothermal treatment to the same upper temperature. This cycling effect may only hold for non-cubic materials with reasonably high expansion anisotropy as listed above in Table 2, or for materials with more than one phase such as the composite materials shown in Figure 2. There is conflicting evidence in the literature on the effect of thermal cycling on stress relief of homogeneous cubic or amorphous materials.

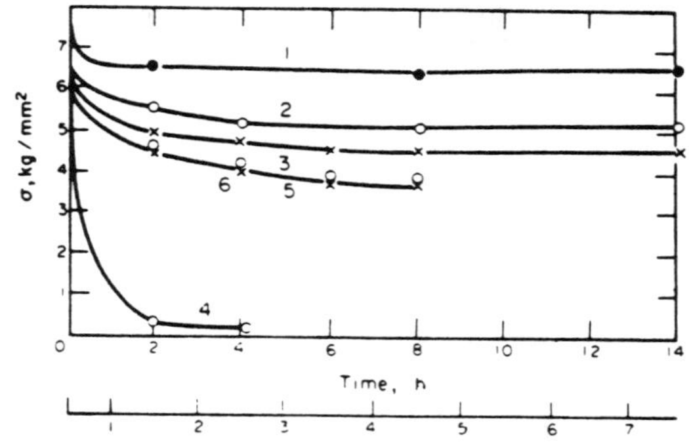

Curve	Thermal conditions
1	Isothermal; 100°C
2	Isothermal; 190°C
3	Isothermal; 400°C
4	Isothermal; 600°C
5	Cycled from -70° to +400°C
6	Cycled from -196° to +400°C

Figure 13. Comparison of the effects of thermal cycling and isothermal exposure on stress relaxation of pure beryllium[21].

For those materials where thermal cycling is more effective than isothermal exposure at the same elevated temperature for reduction of residual stress, the proposed mechanism is as follows. These materials develop short range internal stress when the temperature is changed due to the expansion mismatch between adjacent grains and/or phases. When this stress is added to the long range, or residual stress, the yield, or microyield strength is exceeded locally and plastic strain results. When the temperature is returned to room temperature, the plastic strain remains, but the level of the residual stress is reduced. Holding at the elevated temperature provides no further benefit unless it is close to the annealing temperature. When the temperature is then reduced, the sign of the short range stress is reversed, exercising the material further and providing more stress relief. In a similar manner, vibration, or mechanical cycling can provide stress relief although, as for thermal cycling, there are resulting dimensional changes.

For composites where the interface stresses can be substantial, thermal cycling is mandatory before dimensional stability to microstrain levels can be achieved. As many as 100 cycles are often required to achieve stability under constant moisture level conditions.

For reducing stress levels in critical components we then have a number of options:

1. Thermal treatments such as isothermal exposure or thermal cycling,

2. Mechanical treatments such as vibration or mechanical working,

3. Removal of surface material by chemical etching, controlled grinding and/or polishing or other stress-free methods and

4. Time.

3.3 Microstructural Changes

Microstructural changes in materials can result in both induced internal stress and dimensional change. The type of response depends on the material and the kind of microstructural change. The changes that can take place in materials used in optical and precision instrument systems include devitrification in glasses and glass ceramics; phase transformations, recrystallization and grain growth in metals and ceramics; and devitrification, phase transformations, recrystallization and grain growth in coatings. This cause of dimensional instability is quite common, but can not be covered adequately here. I will only discuss a few examples to illustrate the principle. The first example, illustrated in Figure 14, is the dimensional change that takes place in heat treatable aluminum alloys during precipitation heat treatment, also called aging or precipitation hardening. This example illustrates that a component which is aged from the solution treated and quenched condition to obtain maximum mechanical properties, normally 4 to 8 hours, will undergo a dimensional change due to the precipitation of the second phase. The change is small for 6061, a significant shrinkage for 7075 and a significant expansion for 2014. But notice that additional hours of aging, often performed for stress relief, induces additional significant dimensional change, particularly for the 2014 alloy. This behavior is typical of the 2XXX series Al alloys and must be taken into consideration when they are used. The 6061 alloy, most often used for precision optical structures and mirrors, changes the least, verifying its applicability for these applications.

The second example I previously cited in Figure 1. The change taking place in this material is typical of many tool steels where there is a sluggish phase transformation taking place from retained austenite to martensite, a transformation accompanied by a net expansion. The relatively stable parts, numbers 1-3 in set B, were sub-zero cooled to drive the reaction to completion.

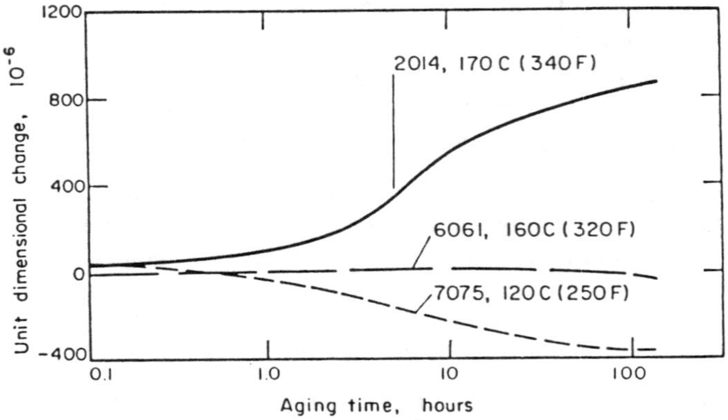

Figure 14. Dimensional change as a function of time at the precipitation heat treating temperature employed to produce the T6 temper for three aluminum alloys[22].

Electroless nickel coatings are used extensively, both for polishable coatings and for corrosion protection of Al, Be and iron alloy components. The coatings are nickel-phosphorous alloys that may be amorphous and are thermodynamically unstable as deposited. The stable form of the coating is nickel phosphide (NiP) in a matrix of Ni, but in this form it is not polishable. Coatings can have significant residual stress depending on phosphorous content and substrate CTE, and annealing changes the stress state[23], as shown in Figure 15.

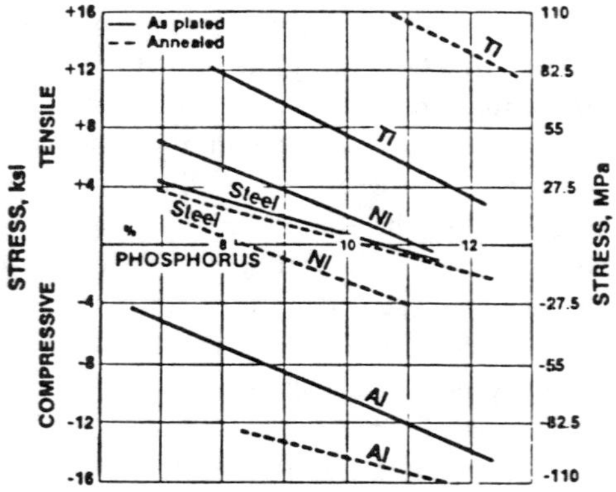

Figure 15. Stress of electroless nickel on aluminum, nickel, titanium and steel as a function of phosphorous content[24].

During thermal treatment (annealing) immediately after plating, hydrogen is driven off, adhesion improves, amorphous coatings become polycrystalline and diffusion of phosphorous toward formation of NiP begins. Since NiP has a lower specific volume and CTE than the quasi-stable Ni-P alloy, there is a shrinkage and a decrease in CTE for the coating during thermal treatment that also changes the stress state; the magnitude of these changes depends on the annealing conditions[25]. For higher temperatures and/or longer times, NiP starts to form as a precipitate in the coating. For machinability by diamond turning and/or polishability, the precipitate should not be allowed to form.

The annealing temperature used in the referenced studies is 190°C, with four hours at temperature for Be[26] and one hour for the other materials[23]. Stress goes from tensile to compressive with increasing phosphorous content and increasing substrate expansion coefficient, while annealing changes the stress toward compression for higher expansion materials and toward tension for lower expansion materials. What this all means is that for a particular substrate material, to obtain zero stress in the coating at room temperature after annealing, the right phosphorous content must be chosen, e.g., 5% on Al and 11.5% for Be. For use at lower or higher temperatures, different phosphorous contents should be selected to minimize the stress.

3.4 Inhomogeneity/Anisotropy of Properties

Most materials, as fabricated, are neither completely isotropic nor homogeneous. By that I mean that they are to some level anisotropic, i.e., having some preferred directionality of properties, and/or inhomogeneous, i.e., having a spatial variation in properties Anisotropy of properties exists in pure single crystals of materials. Inhomogeneity of properties occurs in bulk and is a function of raw material fabrication processes.

Cubic materials generally have anisotropy in their elastic properties. For example, the Young's modulus of elasticity of pure iron varies with crystallographic direction from 19.2 Msi to 41.2 Msi. Similar variations in shear modulus and Poisson's ratio are also present. Comparable variations in these properties occur in other cubic materials such as Cu, Ni, Si, beta SiC, etc. When these materials are used in polycrystalline form, the variations average out and are not noticed. In components that are fabricated from single crystals, or applications that have crystallographic texture such as plated or chemically vapor deposited (CVD) materials, there can be substantial elastic property anisotropy and this should be included in any detailed deformation modeling of such components. Thermal properties such as CTE are isotropic for cubic materials. Thermal conductivity, which is also isotropic, is affected by grain size and grain boundaries so that for plated or CVD materials there is anisotropy. For example, CVD SiC has a deposition texture and both elastic modulus and thermal conductivity have approximately 15% anisotropy.

In a similar manner to the cubic materials, there are variations in elastic properties in non-cubic materials, i.e., hexagonal, rhombohedral, tetragonal, orthorhombic, etc. However the thermal properties of the non-cubic materials are anisotropic. For example, the CTE of Be is 38% higher in the basal plane than it is in the axial direction of a Be crystal or grain. This anisotropy leads to the microstructural strains listed above in Table 2. Some materials such as graphite and quartz have negative CTE in some directions and positive in others. The effects of anisotropy of thermal properties, for the various crystalline materials that exhibit it, can be minimized with a fine-grain, randomly oriented microstructure or a pseudo-isotropic layup of fibers.

Inhomogeneity can be attributed to spatial variations in chemistry, grain size and/or grain orientation, and many other factors. In general, CTE inhomogeneity in glasses and glass ceramics is due either to compositional variations or can be related to annealing history; for metals, ceramics and salts, it is due to compositional or microstructural variations where the latter can be due to crystal orientation differences and/or the presence of other phases. For composites, CTE inhomogeneity, at least on a micro scale, is a given due to the presence of multiple phases. The inhomogeneity can be caused by reinforcement orientation or from variations in the concentration of the reinforcement. Care should exercised in selection of multiphase materials that will be used over a temperature range to insure dimensional stability.

Components having CTE inhomogeneity can behave in the manner shown in Figure 3 at the beginning of this paper. In this particular case, the cause is a combination of Be powder with basal plane cleavage, a poor distribution of grain sizes and inappropriate consolidation parameters. Current Be fabrication processes have virtually eliminated these types of inhomogeneities through the use of impact ground or spherical powder with well controlled chemistries combined with hot isostatic pressing of the powder[27-29].

4. LOW EXPANSION MATERIALS

Low expansion materials have been used for many years for precision instrumentation to minimize thermal dimensional changes in service. For optics, fused silica and fused quartz have replaced borosilicate glasses in many applications, and in recent years ULE and the low expansion glass ceramics have been used extensively for more critical applications. Invar alloys have been used in precision metering applications for many years, especially in combination with fused silica optics because of the excellent CTE match at room temperature. More recently composite materials, particularly graphite-epoxy composites, have been used for precision structures since they can be tailored for near zero expansion over a fairly wide temperature range. However each of these material types has distinct advantages and disadvantages. S.F Jacobs gave a good history of CTE measurements including inhomogeneity, hysteresis and temporal stability[30], and his paper in this volume[31] goes into much more detail in the coverage of this subject.

One very important issue that has been raised[32] is the ultimate stability that can be obtained in precision structures, particularly those fabricated from low expansion materials. In many cases the temporal instabilities far exceed the thermally induced distortions. Another aspect of this problem relates to instabilities in composites, where there is a substantial spatial variation in properties and residual stress within any component. Stabilization by thermal cycling is critical and temperature limitations for interfacial stresses exceeding strength limitations must also be considered.

Borosilicate glasses have been used for large mirror substrates, e.g., the 200 inch Hale telescope primary on Mt. Palomar, but these glasses have relatively high CTE, 3.0 ppm/°C and are no longer considered low expansion glasses. Lightweight mirror blanks continue to be fabricated from these glasses due to the relatively low fabrication temperature compared to fused silica and the utilization of more sophisticated blank fabrication methods[33,34]. Most glasses and glass ceramics are chemically inert, can be highly dimensionally stable and can be used over fairly wide temperature ranges, although each material has a range where it performs best. ULE and the glass ceramics have essentially zero expansion at room temperature and fused silica has zero CTE at approximately 160K. All of these materials are brittle and have very low fracture toughness and fracture strength in tension, have low Young's modulus and relatively high density. Fabrication can be difficult for complex shapes, and attachments require clever methods and thermal compensation. Zerodur can have an instability in the range of 150°C to 300°C[8], but use of Zerodur M or careful control of heating and cooling in this temperature range eliminates any change in dimensions or CTE. Zerodur has been used extensively for metering bar/optical bench applications.

The invar alloys can have near zero CTE over a limited temperature range when given a three step heat treatment[35]. Components can be readily fabricated using conventional metalworking processes and have high fracture toughness and moderate strength. Almost all invar specimens measured have temporal dimensional instabilities ranging from 1ppm/yr to as much as 50ppm/yr, depending on thermo-mechanical history. Super invar has a lower CTE than invar and very high dimensional stability[31,36,37] but can not be used at low temperatures since it undergoes an irreversible phase transition at temperatures below -55°C that substantially changes its dimensions and properties[31].

Composite materials, by their very nature have the disadvantage of having built in residual stress at the phase boundaries which usually lead to temporal dimensional instability. Graphite-epoxy composites can be designed and fabricated with near zero CTE, low density and high modulus. The properties can be tailored to the application, but are anisotropic due to the fabrication methods which require a preferred fiber direction or at best, a pseudoisotropic layup. The biggest drawback of graphite-epoxy is its moisture sensitivity which effects dimensions, CTE and mechanical properties. An excellent review of moisture and viscoelastic effects on the stability of composites has been given by Wolff[10], and the recent SPIE conference on Design of Optical Instruments[38] contained three sessions on composites where much new information was presented on ways to minimize moisture effects.

While silicon carbide (SiC) is not as low expansion a material as those discussed above, its CTE is lower than the borosilicates, and its combined properties, particularly E/ρ and CTE/D give it a unique combination of properties that should be considered for precision optical instruments. As shown in Table 1, there are several kinds of SiC whose properties are similar, but the fabrication methods and composition can be quite different. The single phase CVD SiC[39] is nominally 100% β SiC and since it is made by a deposition method, is well suited to forming faceplates, shells, tubes and other forms that can be replicated from a mandrel. There is normally some locked in stress in CVD SiC as evidenced by a slight springback from the mandrel. The other common form of SiC is a two phase form consisting of 60% to 88% α SiC (the hexagonal form) with the balance pure silicon that is called reaction bonded (RB), reaction sintered or siliconized SiC[40,41]. This material can be cast into virtually any shape and depending on the processing methods can be formed to net shape. Components such as lightweight mirrors and optical benches have been fabricated from RB SiC, and limited testing has shown the material to be dimensionally stable. Much more testing is required before this developmental material can be qualified for nanospec applications.

5. PROMOTING DIMENSIONAL STABILITY

Now that I have given you some insight into the causes of dimensional instability, you should see that there are many potential pitfalls in the design and fabrication of dimensionally stable components. In order to avoid these pitfalls there is a sequence of actions that can be taken that should lead to stable components.

The first step is to establish a budget for the allowable dimensional change for each component in the system and allocate a tolerance to each element. Structural components usually do not have as tight a tolerance as optical components and all components of each type will not necessarily have the same requirements. Next consider the sources of dimensional change as they relate to the components to be fabricated. For example, if the system is to operate at reduced or elevated temperatures, then both thermal instability and thermal cycling instability are potential types of dimensional instabilities to consider, and the sources for such behavior are changes in internal stress, inhomogeneity and/or anisotropy, and microstructural changes. Consider the fabrication options and how they relate to the dimensional instability sources and the component performance requirements. With all these factors in hand, select the candidate materials and reevaluate the sources and fabrication methods for each material with respect to meeting the budgeted dimensional tolerances in the specified use environment.

In order to make the final materials and fabrication method selection, you need to demonstrate that external stresses will not cause excessive strain in the component; that internal stresses in the component due to fabrication methods or inhomogeneities and/or anisotropies will not change excessively; and that any microstructural changes will not cause excessive strains. While this method may sound complicated, once you understand the basic sources of potential instability and the magnitude of the possible changes for each of the common candidate materials and their respective fabrication methods, the selection process becomes almost intuitive. The difficult part is when you must produce a component that operates in an environment for which properties information is limited, or when designing components to nanotolerances.

A good example of how this process can be implemented is given by Hassler, et al[42], for the design of a dimensionally stable composite position encoder scale. The stability requirements are defined, the conventional materials are reviewed and consideration is given to composites. Both the CTE and coefficient of moisture expansion (CME), and creep (temporal stability) are taken into account in the design. Fabrication methods were considered, test pieces were fabricated and tests confirmed the efficacy of the final design choices.

The bottom line is that dimensional <u>in</u>stability of materials *is* critical in the design of optical instruments and must be considered early in the design process.

6. REFERENCES

1. C.W. Marschall and R.E. Maringer, <u>Dimensional Instability, An Introduction</u>, Pergamon Press, New York (1977)
2. R.A. Paquin, ed., <u>Dimensional Stability</u>, *SPIE Proc.* **1335** (1990)
3. D.A. Vukobratovich and R.A. Paquin, eds., <u>Optomechanics and Dimensional Stability</u>, *SPIE Proc.* **1533** (1991)
4. W.B. Emerson, "Secular length changes of gage blocks during twenty-five years", in <u>Metrology of Gage Blocks</u>, U.S. National Bureau of Standards, Circular 581, p. 71 (1957)
5. Ref. 1, p. 187
6. R. Dawson and D.G. Moffat, "Vibratory Stress Relief: A Fundamental Study of Its Effectiveness", *J. Eng. Matls. and Tech.* **102**, p. 169 (April 1980)

7. R.A. Paquin, "Processing Metal Mirrors for Dimensional Stability", Wkshop on Optical Fab. and Test, Tech. Digest, Opt. Soc. of Amer., p. TB-1 (Dec 1981)

8. O. Lindig and W. Pannhorst, "Thermal Expansion and Length Stability of Zerodur in Dependence on Temperature and Time", *Appl. Optics* **24**, p. 3330 (1985)

9. Ref. 1, p. 23

10. E.G. Wolff, "Moisture and viscoelastic effects on the dimensional stability of composites", in Dimensional Stability, R.A. Paquin, ed., *SPIE Proc.* **1335** p. 70 (1990)

11. Ref. 1, p. 13

12. R.A. Paquin, moderator, "Dimensional stability workshop", in Dimensional Stability, R.A. Paquin, ed., *SPIE Proc.* **1335**, p. 217 (1990)

13. A.G. Imgram, M.E. Hoskins, J.H. Sovik, R.E. Maringer and F.C. Holden, "Study of microplastic properties and dimensional stability of materials", U.S. Air Force Materials Laboratory Report AFML-TR-67-232, Part II, Battelle Columbus Laboratories (1968)

14. V.A. Likhachev, "Microstructural strains due to thermal anisotropy", *Sov. Phys. Solid State* **3** p. 1330 (1961)

15. N.N. Davidenkov, V.A. Likhachev and G.A. Malygin, "Investigation of the irreversible thermal change in shape of zinc", *Phys. Metals Metallogr. (USSR)* **10** p. 95 (1960)

16. R.S. Barker and J.C. Sutton, "Stress relieving and stress control", Ch. 10 in Aluminum, Vol. III, Fabrication and Finishing, K.R. Van Horn, ed., Amer. Soc. for Metals, Metals Park, OH (1967)

17. W.P. Koster, M. Field, L.J. Fritz, L.R. Gatto and J.R. Kahles, "Surface integrity of machined structural components", U.S. Materials Laboratory Report AFML-TR-70-11, Metcut Research Associates, Inc. (1970)

18. M. Field and W.P. Koster, "Surface integrity in conventional machining chip removal processes", Society of Manufacturing Engineers Paper EM 68-516 (1968)

19. C.W. Marschall and R.E. Maringer, "Stress relaxation as a source of dimensional instability", *J. Mater.* **6** p. 374 (1971)

20. J.W. Bender and R.L. Wahl, "Work-induced stress and long-term stability in optically polished silicon", in Optomechanics and Dimensional Stability, D.A. Vukobratovich and R.A. Paquin, eds., *SPIE Proc.* **1533** p. 264 (1991)

21. I.Kh. Lokshin, "Heat treatment to reduce internal stresses in beryllium", *Metal Sci. Heat Treat. (USSR)* p. 426 (May, 1970)

22. H.Y. Hunsicker, "The Metallurgy of heat treatment", Ch. 5 in Aluminum, Vol. I: Properties, Physical Metallurgy and Phase Diagrams, K.R. Van Horn, ed., Amer. Soc. for Metals, Metals Park, OH (1967)

23. K. Parker and H. Shah, "Residual Stresses in Electroless Nickel Plating", *Plating* **58** p. 230 (Mar. 1971)

24. K. Parker, "Effects of Heat Treatment on the Properties of Electroless Nickel Deposits", *Plating and Surface Finishing* p. 71 (Dec. 1981)

25. D.L. Hibbard, "Dimensional stability of electroless nickel coatings", in Dimensional Stability, R.A. Paquin, ed., *SPIE Proc.* **1335** p. 180 (1990)

26. K. Parker and H. Shah, "The Stress of Electroless Nickel Deposits on Beryllium", *J. Electrochem. Soc.* **117** 8 p. 1091 (1970)

27. R.A. Paquin, "Hot Isostatic Pressed Beryllium for Large Optics", *Opt. Eng.* **25** 9 p. 1003 (1986)

28. R.A. Paquin, "New technology for beryllium mirror production", in Current Developments in Optical Engineering and Commercial Optics, R.E. Fisher, H.M. Pollicove, W.J. Smith, eds., *SPIE Proc.* **1168** (1989)

29. R.A. Paquin and G.J. Gardopée, "Fabrication of a fast aspheric beryllium mirror", in Reflective and Refractive Optical Materials for Earth and Space Applications, M.J. Riedl, R.R. Hale and T.B. Parsonage, eds., *SPIE Proc.* **1485** p. 39 (1991)

30. S.F. Jacobs, "Unstable Optics", in Dimensional Stability, R.A. Paquin, ed., *SPIE Proc.* **1335** p. 20 (1990)

31. S.F. Jacobs, "Variable invariables - dimensional instability with time and temperature", in Optomechanical Design, P.R. Yoder, ed., *SPIE Proc.* **CR43** (1992)

32. B.P. Dolgin, J. Moacanin and T.P. O'Donnell, "Theoretical limits of dimensional stability for space structures", in Optomechanics and Dimensional Stability, R.A. Paquin and D. Vukobratovich, eds., *SPIE Proc.* **1533** p. 229 (1991)

33. L. Goble, J.R.P. Angel and J.M. Hill, "Steps toward 8m honeycomb mirrors VII. Spin casting an experimental f/1 1.8m honeycomb blank of borosilicate glass", in Large Optics Technology, G.M. Sanger, ed., *SPIE Proc.* **571** p. 92 (1985)

34. R.K. Melugin, J.H. Miller, J.R.P. Angel, P.A.A. Wangsness, R.E. Parks and D.A. Ketelson, "Development of lightweight, glass mirror segments for the Large Deployable Reflector", in Large Optics Technology, G.M. Sanger, ed., *SPIE Proc.* **571** p. 101 (1985)

35. B.S. Lement, B.L. Averbach and M. Cohen, "The dimensional behavior of invar", *Trans. ASM* **43** p. 1072 (1951)

36. S.F. Jacobs, J.N. Bradford and J.W. Berthold, "Ultraprecise Measurement of Thermal Coefficients of Expansion", *Appl. Optics* **9**, 11, p. 2477 (1970)

37. S.R. Patterson, "Dimensional Stability of Superinvar", in Dimensional Stability, R.A. Paquin, ed., *SPIE Proc.* **1335** p. 53 (1990)

38. Design of Optical Instruments, D.M. Aikens, V.L. Genberg, G.C. Krumweide and M.J. Thomas, chairs, *SPIE Tech. Conf.* **1690**, Orlando (1992)

39. M.A. Pickering, R.L. Taylor, J.T. Keeley and G.A. Graves, "Chemically Vapor Deposited Silicon Carbide (SiC) for Optical Applications", in Space Optical Materials and Space Qualification of Optics, R.R. Hale, ed., *SPIE Proc.* **1118** p.2 (1989)

40. R.A. Paquin and M.B. Magida, "Low scatter surfaces on silicon carbide", in Laser Induced Damage in Optical Materials: 1989, H.E. Bennett, A.H. Guenther, L.L. Chase, B.E. Newnam and M.J. Soileau, eds., *NIST Spec. Publ.* **801** p. 256 (1990)

41. M.B. Magida, R.A. Paquin and J.J. Richmond, "Dimensional stability of bare and coated reaction-bonded silicon carbide", in Dimensional Stability, R.A. Paquin, ed., *SPIE Proc.* **1335** p. 60 (1990)

42. W.L. Hassler, Jr. and R.K. Nakayama, "New low-cost dimensionally stable composite linear absolute position encoder scale", in Dimensional Stability, R.A. Paquin, ed., *SPIE Proc.* **1335** p. 80 (1990)

Variable invariables - dimensional instability with time and temperature

S.F. Jacobs

Optical Sciences Center, University of Arizona
Tucson, Arizona 85721

ABSTRACT

How stable are structures made of fused silica, Zerodur, ULE, Invar or Superinvar? We review what is known about dimensional changes with time or temperature for these important materials, including some recent findings.

What are the variations in commercially available materials? How does one go about *obtaining* the best Invar and Superinvar?

1. INTRODUCTION

Recent advances in astrometry and ultrastable lasers are imposing very stringent demands for dimensional stability of precision structures and science instrument components. As a result, it has become necessary to critically reexamine much of our old reliable data to see whether it truly applies to real-world conditions. We describe here some of the variations that exist in what we may have considered invariable expectations. The discussion is restricted to temperature and time. Other parameters, such as refractive index, may require consideration as well.

When it comes to real-world performance, there are many parameters to consider. For example, fused silica is a very important and well refined material, excellent both for its structural and its optical properties. Most manufacturers provide detailed information concerning its spectral transmittance, but how much variation is there in the instantaneous[1] linear coefficient of thermal expansion, α', vs temperature for **different** fused silicas, including behavior at cryogenic temperatures? What about **hysteresis** (return to original length after thermal cycling)? What about Invar's thermal expansivity **vs. temperature**? How can one reliably *obtain* Invar with the lowest thermal expansivity? What can be meaningfully said about thermal expansivity vs. temperature of composite materials? What about **temporal instability** (changes in length at constant temperature)? For example, Invar generally elongates as it ages at room temperature. What can we expect it if is used at an **elevated** temperature? What is known about **temporal** instability of other low thermal expansivity materials? We are finding that any synthetic material whose dimensions **can** change, **will** change. And that we can observe this change, given enough time or measurement sensitivity.

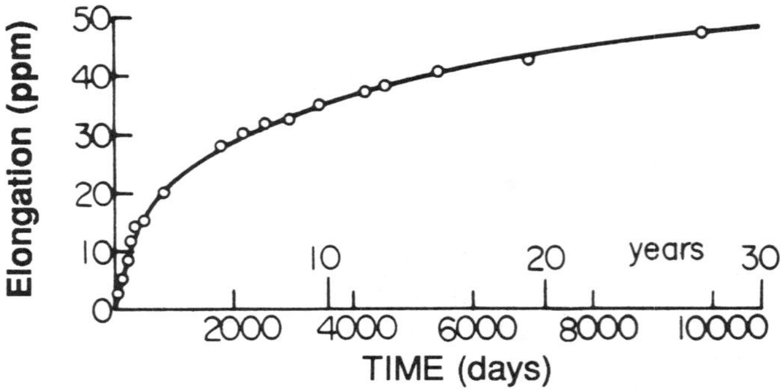

Fig. 1 Elongation of Invar at room temperature.[2]

In the late 1920's, Guillaume described Invar's temporal instability, which he measured at constant temperature, by mechanical means, over a period of 30 years! (See Fig. 1.) In the early 1970's, Justice[3] studied temporal instability of fused silica and ULE over a period of three years, using optical interferometry. More recently, lasers have made possible the technique of optical heterodyning, which faithfully compares a stable laser's frequency with that of another laser, which may be locked to the resonant frequency of an optical cavity with mirrors spaced by a material whose length change is being measured. Initially, HeNe lasers were used, stabilized with respect to neon atoms ($\Delta\nu/\nu \sim 10^{-9}$), or molecules such as Iodine ($\Delta\nu/\nu \sim 10^{-11}$). Later Hall et al. showed[4] that even better stability can be achieved by stabilizing a laser with respect to the length of an empty cavity which is superbly temperature controlled. For example, if the cavity spacer material's thermal expansivity is .01 ppm K^{-1}, then $\Delta\nu/\nu = \Delta L/L = \alpha\Delta T \approx 10^{-14}$.

Renewed interest in highly stable lasers has been motivated by such challenges as gravity wave detection and the availability of trapped and frozen atoms. Laser linewidths approaching a few milliHertz have been demonstrated, which makes them comparable in stability to the best hydrogen masers. Thus the search for dimensionally stable materials has come full circle: stable lasers helping to evaluate stable materials which in turn will be used to make stabler lasers.

2. THERMAL EXPANSIVITY

2.1 Fused silica

The term "quartz" is used here to mean the crystalline form of SiO_2. It occurs in nature, but can be made synthetically. Vitreous silica is SiO_2 in the glassy state, which can be made, starting either from quartz (fused quartz) or from compounds such as $SiCl_4$, e.g. by flame hydrolysis (fused silica).

Manufacturers' data today contains little information about the instantaneous thermal expansivity α'. Often α' is stated, but the temperature intervals are not small, and just as often a single α is quoted without mention of any temperature at all. To clarify the situation, we present Fig. 2, which includes G.K. White's low temperature measurements and ours at higher temperatures. I was surprised to see how little variation there is in commercially available fused silicas, despite the variations possible by varying heat treatments.[6] One thing that **does** vary a little is the α′ zero-crossing temperature for fused silicas: from 137K (for specially annealed fused silica tubing[7]) to 175K (for fused quartz[8]). It is interesting to note that although crystalline quartz has very large thermal expansivity at room temperature, its expansivity becomes smaller in magnitude than that of fused silica at temperatures below 20K[5] (See Fig. 3; Table I).

Fig. 2 Thermal expansivity vs. temperature of quartz, fused silica, Invar, Superinvar, and ULE.

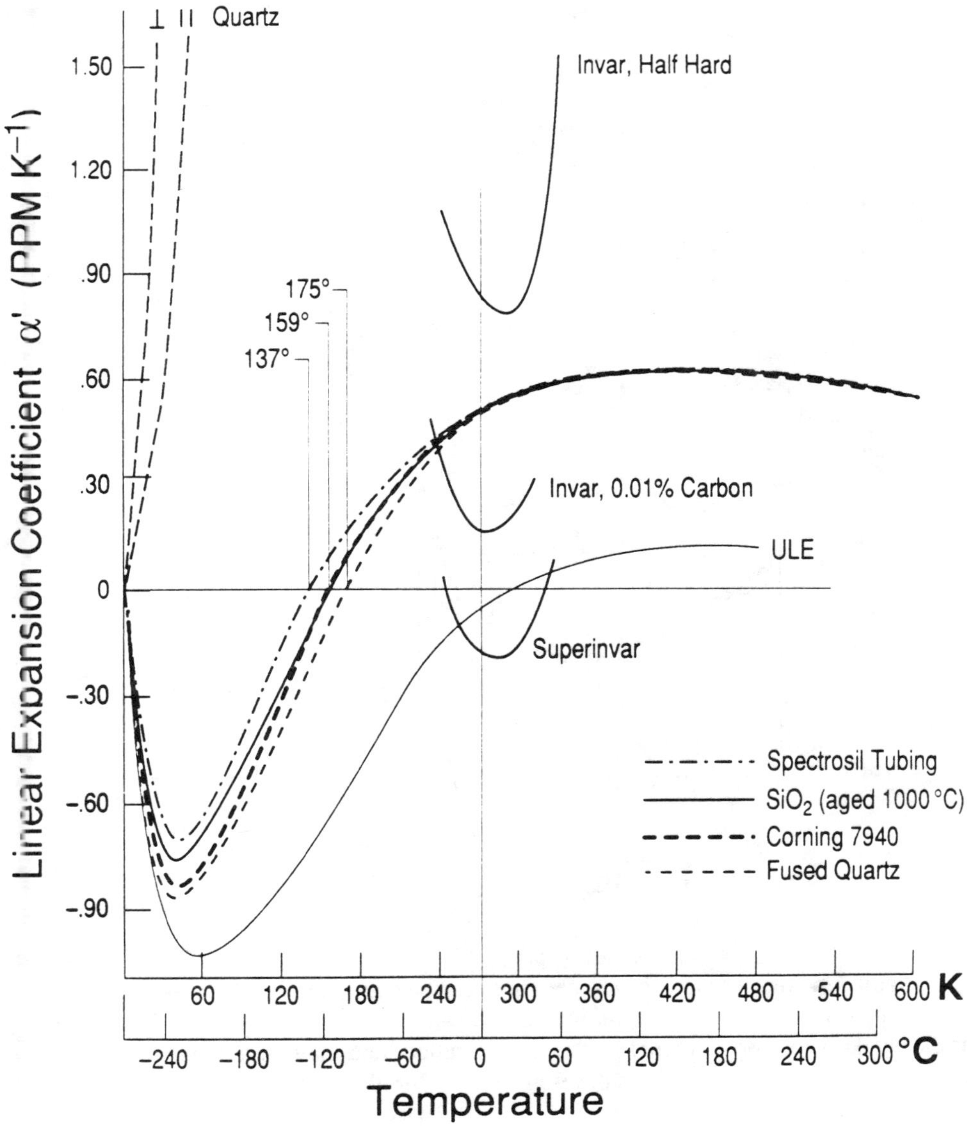

Fig. 3 Thermal expansivity of glasses and quartz; curve for copper is included for comparison. (Reproduced by permission of G.K. White and the publishers, Butterworth Heinemann Ltd. ©)

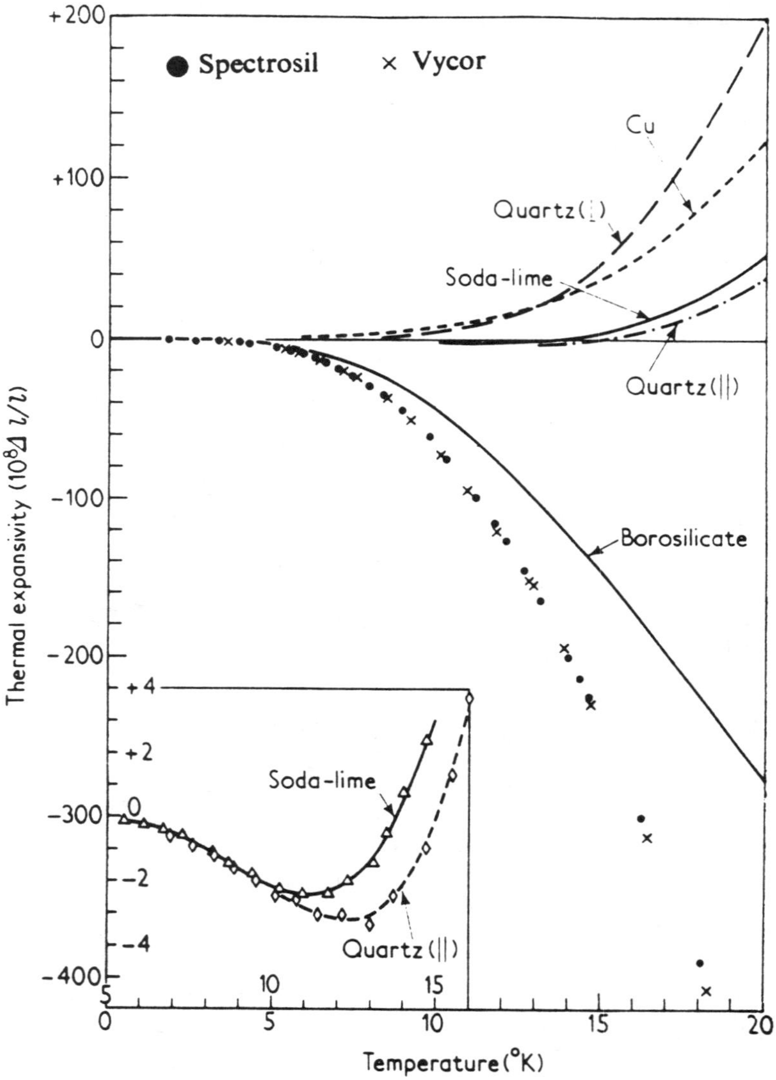

Again in the cryogenic temperature region, around 235K, we have often encountered hysteresis-like effects in well annealed fused silica (see Fig. 4), Zerodur, and Solgel. These effects need further investigation, and may be avoidable, as Schott Glaswerke has shown with their development of Zerodur M (see next section).

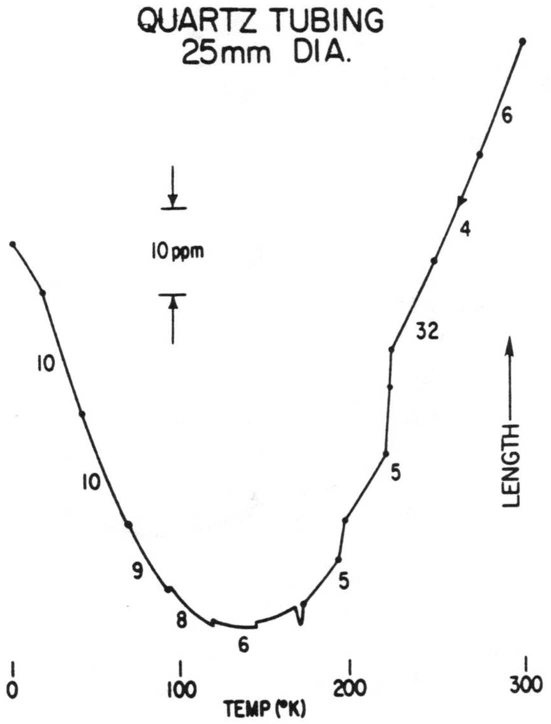

Fig. 4 Thermal expansion of Spectrosil grade tubing annealed 1050 to 850°C at 15°/hour. Numbers show waiting times in hours. (Similar behavior was observed for 18 and 14mm tubing).

2.2 Zerodur

Zerodur is a glass-ceramic made by Schott which achieves near-zero thermal expansivity over a wide temperature range by balancing a positive expanding glassy phase component with a negative expanding crystalline phase component. Fig. 5 shows α vs temperature of Zerodur.[8,9,10] What is not shown in this curve is the hysteresis behavior[11,12] shown in Fig. 6. Here the sample is thermally cycled at 60 K/hr, waiting every 25° for the sample length to stop changing. We see that after reaching thermal equilibrium, the sample sometimes undergoes (often very time-consuming) relaxation effects. To avoid this problem, Schott developed Zerodur M.[13] Note that the general shape of the curve α' vs temperature is greatly changed from that of standard Zerodur. A comparison is shown in Fig. 5.

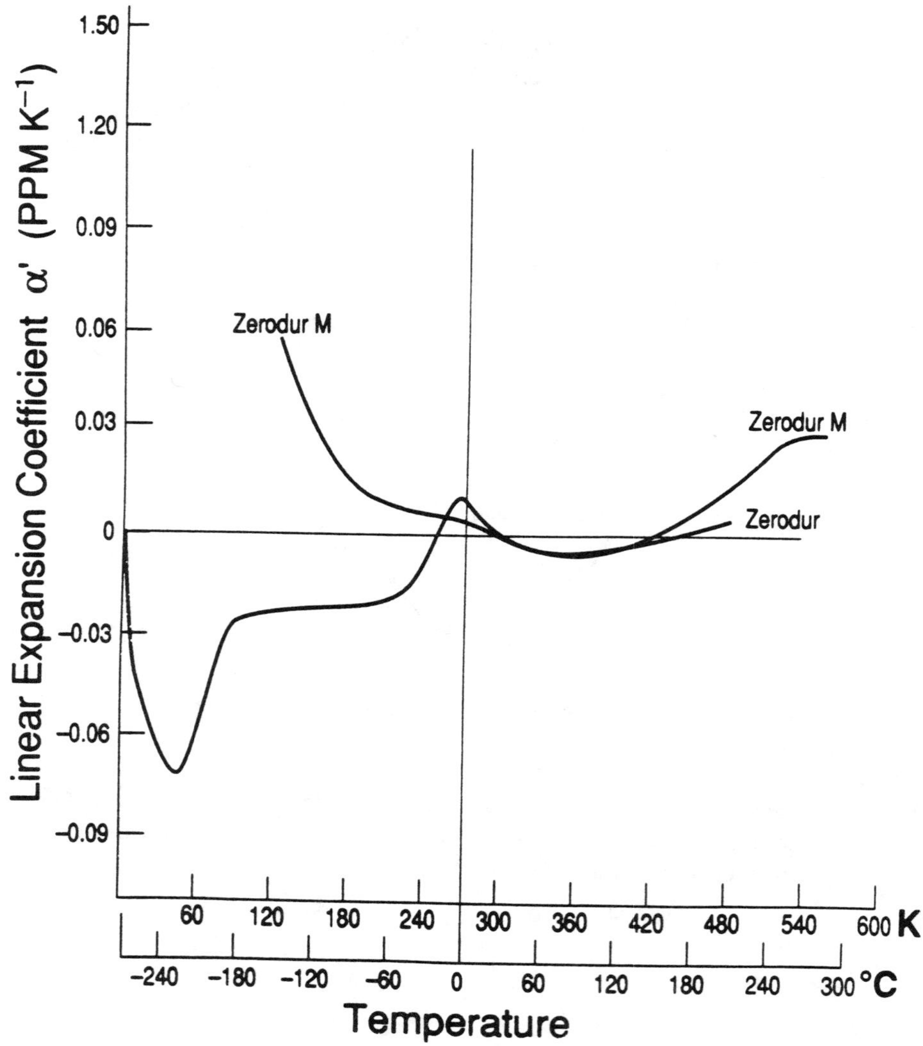

Fig. 5 α' **vs temperature for Zerodur and Zerodur M.**

T°K	*Spectro-sil*	*Vycor*	*Boro-silicate*	*Soda*	*Quartz* (⊥)	*Quartz* (∥)	*Quartz Average*
283	(50)	(75)	(300)	(900)	1370	750	1160
85	—	—	—	—	640	260	510
75	—	—	—	—	570	225	455
30	−72*	−68	—	+62	160	49	123
25	−66	−64	−20	+37	103	30·5	79
20	−59	−58	−27	+17	53	13·3	40
15	−45	−45	−24	+4·0	17·5	3·0	13
12	−33	−33	−19	+0·5	6·9	0·0	4·6
10	−23·5	−24	−15	−0·5	2·7	−0·6	1·6
8	−14·3	−14·5	−9·5	−0·7	1·0	−0·5	0·5
6	−6·8	−6·4	−4·5	−0·2	0·4	−0·25	0·2
4	−2·0	−1·8	−1·3	—	—	—	—

Table I. Linear thermal Expansion Coefficients (x 10^8 deg. K^{-1}). Value marked * is extrapolated. Those in brackets have been quoted in commercial literature of tables for glasses of similar composition.[5] (Reproduced by permission of the publishers, Butterworth Heinemann Ltd. ©)

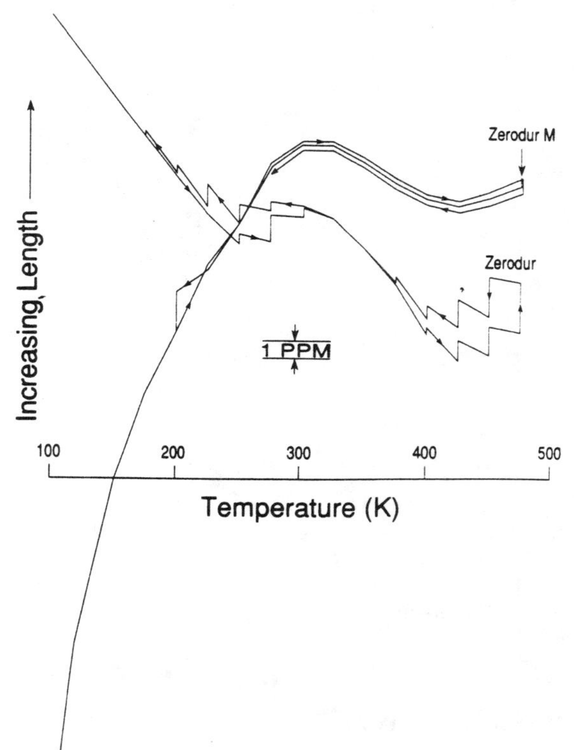

Fig. 6 Thermal expansion ΔL/L vs temperature of Zerodur and Zerodur M.

2.3 ULE

ULE (Corning Code 7971) is a vitreous mixture of SiO_2 + nominal 7.4 wt% of TiO_2. α′ vs temperature is shown in Fig. 2.[14,15] Note the similarity between the shapes of α′ vs temperature curves for ULE and fused silica. This is because TiO_2 has a negative expansivity near room temperature. Of special interest is the fact that ULE does not exhibit the hysteresis problem that can occur in glass ceramics. Fig. 7 shows the results of repeated thermal cycling of ULE.

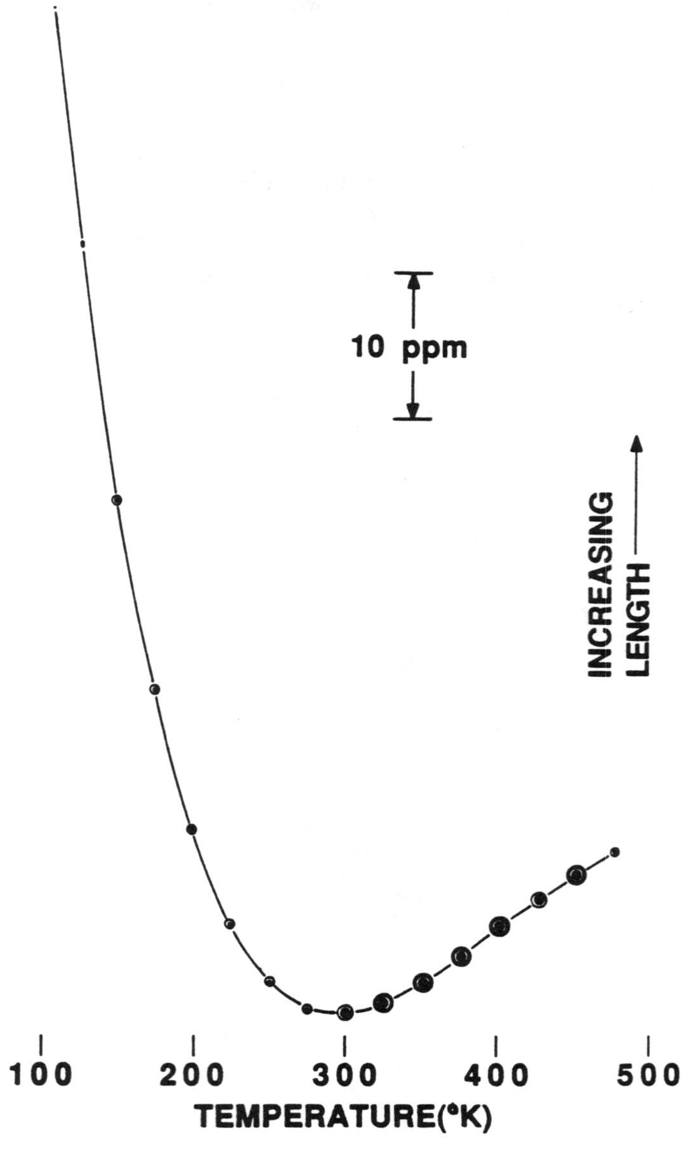

Fig. 7 Repeated thermal cycling of ULE.

2.4 Invar

When a low thermal expansivity **metal** is required, usually for reasons of strength, there is little alternative but to use some form of Invar, an iron-nickel alloy containing about 36% nickel. Discovered by Guillaume in 1886,[16] Invar has a low thermal expansivity near room temperature which arises from the interplay of two phenomena. Although Invar is face-centered cubic at room temperature, it is ferromagnetic, with a Curie temperature of 260°C. On heating, at temperatures below the Curie temperature, the magnetic order decreases, with a corresponding decrease in volume. The normal lattice expansion occurs simultaneously, however, and for a temperature interval below the magnetic transition, there is a temperature at which the net expansion can be very close to zero, and in some instances[17] even less than zero. Referring to Fig. 2, the lower Invar curve describes the lowest expansivity Invar we have measured. In general, we find U-shaped curves with various magnitudes, such as that shown for half-hard[18] Invar. G. Girard at the Bureau International des Poids & Mesures has done excellent work on treating Invar **tapes** and **wires,** by a combination of thermal and mechanical working. Expansivities < 1 ppm K^{-1} are reported, along with temporal stability < 1 ppm/yr.[19]

For certain applications, the smallest expansivity possible is not desired, but rather an expansivity *match* with vitreous silica. Fig. 2 shows the temperature range over which this match is possible.

Lement and Averbach showed[17] that Invar's low expansivity is adversely affected by the presence of carbon, manganese, silicon, sulfur, phosphorus, and virtually any other impurity. We describe in Appendix I how Jet Propulsion Laboratory obtained the Invar shown in the lower curve of Fig. 2.

2.5 Superinvar

Several variations on the Invar theme have been produced. These include Superinvar (Fe, Ni, Co) and Stainless Invar (Fe, Co, Cr).[16] It has been possible to obtain thermal expansivity zero crossings with such materials (see Fig. 2), as well as good temporal stability. (See Section II.) Appendix II shows the composition of the Superinvar sample shown in Fig. 2, as well as its heat treatment. Unfortunately, with Superinvar there appears to be a delicate balance not only in the fabrication (composition and heat treatment) but in post-fabrication thermal, magnetic and mechanical treatment. E.g., Don't heat it, magnetize it, or drop it! In addition, with Superinvar (but not Stainless Invar!) there is an irreversible phase transition (from face-centered cubic to body-centered cubic) which occurs at temperatures ranging from about 270K to 100K, depending on composition.[16] We deliberately cycled Superinvar through this phase transition by immersing a sample in liquid nitrogen. The material promptly elongated by 3/4 of one percent, and its temporal stability worsened. This will be discussed further in Part 3.

3. TEMPORAL INSTABILITY - LENGTH CHANGE AT CONSTANT TEMPERATURE

3.1 Fused silica

We have always considered vitreous silica to be dimensionally stable with time. In the early 1970's, we measured the temporal stability of different fused silicas, Heraeus-Amersil's Homosil and Corning Code 7940, ULE, Zerodur, Invar and Superinvar using an iodine-stabilized HeNe laser.[20] Since that time, we have included the same optically contacted Homosil sample (cavity) in all our temporal stability measurements to serve as a secondary standard; the primary standard was a Zeeman stabilized HeNe laser.[21] Recently, Ben Dolgin, noting the excellent agreement between our early measurements of Homosil and those of Ben Justice, urged us to examine our subsequent data obtained with the same sample, over various time periods ranging from two to four months. Fig. 8 shows this. The large error bars are due chiefly to temperature fluctuations. For fused silica near room temperature, $\Delta T = .04°C$ corresponds to $\Delta L/L = .02ppm$. These results suggest that perhaps fused silica shrinks with time, with a time constant of about 10 years. Because of the importance of this material, further work, with better temperature control, seems indicated.

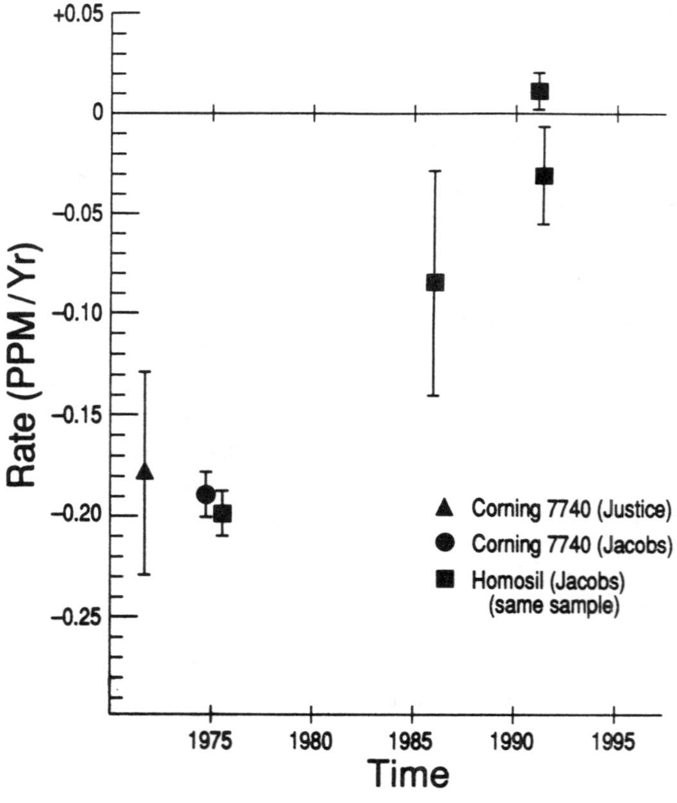

Fig. 8 Fused silica shrinkage with time.

3.2 Zerodur

In 1975 we reported measurements of temporal stability of Zerodur[20]. Over a six-month period, we found less than .01 ppm change. However, Schott only claims <0.1 ppm/yr stability. Recently, John Hall evaluated Zerodur for use as an ultrastable cavity for stabilizing lasers. Fig. 9 shows his unpublished results, which were obtained using an iodine-stabilized laser and (several) microdegree temperature control. These results imply a shrinking of 0.1 ppm/yr. This stability is more than adequate for most purposes, however for the ultimate in stability there remains some concern about the so-called "thermal memory" of glass ceramics.

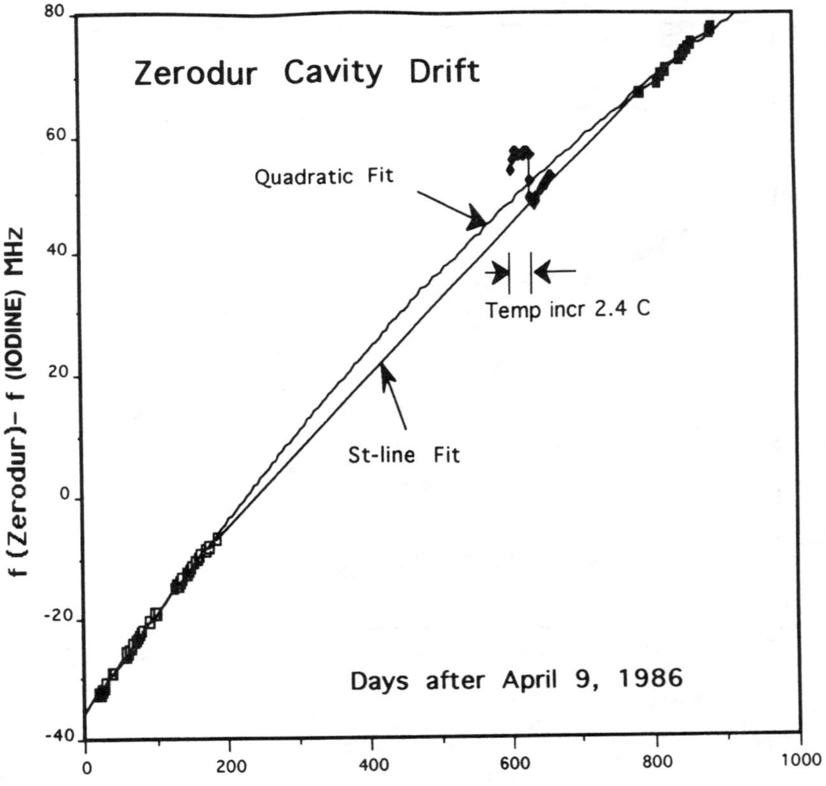

Fig. 9 **Heterodyne beat of Zerodur cavity stabilized laser vs iodine stabilized laser. (Adapted from J.L. Hall unpublished figure.)**

3.3 ULE

Our 1975 results[20] showed ULE shrinking .06 ppm/yr. John Hall's recent results with ULE show shrinkage of .025 ppm/yr., (drift of .2548 Hz/sec.). Fig. 10 shows the laser stability he achieved by locking to a cavity made of this material, with the material's length drift removed.[22]

Fig. 10 Relative beat frequency between ULE cavity stabilized laser and iodine stabilized laser. (Adapted from J.L. Hall unpublished figure.)

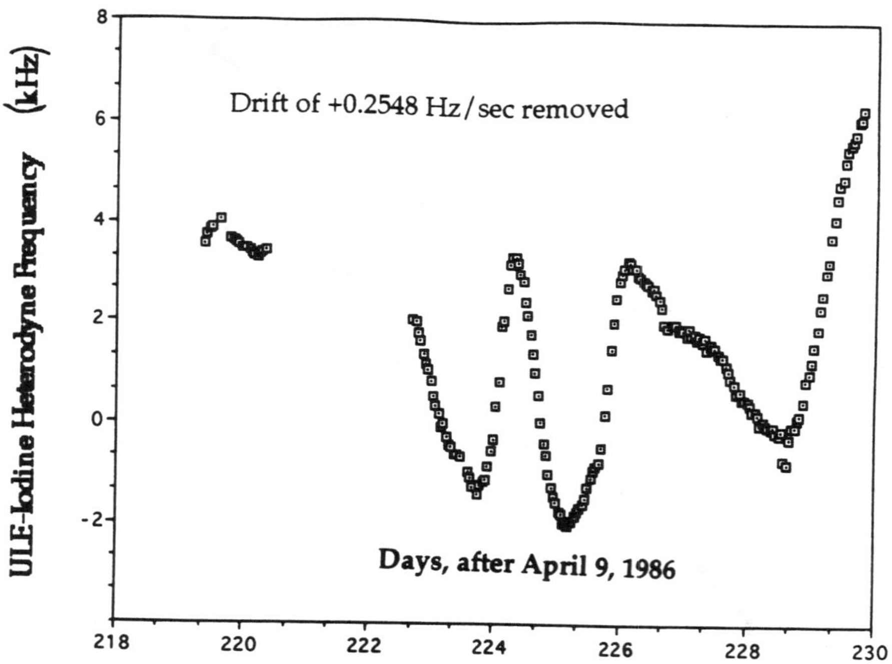

3.4 Invar

The temporal instability of Invar has long been recognized, yet its desirability as a low thermal expansivity metal often makes Invar the material of choice for applications near room temperature. One approach to the temporal instability problem has been to stockpile and age the material. Another has been to try various heat treatments in order to accelerate the aging process.

In 1983 a study was designed by David Schwab[23] to investigate the aging effectiveness of four different heat treatments. Fig. 11 shows the results at 38°C. It was found that the simplest of the treatments tried[24] proved the most effective.

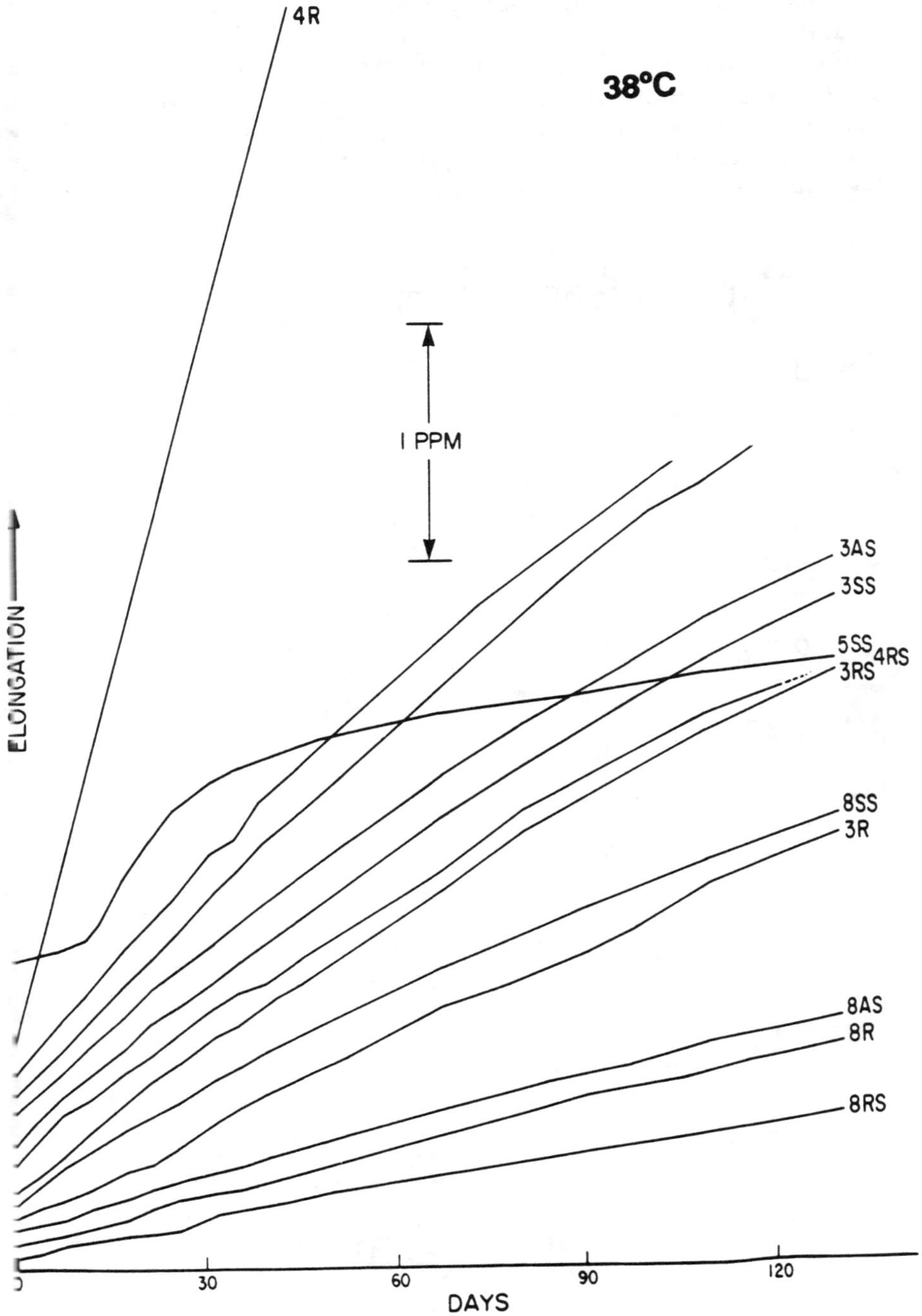

Fig. 11 Elongation with time of Invar samples with various heat treatments. R = untreated; RS = (see reference 23).

Recently, an unusual problem came up: How to predict accurately the future length change with time for well characterized Invar, to be used **at an elevated temperature**. The literature contained almost no information, and so an experiment was performed[25] to compare temporal stability, first at 25°C, then at 35°C, then back at 25°C for variously treated Invars. In addition to obtaining excellent time constant measurements for this material, it was concluded that most of the material variables (see Table II) made little difference to the aging curves (Fig. 12), but low carbon content notably improved the temporal stability. This is consistent with the findings of Lement et al.[17]

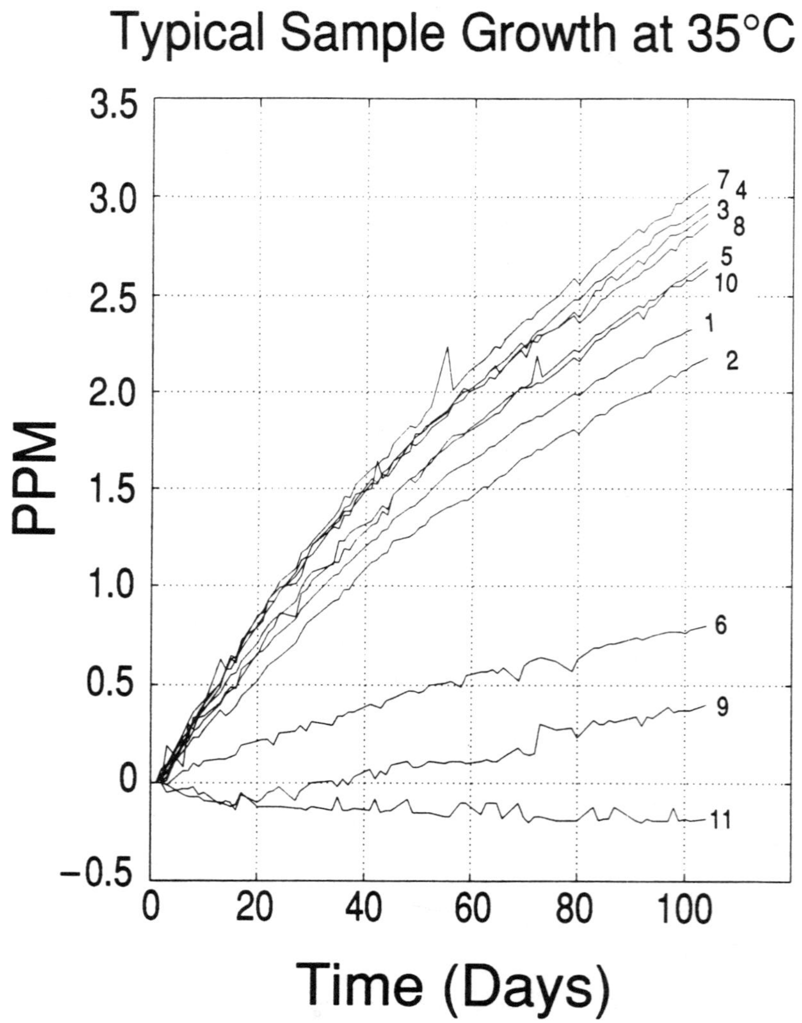

Fig. 12 Invar temporal length change with time. Numbers refer to sample variables shown in Table II.

Despite the many studies of temporal instability of Invar, today's user is faced with a problem: What can I expect of material I obtain *now*? We are currently measuring the temporal stability of the material described in the lower Invar curve of Fig. 2. This appears to be exceptionally stable material, but unfortunately it has not yet proven cost effective for a steel manufacturer to produce, characterize and make available for sale a large batch of material. As a result, every special job remains a special project for a metallurgical specialist.

3.5 Superinvar

We discussed earlier what happened when Superinvar underwent the infamous phase transition. After we deliberately immersed two of three samples of Superinvar in liquid nitrogen, their subsequent temporal instability, compared with that of the pristine third sample, was considerably worsened, as shown in Fig. 13.

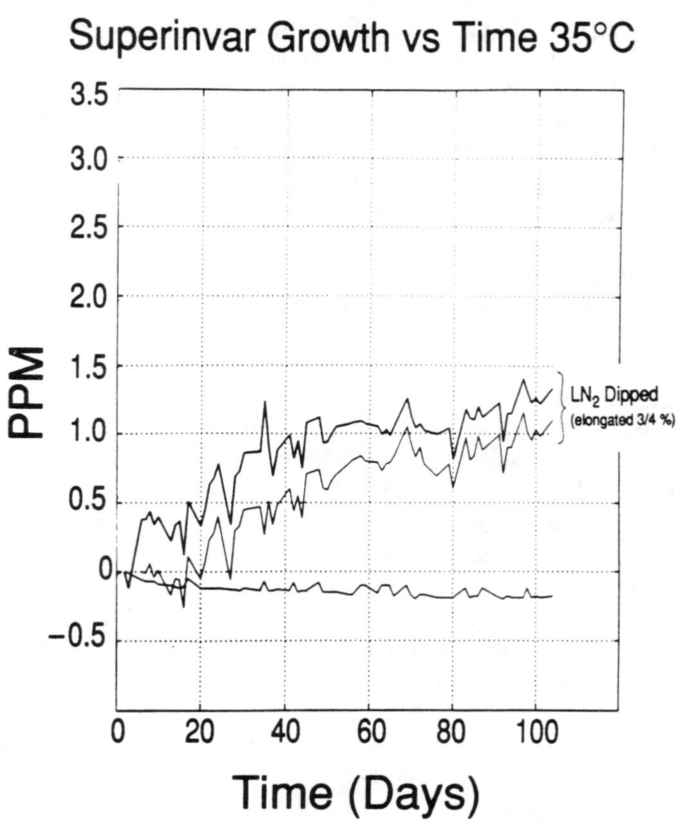

Fig. 13 Effect of phase transition on temporal stability of Superinvar.

The best Superinvar we have encountered is that whose expansivity is shown in Fig. 2. Its composition has been verified several times, and is shown in Appendix II along with its heat treatment. Because of its low thermal expansivity, we have studied the temporal stability of this sample frequently over the past 20 years, with varying results (see Table III). We find that the superb temporal stability, initially obtained through the heat treatment described, is easily degraded by rather small changes in temperature. It is undoubtedly because of this that Averbach recommends[26] that the most reliable route to development of a very stable Invar is ultrahigh purity Invar, rather than through the delicate balances involved in Superinvar.

4. MEASUREMENT METHOD

The most prevalent expansivity measurement method in use today is the dilatometer,[27,28] employing for example a linear differential voltage transformer to sense differential length changes with respect of a known, calibrated sample. These instruments are very useful for general studies of expansivity, providing high sensitivity often exceeding 0.1 ppm. The ΔT part of the problem is often the weak link, especially when expansivity is not small. Reaching thermal equilibrium takes time, and time costs money. As a result, most measurements are made without waiting for true thermal equilibrium, but rather with a slow temperature scan. What is lost by this omission, in addition to precision, is information about the waiting time, which gives us a clue concerning what is going on inside the material. For example, Zerodur in some temperature ranges can take over 24 hours to equilibrate to a criterion of $\Delta L/L \leq .01$ ppm over an hour's time.[29]

The laser/interferometric method we use was described by A.D. White,[30] who frequency locked a tunable HeNe laser to a Cervit cavity of length L and then compared the tunable laser's frequency ν with that of a HeNe laser stabilized with respect to a Neon atom. He then varied the temperature of the Cervit cavity and deduced the change in cavity length $\Delta\nu$ in the locked tunable laser's frequency, using the relationship $\Delta L/L = \Delta\nu/\nu$. The precision of this method is essentially that of the stable laser's stability, excluding temperature errors. Thus, for low expansivity materials, where temperature errors are generally unimportant, laser stability $\Delta\nu/\nu = 10^{-9}$ allows $\Delta L/L$ precision of .001 ppm. Parks recently noted[31] that the Japanese have written a National Industrial Standard based on this technique.

Figs. 14 and 15 show how we utilize the above principal for thermal expansivity and temporal stability measurements respectively. Fig. 15 also shows our typical sample configuration. For very high precision work, we optically contact the mirrors to the sample ends, however it is more convenient to use a vertical optical axis, with gravity holding the cavity together.

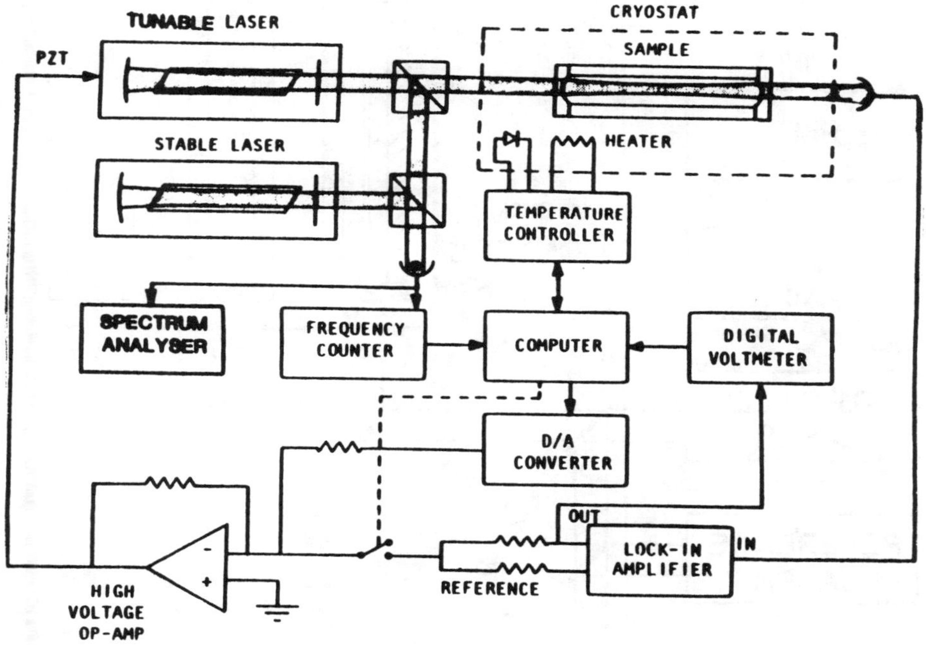

Fig. 14 Arrangement used for laser/interferometric measurement of length changes with temperature.

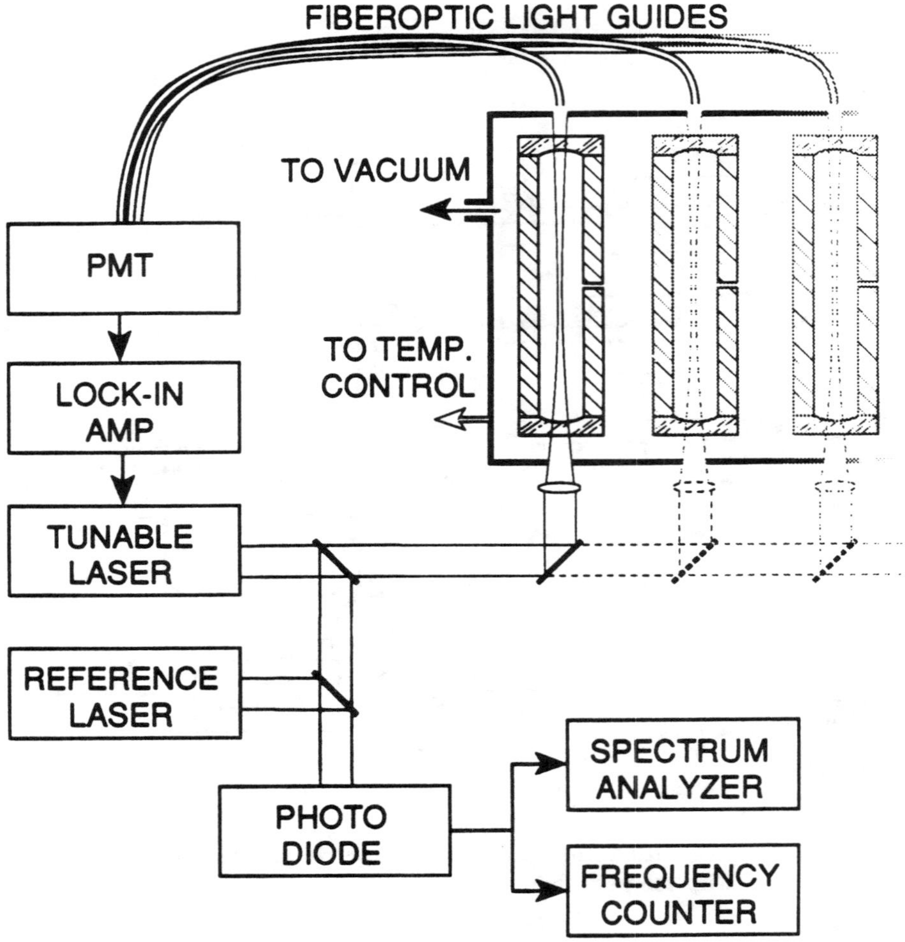

Fig. 15 **Arrangement used to study changes of length at constant temperature. Present chamber capacity: 37 samples.**

5. CONCLUSIONS

We have shown that

1. Regarding thermal expansivity of different fused silicas, the few differences there are, lie at **cryogenic** temperatures.

2. Regarding Zerodur and Zerodur M, both have extremely low thermal expansivity over a wide temperature range. One should be aware of the hysteresis problem and the different shapes of the curve of α' vs temperature.

3. ULE has no hysteresis problem.

4. Regarding Invar, indications are that $\alpha' = 0$ can be obtained from Invar made extremely free from impurities. Especially carbon. It is then reasonable to expect temporal stability better than 1 ppm/yr. (See acknowledgements.)

5. Regarding Superinvar, detailed information is presented describing the best Superinvar we have seen. While the thermal expansivity is superb, the temporal stability is shown to vary with temperature.

Work is in progress in our laboratory to characterize graphite/epoxy, PEEK, and carbon/carbon with respect to thermal expansivity and temporal stability. These materials can be tailored to meet various requirements, however their technology has not yet matured to the point where detailed characterization is useful. Graphite/epoxy is especially attractive for its shape flexibility as an optics mounting material. Unfortunately, unlike carbon/carbon, graphite/epoxy has a water absorption problem which hampers characterization. A current approach is to seal graphite/epoxy and hope to prevent water absorption in this way. Two ways of evaluating this are:

1. Test different sealants.
2. Correlate weight change with temporal instability.

6. ACKNOWLEDGEMENTS

We gratefully acknowledge the encouragement and assistance of JPL personnel Tim O'Donnell, Cheng Hsieh, Mark Lane and Witold Sokolowski, who have succeeded in obtaining the strong, low expansivity Invar described in Appendix I and are willing to advise others. Many thanks to John Hall for useful discussions and permission to use his unpublished data. Also thanks to Henry Hagy for background concerning ULE and fused silica, and to Guy White for permission to use his excellent low temperature expansivity data. Finally, thanks to Dan Bass, who made most of our measurements.

7. APPENDICES

Appendix I. Preparation of the Invar shown in Fig. 2, lower curve.

This material was produced through powder metallurgy by Spang Specialty Metals in July, 1990, with the request for very high purity material. Essentially, Fe 64%, Ni 36%, and nothing else. The material tests showed

Al	<	.001%
C		.002
Ni		36.0
Mn	<	.001
P	<	.01
S		.001
Fe		(balance)

Spang produced a 500 lb. sintered billet, 4 x 8 x 54". To increase the material's strength, the billet was then sent to Scientific Alloys, Inc. for hot forming (hammering to increase density), and finally extrusion into rods and other shapes.

Appendix II. The Superinvar sample shown in Fig. 2 was manufactured by Simond Steel Division of Wallace-Murray Corporation in the mid-1960's. It was Heat J-3058, purchased by Charles Freed of MIT. Its composition and heat treament is as follows:

Al	.01%
C	.03
Co	5.10
Cr	.05
Cu	.04
Fe	(balance)
Mn	.46
Mo	.024
Ni	31.2
P	.005
Si	.06
S	.013
Ti	.01
V	.01
Zr	.001

Lement, et al.[17] developed a heat treatment that is often referred to as the *MIT triple treat*. The first step is an 830°C soak and then water quench, typically followed by rough machining. The second step is soak at 315° for one hour and then air cool, followed by a third soak at 93° for 48 hours and air cool or furnace cool. The Superinvar part is then finish-machined as required. A 780 - 830° soak with air cooled or controlled oven cool heat treatment can be substituted for Step 1. This results in higher expansivity, but greater temporal stability compared with the water quenched material.

8. REFERENCES

1. The ASTM E338 Committee recommends the following nomenclature to describe thermal expansion:

Coefficient of linear thermal expansion (or thermal expansivity)

$$\alpha \equiv \frac{1}{L}\frac{\Delta L}{\Delta T}$$

Instantaneous coefficient of linear thermal expansion

$$\alpha' \equiv \lim_{\Delta T \to o}\left(\frac{1}{L}\frac{\Delta L}{\Delta T}\right)$$

Mean coefficient of linear thermal expansion

$$\overline{\alpha} \equiv \frac{1}{T_2 - T_1}\int_{T_1}^{T_2} \alpha' dT, \text{ stating } T_1 \text{ and } T_2$$

2. Ch. Ed. Guillaume, *Trav. Mém. Bur. Int. Poids Més.* **17**, 1, (1927).

3. Justice, B., Corning Glass Works, Corning, N.Y., Private Communication, 1973.

4. Dieter Hils and J.L. Hall, *Phys. Rev. Letters* **64**, 1697-1700, (1990).

5. G.K. White, "Thermal expansion of silica at low temperatures," *Cryogenics* **4**, 2-7, (1964). Fig. 3 and Table I

6. R. Brückner, "Properties & structure of vitreous silica I," *J. Noncrystalline Solids* **5**, 139, (1970).

7. Material described in Fig 4.

8. Heraeus-Amersil TO8E - see S.F. Jacobs, Dean Shough, Cliff Connors, "Thermal expansion uniformity of materials for large telescope mirrors," *Appl. Opt.* **23**, 4237-4244 (1984).

9. R.B. Roberts, R.J. Taish and G.K. White, "Thermal properties of Zerodur at low temperature," *Cryogenics* **22**, 566-568 (1982).

10. See Schott Product Brochure No. 10008.

11. Stephen F. Jacobs, Steve C. Johnston and Glenn A. Hansen, "Expansion hysteresis upon thermal cycling of Zerodur, *Appl. Opt.* **23**, 3014-3016 (1984).

12. O. Lindig and W. Pannhorst, "Thermal expansion and length stability of Zerodur in dependence on temperature and time," *Appl. Opt.* **24**, (1985).

13. See Schott Product Brochure No. 10041.

14. G.K. White, "Thermal expansion at low-temperatures of glass-ceramics and glasses," *Cryogenics*, **16**, 488 (1976).

15. J.W. Berthold and S.F. Jacobs, "Ultraprecise thermal expansion measurments of sevel low expansion materials," *Appl. Opt.* **15**, (1976); S. F. Jacobs, "Dimensional stability of materials useful in optical engineering," *Optica Acta* **33**, 1377-1388 (1986).

16. "Physics & Applications of Invar Alloys," Maruzen Co. Ltd., Tokyo, 1978.

17. B.S. Lement, B.L. Averbach and Morris Cohen, "Dimensional behavior of Invar," *Trans. Amer. Society for Metals*, **43**, 1072 (1951).

18. Work-hardened by drawing to 50% of its initial cross section.

19. G. Girard, "Sur le mesure des etalons geodesiques au BIPM," Proc. IAG-Symposium "High Precision Geodetic Length Measurements," June 19-22, 1978, Helsinki, Finland. Ed. by Teuvo Parm, Helsinki, 1980. ISBN 951-711-063-4, ISSN -0355-1962.

20. J.W. Berthold III, S.F. Jacobs, M.A. Norton, "Dimensional stability of fused silica, Invar and several ultra-low thermal expansivity materials," *Metrologia*, **13**, 9-16 (1977).

21. T. Baer, F.V. Kowalski and J.L. Hall, "Frequency stabilization of a 0.633μm HcNe longitudinal Zeeman laser," *Appl. Opt.* **10**, 3173-3177 (1980).

22. Two years later, Halls reports an inexplicable increase of a factor of two in shrink rate.

23. D.E. Schwab, S.F. Jacobs and S.C. Johnston, "Isothermal dimensional instability of Invar," SAMPE Proceedings, Reno, Nevada (1984); S.F. Jacobs, C. Johnston, D.E. Schwab, "Dimensional instability of Invar," *Appl. Optics*, **23**, 3500 (1984).

24. Heat at 315°C for 60 min. in air; then cool to room temperature; reheat at 95° for 40 hours in air; cool to room temperature again.

25. Jeffrey M. Steele, Dennis A. Thompson, Stephen F. Jacobs and Dan Bass, "Temperature and aging effects on the temporal stability of Invar," Proc. SPIE, **1752** (in publication).

26. Private communication 1984

27. AIP Conference Proceedings, No. 17, Thermal Expansion - 1973, Ed. by R.E. Taylor and G.L. Dennan
AIP, N.Y. (1974)

28. C.W. Marshall, R.E. Maringer, "Dimensional Instability - an Introduction," Pergamon Press, NY (1977).

29. S.F. Jacobs, S.C. Johnston, J.M. Sasian, M. Watson, J.D. Targove and D. Bass, "Surface figure changes due to thermal cycling hysteresis," *App. Opt.* **26**, 4438-4442 (1987).

30. A.D. White, *App. Opt.* **6**, 1138 (1967).

31. R.E. Parks, *Optics and Photonics News* p. 45, Sept. (1991).

Table II.

Invar variables tried in elevated temperature stability tests.

	INVAR										SUPERINVAR
SHEET											
.040 thick			x	x	x	x	x		x	x	x
.060 thick								x			
Bent			x	x	x	x		x	x	x	x
e-beam Welded							x				
48 hrs Stabilization			x	x		x	x	x	x	x	x
96 hrs Stabilization					x						
Soln Anneal-Slow Cool										x	
Decarburized-Quenched						x					
Decarburized-Slow Cool									x		
⊥ rolling direction			x		x	x	x	x	x	x	x
∥ rolling direction				x							
PLATE											
24 hr Stabilization		x									
48 hr Stabilization	x										
Thermal Cycled											x
NO. SAMPLES	3	3	3	3	3	3	3	3	3	3	3
GROUP NO. (see Figure)	1	2	3	4	5	6	7	8	9	10	11

Table III.

Temperature history and temporal instability of the Superinvar sample described in Fig. 2. (Sample was stored at room temperature between measurement periods.)

Date	Temperature (C)	Instability (ppm/yr)
1975 (6 months)	27	< .01
1983 (4 months)	60	+ 1.2
1991 (3 months)	25.8	+ 1.0
1991 (3 months)	37.8	- 0.5
1991 (3 weeks)	25.8	+ 1.2

Optomechanical Component and Mechanism Design

Chair
Daniel Vukobratovich
Optical Sciences Center/University of Arizona

Design Considerations for Mirrors with Large Diameter to Thickness Ratios

John W. Pepi
Itek Optical Systems
A Division of Litton Systems, Inc.
10 Maguire Road
Lexington, MA 02173

ABSTRACT

As optics increase in size, so increases the difficulty in achieving optical performance goals. Such challenges, however, can be accomplished by careful consideration of design, mounting, and material characteristics to lead to a selected concept of large ground or space based optics with high aspect ratios.

This paper outlines some of the large mirror design methodology and understanding required to meet specified performance criteria for both monolithic lightweight and thin solid meniscus design approaches. These criteria are many. The optic must be shown capable of meeting performance goals during a changing gravity environment, for ground based telescopes, often looking from horizon to zenith, or be tested for the zero-g environment of space. It must further be insensitive to, or provide accommodation for, thermal swings about the nominal fabrication temperature. At temperature, it must preclude excessive error causes by real time thermal gradients. It must resist vibratory and wind loadings as well, and be of adequate strength to withstand all phases of handling and transportation in various stages of completion.

The mirror must be generated, cut, ground, polished, tested, mounted, and coated. During these stages, the effects of residual stress, temperature, assembly tolerances, mount force errors, tool pressure, bimetallic distortion, surface flaws, and humidity must be duly considered to meet the stringent criteria of fractional visible wavelength performance and fractional arc second encircled energy requirements.

The body of this paper presents an overview of the detailed analyses required to meet such criteria. Discussed are methods of supporting, mounting, and controlling such optics. The analytical tools of finite element math modeling are presented which assist in determining effects of material inhomogeneities, thermal strain, initial stress conditions, lightweighting, and mirror curvature, which play an important role as the diameter to thickness ratio increases. Material phenomenon and comparative trades are developed to aid in the choice of ideal candidates. Design examples from experience are finally given to illustrate the sensitivity of proven optics to the design criteria which have met stringent performance goals. These include lightweight monolithic optics with diameter to thickness ratios in excess of 15:1 and thin meniscus designs with ratios in excess of 100:1. In the latter case, active control is demanded, and correctability to both fabrication and operational errors is discussed, including actuator count determination.

1.0 INTRODUCTION

The laws of physics which govern the design of mirrors with high aspect (diameter to thickness) ıatios are, or course, the same as for those of low values. The difference, however, lies in the magnitude of distortion induced in the high aspect ratio designs by the operational, fabricating, or test environments. Depending on the particular environmental conditions, performance errors can be orders of magnitude higher for these optics than those of the more common and lower ratios. What is "in the noise" or unmeasurable for that latter class suddenly becomes a design driver.

It is in this vain that the subject paper is presented. Design, material, and mounting considerations for large diameter to thickness ratio optics will be discussed from an analytical standpoint to point out difficulties which are often unseen in conventional optics. The presentation to follow however, is not meant to be an esoteric theoretical discussion involving advanced mathematics. Rather, a critical review, or overview, of some of the issues which must be addressed is made, as outlined in the abstract.

2.0 DEFINITION

A definition of large diameter and large diameter to thickness ratio is relative. Large diameter to some may be small to others. In turn, large diameter to thickness ratio is a subjective, relative quantity. In this latter regard, it is noted that aspect ratios of less than 6:1 are generally considered as conventional. For the purpose of this paper, aspect ratio in excess of 15:1 will be considered large. Indeed, designs will be discussed with ratios in excess of 200:1. It is noted further that environmental effects on optical distortion usually scale as the power of diameter in direct proportion. To this end, performance of mirrors with high aspect ratio will be discussed for mirrors with diameters in excess of one meter. With this arbitrary definition, issues can be address for a certain class of optics where not only analysis has been made but also facts have been revealed through interpretation verified by the data of actual test results and performance measurements.

3.0 BACKGROUND

For the class of optics being presented, two questions need to be addressed. The first is: how will the mirror meet performance goals in its operational environment? And the second is: how will you make it in the first place?

The first question concerns itself with the site, whether ground, aircraft, or space based. In any event common to those environments are gravity (or the lack of it) and temperature. It is those two environments to which questions must be largely addressed. In the former, analysis and design must be accomplished to maintain performance for varying angular orientations ground or in the air, from zenith or nadir to horizon. In space, the dilemma of performing a meaningful ground test must be assessed. In the latter, temperature soak changes even of a few degrees only, must be considered, as well as the potential for variations and gradients about the operating temperature as may be induced by sun, wind, or on station heat sources.

The second question concerns itself with whether one can get there in the first place. Can an optic be so large in aspect ratio that one cannot fabricate it to the required precision? Can an optic be so large in aspect ratio that you cannot test it? These questions address the concerns that, although critical, are less so in the more massive designs of low aspect ratio. Among those concerns are the effects of grinding, polishing, cutting, testing, mounting, coating, temperature, residual stress, handling, assembly, metrology, humidity, surface flaws, tolerances, mass uncertainties, force uncertainties, hysteresis, and peculiar material properties. These effects, along with those of the gravity and operational environments are discussed in the body of this report.

4.0 LIGHTWEIGHT CONSIDERATIONS

Along with the criteria for risk, schedule, cost, and performance is the requirement for lightweight, a feature of large diameter of thickness ratio optics. These can be classified as either thin solid optics (thin meniscus) or somewhat thicker but lightweighted, monolithic optics. This latter class is comprised of classical sandwich construction, i.e. a lightweighted core with top facesheet and bottom facesheet with or without holes. For very large diameter optics made of glass or glass ceramic, this type of construction often becomes impractical due to fusion or frit bonding requirements, particularly for the near zero thermal expansion candidates. (The Steward Observatory Mirror Laboratory has had much success in spin casting very large optics using borosilicate glass of relatively high thermal expansion.) To reduce risk, schedule, and cost to fabricate monolithic and lightweight optics, the use of "open back" construction is made, whereby the optic can be formed by machining away glass from an initially solid blank, where over 90 percent of the material can be removed. Such designs, in order to optimize stiffness, should consist of triangular cell cores. In the further interest of weight reduction, design optimization programs have been developed to minimize weight while maximizing stiffness, yet resist manufacturing pressures with no undue quilting, or "print-thru", during polish. While this technique is routinely used, advances using "fragile polish" removal techniques, in which nearly zero not pressure is applied while maintaining equal or greater removal rates as conventional methods allow, have obviated quilting concerns. This approach has been used with much success at Litton Itek Optical Systems.

An example of such a lightweight open back monolith is shown in Figure 1(a) which exhibits a diameter to depth ratio of 15:1. The equivalent solid thickness of this blank, i.e.. one giving the same stiffness, results in a effective aspect ratio of 30:1. Shown in Figure 1(b) is a thin meniscus solid, lightweight only by virtue of its small thickness. This optic has a diameter to thickness ratio in excess of 100:1. Performance comparisons of these optics are later given. It is noted that the mounting for each of the optics is different, the former performing on three points and the latter requiring multipoint. It is important to note that for the thin meniscus designs, and in may cases for the monolithic ones of high aspect ratio, some form of active control is generally demanded to meet the stringent tolerances specified for visible wavelength imagery. The need for multipoint mounting and active control is now presented.

5.0 MOUNT CONSIDERATIONS

The mounts which support the optic to its bezel or reaction structure frame must serve several purposes. They must be stiff in all directions to maximize rigid body frequencies, rigid enough to preclude buckling, and strong enough to resist operational and non operational stresses under acceleration loadings. They must also be flexible (in the appropriate degrees of freedom) to minimize mount induced assembly loading mirror distortion, to which the high aspect ratio optics are most susceptible. They must be flexible enough as well to minimize deflection caused by thermal expansion mismatch. To achieve such a design, the mounts must be near kinematic in nature. Truly kinematic mounts which rely on bearings, universal joints, pins, slots, rollers, or the like are to be avoided generally due to the unpredictable effects of stiction and friction. A near kinematic mount usually takes advantage of flexible parts in the appropriate degrees of freedom which are both predictable and accountable in the design.

An example of a three point near kinematic mount is shown in Figure 2(a) and 2(b). The former is a bipod design in which the flexibility requirements are met by necking down the members at each end. The stiffness requirements are met by the inherent axial member stiffness of the bipod legs themselves. The latter design is a three point leaf flexure mount with rotational pivot, the flexural providing the required stiffness and flexibility in all degrees of freedom except radial rotation which is provided by the pivot.

The ideal three point mount system often becomes impractical for mirrors of large diameter to thickness ratio, largely due one of the main enemies of those optics - gravity. Gravitational errors in conjunction with mount design are given in the section below.

6.0 GRAVITATIONAL CONSIDERATIONS

It is difficult to address mounting issues without addressing the gravity issue. This section will address some of the ways of mounting to resist gravity deflection in both optical axis vertical and horizontal configurations, particularly as the aspect ratio increases. A rather extensive discussion on mount configurations is given by Yoder[1], which provides an excellent reference. It is not the intent here to review detail but rather, provide an overview of the issues.

For mirrors with optical axis vertical, deflection under gravity on three points at the mirror periphery is given by

$$y = \frac{k_1 \rho D^4}{Et^2} \tag{1}$$

for a thin meniscus solid, in which ρ is the material density, D the diameter, t the thickness, E the modulus, and k_1 a geometry/mount constant. (Nomenclature for this and all equations is given in Table I). When the optical axis is horizontal, deflection is given by

$$y = \frac{k_2 \rho D^5}{R_s E t^2}, \tag{2}$$

in which R_S is the mirror radius of curvature

Note the rapid increase as the diameter increases and thickness decreases. As the allowable performance is exceeded, the mount locations can be moved inboard to reduce distortions by up to a factor of four in both orientations. For the horizontal axis, however, care must be given to locate the mount intersection near the mirror center of gravity. As the mirror aspect ratio increases, the designer may be forced into a multipoint mount of six, nine, or more points. To maintain kinematics, a hindle, or whiffle tree, mount is then required for the optical axis vertical direction. One such mount is shown in Figure 3 for segment of the Keck telescope mirror, with aspect ratio of 25:1. Note here that the mount attachment is not truly kinematic, but the necessary compliance is achieved by flexured posts.

Other vertical axis mounts at many points include pneumatics, hydraulics, and counterweight systems, which serve to float the optic, leaving only gravitational residual error in between the mount points. Extreme care is needed in design to avoid problems of stiction and friction, however, and such designs will most often need some type of force sensing or active control.

For horizontal axis loading, a different scheme is required for multipoint loading. This can take the form of edge mounting with sinusoidally distributed forces, or mercury tubes. The sinusoidal forces can be achieved by lateral edge counterweights, as an example. Schewesinger[2] has provided an excellent theoretical treatise on the subject in several papers. This approach, too, has its limitations for high aspect ratio optics. It has been used successfully for mirrors up to four meters. But for the aspect ratios being considered here, distortion may still be too high for high acuity optical systems. In this case, supporting the optic edge tangentially can provide a good support. One such optic is shown in Figure 4, in which tangent rods provide the necessary lateral restraint yet maintain compliance in the axial direction. Angling the rods out of plane and other combinations with radial load distributions can further minimize distortion. As mirrors increase to the 8 meter class of aspect ratios in excess of 25:1, again some form of active control is still demanded.

The next level of support in terms of minimizing passive errors of gravity in the horizontal direction consists of multipoint restraints interior to the mirror edge. Again counterweights or hydraulics can be used, and combined to provide axial restraint as well. In fact, collocation of both axial support and lateral support systems can optimize the design in all orientations. The Hale telescope uses this concept. For thin meniscus designs, this approach minimizes distortion, but active control is usually demanded. Also, extreme care need again be taken to minimize mount induced rotational moments which may yield uncorrectable surface errors. A schematic example of a dual acting interior mount and rotationally compliant attachment scheme is shown in Figure 5, with provision for active control. Attachment to the optic back surface is achieved through epoxy bonding in the case of glass, to an attachment button matched in thermal expansion coefficient.

An example of a completed optic of aspect ratio greater than 100:1 is shown in Figure 6. Here, the optimized back internal multipoint support is used to resist gravity in all orientations. Each support also is under active control to correct errors from sources other than gravity. In this case no counterweights or hydraulics are required, since the weight per support is quite low. Lateral support is achieved through rotationally compliant bipods and tripods attached to the mirror back which point to the neutral surface of the mirror (i.e. the midpoint of its thickness), as seen in Figure 7.

7.0 MATERIAL CONSIDERATIONS

Properties of some of the potential common mirror materials are, in general readily available. When discussing large aspect ratio aspects, it is important to understand all the design criteria which might drive a design. Figures of merit in themselves are nice selling points for the manufacturer, but may have no bearing on the design. For example, common mirror choice materials include glass ceramics such as Zerodur, glasses such as borosilicate, fused silica, and ULE, metals such as beryllium and aluminum, and silicon carbide. A combined thermal mechanical figure of merit may look like

$$\frac{EKS}{\rho\alpha\Delta\alpha} \qquad (3)$$

which E is the stiffness (Young's modulus) K the thermal conductivity, S the strength, ρ the density, α the coefficient of thermal expansion, and $\Delta\alpha$ the expansion homogeneity. Materials such as glasses have poor conductivity and hence do not resist thermal gradients, but the low expansion glasses make up for this shortfall. Beryllium has excellent conductivity but its benefit is negated relative to glass at room temperature due to relatively high expansion coefficient. At colder temperatures, however, conductivity to expansion ratios are better than the glasses. In the absence of heat flux, such as solar exposure, the figures of merit may have no bearing. Gravitational sag and weight may now drive the design (E/ρ). Strength may be insignificant if a design is stiffness driven. In the presence of soaks, expansion homogeneity is important, as will be discussed, particularly for large optics. Not addressed by usual figures of merit are the effects of coatings, residual stress, curvature, surface flaws, hysteresis, etc., all of which may play an active role in material selection for the high aspect ratio optics being considered. Some of the peculiar behavior of those optics are addressed in the remainder of this paper.

8.0 TEMPERATURE CONSIDERATIONS

In order to understand the behavior of the large optics detailed mathematical models are required to determine distortion during thermal soak and gradient conditions. This is important not only for the operational environment but for the fabrication environment as well, where active control is not as readily invoked. Figure 8 shows a typical method used to determine the effects of expansion coefficient inhomogeneity during thermal soak. In that case, a finite element model is prepared which maps, one for one, measured CTE data from a Corning ULE glass boule, where CTE values are expressed in parts per million. Although near zero in thermal expansion, it is the variability about the nominal which is often of concern in the approach to large optics design. Errors due to expansion inhomogeneity are found approximately proportional to the relation.

$$y = \frac{k_3 T \Delta \alpha D^2}{t}. \tag{4}$$

where T is the soak temperature.

As an example, an optic exhibiting a D/t ratio of 60:1 and presented inhomogeneity is shown in Table 2, using the finite element modeling technique for a proposed 7 meter diameter optic. Notice the effects even after power (focus) removal of the relatively small variation in expansion coefficient for only 1°C change.

Thermal gradients through the thickness can also be a driver in design of thin meniscus mirrors. Due to curvature (shell effects), errors so realized are not entirely focusable. The same example for the 60:1 optic shows the results before and after focus in the same table. The radius of curvature is 25 meters. For low aspect ratios, error is related to

$$y = \frac{k_4 \alpha \Delta T D^2}{t} \tag{5}$$

where ΔT is the axial thermal gradient. The error is nearly all focusable, for conventional optics.

For high aspect ratios, with finite radius of curvature, however, the mirror shall under the described gradient no longer remains in the stress free state afforded by the flat condition, so that the above equation no longer applies. In this case, the thermal bending of a circular segment of a thin spherical shell subjected to a constraint temperature gradient through its thickness is examined. The physical parameters and governing equations, which use complex Bessel (kelvin) functions, are described by Barnes.[3] A significant amount of spherical aberration is found as the aspect ratio increases.

To illustrate what happens to the residual error before and after focus correction, a detailed NASTRAN model of a 40 inch diameter mirror was made, with the mirror thickness varied accordingly. The model was initially run as a flat plate, and the results found to agree with the theoretical solution to within one-half of a percent. A radius of curvature of 175 inches was then included to represent a typical curved optic.

Utilizing a constant unit gradient through the mirror thickness, the resulting displacement from the math model were then input to a post-processor to determine residual error both before and after focus. The results are shown in Figure 9 along side the theoretical solution afforded in the reference above. The correlation is markedly good.

As evidenced by the curves, the departure from the flat plate error is most pronounced as the diameter to thickness ratio increases. For a D/t of 50:1, the error is about 2.5 times less than the flat plate without correction. After focus, however, about 16 percent residual still remains, whereas the flat plate error is entirely focusable. For a D/t of 10:1, the uncorrected error is very close to that of the flat plate, while the residual error remaining after focus is less than 1 percent.

We conclude that the shell effect reduces the residual error before a focus fit, under a thermal axial gradient. After focus, the residual error may be significant.

9.0 RESIDUAL STRESS

It is of extreme importance to understand the nature of residual stresses in the optic to be fabricated. For small diameter optics with relatively low D/t ratios (<10:1), residual stress effects are often negligible. For large diameter optics with increasing aspect ratios, the effect could be significant in understanding behavior during cutting, lightweighting, or edging operations. Stresses due to cooling processes for Zerodur, for example, while low, could cause excessive warping after release. Stress residual magnitude for ULE and other glassy materials are of similar magnitude, and could be significantly higher for metal materials. Finite element analyses have correlated markedly well with measurements made on several large optics programs.

Consider for example, a circular disc subjected to hoop compressive stresses at its outer periphery and tensile stresses toward its center. This is a likely scenario in the forming process of Zerodur glass ceramic, in which the outer periphery, due to heat transfer laws, is cooled more rapidly than the center. As the center continues to shrink (positive expansion coefficient) during cooling after ceramizing and anneal against the already cooled, more rigid outer edges, tensile stresses are set up near the center with hoop compression near the edge as shown in Figure 10. The locked-in stresses are a function of the expansion characteristic of the material, and since this value is small, so are the stress levels. Similar effects can be found in ULE, where low levels of residual stress are present due to small but finite differences in expansion coefficients throughout a boule due in part to variations in composition. Generally outer zones of the boules are in residual tension. For other materials, residual stress levels can be expected to be of greater magnitude. For the glasses, birefringence test can be made to correlate results with analysis. As with any polarized measurement, only the difference in principal stresses are readily made, with more complex techniques requiring oblique incidence viewing and shear integration required to isolate the stresses. Thus, an element in pure (bi-directional) tension gives no indication of stress if the usual technique is applied. Nonetheless, measurements near the edge, where radial stress is zero, give a good indication of the magnitude and direction of internal edge stress.

Again finite element modeling can be used to predict the magnitude of springing for various operations. Table 3, for example, is an example of warping after edge cutting and central hole cutting on a Keck telescope segment. These analytical results correlated very well with test data.[4] Stress level residuals are on the order of 30 psi only. Note that power is not the only anticipated change, but higher order symmetrical aberrations as well are realized, due to the shallow shell effects. The segment has a D/t ratio of about 25:1.

Springing effects are related to

$$y = \frac{k_5 \sigma_r D^2}{E t^2} \tag{6}$$

in which σr is the residual stress level.

10.0 COATING CONSIDERATIONS

Coating of optics are subject to residual stresses during deposition and differential stresses during thermal excursions due to CTE mismatch. However, for normal reflective coatings measured in Angstroms of thickness, distortion to even large optics is of minimal concern. On the other hand, the use of much thicker metallic coatings for use on aluminum or beryllium for example, required for fine polish, can result in significant distortion. Over normal temperature ranges and accounting for backside coating as well to minimize this bimetallic effect, errors in excess of several hundred micro inches (tens of waves) are easily realized for an 8 meter optic with D/t of 40:1. While active control can alleviate this problem, correctability must be invoked continually and actuator density must be high. During fabrication, thermal changes could mask out what is trying to be interpreted.

Under thermal soak or residual coating stress, distortion is related to

$$y = \frac{k_6 \Delta \varepsilon h D^2}{t^2} \quad (7)$$

$\Delta\varepsilon$ is the strain residual in the coating ($T\Delta\alpha$ or σ_c/E) are h its thickness.

11.0 SURFACE FLAW CONSIDERATIONS

While surface flaws must be addressed due to slow crack growth phenomena, additionally during grinding large changes can be observed due to the Twymann effect, which is related to the residual stress discussion above and follows the same relationship. A material is removed, subsurface flaws and hence locked in stress will cause radius changes and higher order aberrations to a lesser degree. These will disappear after final polish, but must often be accounted for in the grind to understand what is happening. These effects could exceed several microns, for example, in grind of a glass optic with D/t of 40:1.

Another phenomenon related to surface flaws is humidity, which can cause chemical reactions and residual stress if moisture is allowed to propagate in flaws. Again, a few microns of power change can be realized with marked humidity changes for optics with aspect ratios in excess of 50:1 if surface flaws are left unchecked.

12.0 METROLOGY CONSIDERATIONS

Of prime importance to the large aspect ratio optics is the ability to fabricate and test it. A metrology mount is usually required to "float" the thin meniscus in order to perform a test of the mirror surface quality in the "absence" of gravity. Such mounts can take on several forms, including air bag, discrete piston, a combination of air and pistons, or perhaps the final mount itself. Mount spacing is chosen to minimize gravity sag in between supports, and often the use of active control correctability can be invoked, since what the mount puts in can be largely taken out by actuators. For these optics, however, a rather large sensitivity to force errors is possible. These errors can be caused by pressure variations stiction, mass uncertainties, force hysteresis, friction, and the like. Errors due to force uncertainty propagate as

$$y = \frac{k_7 F D^2}{E t^3} \quad (8)$$

in which F is the force.

Figure 11 gives an example of the sensitivity to force error for a four meter optic with D/t ratio in excess of 200:1. Note that the mirror is gram sensitive with up to a micron of error realized for randomly distributed forces of ±1 gram maximum. This indicates the care which must be taken in understanding the subtle effects of metrology for such optics. For the subject case, a fluid, single zone, multi piston mount arrangement was chosen which utilized nearly hysteresis free diaphragm assemblies. Accounting for force uncertainties and active control correctability which would be collocated with some (but only a portion) of the pistons, high performance goals were analytically achieved and validated by test of a smaller optic with D/t of 100:1.

13.0 TEMPORAL INSTABILITY

What is considered stable with time for optics of small D/t ratio and small deflections may not be so for those with high D/t ratio and the resulting larger deflections which can be accommodated due to the flexible nature of the piece. The allowable deflection an optic can undergo within its stress limits is proportioned to

$$y = \frac{k_8 \sigma_a D^2}{t} \tag{9}$$

where σ_a is the allowable stress.

Thus, as D/t increases, capabilities exist in such areas as active control, as is planned for the Very Large Telescope program (VLT), for example, in which distortions of up to 25μm may be required for multiple use of the system.[5] The Keck Telescope program utilizes a stress polish technique[6], in which deflection of up to 100μm are required during fabrication. In other cases, large operational deflection under gravity or thermal loading, for example, may occur and be compensated for using active control.

One such phenomena that can occur and be noticed with large deflections is that of delayed elasticity, defined as that portion of strain under stress beyond the instantaneous value under load, and not immediately recovered upon unloading.

Much has been written about structural relaxation, viscous flow, delayed elasticity, hysteresis, and other dimensional instability phenomena of glass and ceramics at elevated temperatures. Less has been documented about similar effects at room temperature. The time-dependent phenomenon of delayed elastic exhibited by Zerodur, a low-expansion glass ceramic manufactured by Schott, has been studied at room temperature.[7] The effect is related to the alkali oxide content of the glass ceramic and to rearrangement of the ion groups within the structure during stress. The effect, apparent under externally applied load, is elastic and repeatable, that is, no hysteresis of permanent set, as measured at elevated temperature, is evident within measurement capabilities. Nonetheless, it must be accounted for in determining the magnitude of distortion under load (delayed elastic creep) and on load removal (delayed elastic recovery). This is particularly important for large and lightweight optics, which might undergo large strain during fabrication and environmental loading, such as experienced in gravity release or in dynamic control of active components.

Itek Optical systems first encountered the delayed elastic effect in Zerodur while fabricating the primary-mirror segments for the Keck 10 meter astronomical telescope. The effect was measured using a high performance mechanical profilometer. To confirm the Keck data for delayed elasticity further, we conducted additional experiments on both glass manufactured by Corning. The samples used were 12.7 cm-diameter stress relieved plates polished flat on one side. The Zerodur was a piece taken from an actual Keck blank. Measurements were made using interferometry to eliminated metrology errors. The load was applied for two weeks at room temperature. The setup was modeled with the NASTRAN finite-element computer program, and the predicted deflections were characterized as Zernike polynomials. The instantaneous delayed elasticity expected was 0.7% of the maximum deflection, whereas the approximate theory would indicate that the figure of the plate would return to its original condition linearly with log time over a period equal to the loading time. The actual data are plotted for Z5 (C 2, 0), Z7 (C3, -3) and Z13 (C 4,0) in Figure 12. When the measured data are compared with the predictions, the correlation is excellent.

The effect for glasses containing alkali oxides is up to 1% or more, in excess of ten times that of fused silica. For usual and most applications, the results are readily accommodated, since the effect is reversible, without hysteresis. Unless strains are large, it will never be noticed. In cases where strains are large, however, it must be understood, to prevent misinterpretations of the source of time changing data during metrology.

14.0 ACTIVE CONTROL CONSIDERATIONS

Most of the class of optics being discussed here will demand some form of active control, since environmental, fabrication, or test deflections may be in excess of allotted requirements, particularly for visible wavelength systems. It is noted that a thin active meniscus design can result in less expensive facesheet cost; is very lightweight, even when actuator and reaction structure weight is included; results in low thermal mass; and exhibits superior performance to the more massive and passive designs. In general, for a given and equally spaced actuator count, correctability for active mirrors is independent of both diameter and thickness, and hence D/t ratio, for a given shape and magnitude of error. The force requirements of course, increase for smaller diameters and higher thickness according to the relations earlier described. Actuator count can be determined by the need for minimizing inter-actuator gravity sag (which is a function of thickness) and the magnitude and shape of aberrations to be corrected. In the former regard, actuators are often made coincident with the gravitational operational mount. In the latter case correctability algorithms must be used to determine count, and detail trades made to size the facesheet within total weight limitations. Residual error after correction is approximately proportional to actuator count. For example, as few as 50 discrete force actuators can correct low order astigmatic error to a residual of 1% of the original aberration; 100 actuators reduces that residual in half. For higher order, residual error increases for a given count of actuators; 100 actuators may reduce residual error to 10% of the input, for example, in the case of spherical aberration. Force requirements along with weight may drive the meniscus thickness trade. A 4 meter optic with D/t of 15:1 may operate passively in a gravity environment, but hydraulics or counterweights may be needed; increasing D/t to 40:1 requires actuator force sensing and local loop closure; increasing D/t to 100:1 will require active control and wavefront sensing, with the benefit of eliminating counterweights or hydraulics, directly attaching the facesheet to its supportive structure. The design of the latter thin meniscus must take into account all of the design principles included herein. Facesheet thickness versus actuator count plots can then be made for the multiple criteria, and an optimum combination of thickness and count found in which all requirements are achieved. These requirements, in addition to the gravity concern, may include thermal gradient, aberrated incoming wavefronts, weight, stress during acceleration or in flexure, fundamental frequency for control stability, manufacturability, metrology error force limitations, power, stroke, resistance to uncorrectable assembly loads, and manufacturing errors, among others.

15. PERFORMANCE COMPARISONS

At the outset of this paper, an arbitrary definition of size and aspect ratio was given for purpose of the above discussions, i.e. diameter in excess of 1 meter and D/t in excess of 15. In this section, a comparison is given in terms of sensitivity to mirror displacements based on the scaling relationships developed. This can give the reader a sense of why design considerations for large high aspect ratio optics are important, since displacements unseen in the smaller optics suddenly increase to significant proportions.

As an example, the summary of Table 4 is presented which compares a 1 meter optic with D/t ratio of 6:1, a low aspect ratio design, to one of 2 meters in diameter with D/t ratio of 15:1, another of 4 meters in diameter with aspect ratio 25:1, still another of 8 meter diameter with D/t of 40:1, and a final of 8 meter diameter of aspect ratio 150:1. The chart is normalized to the 1 meter design of D/t ratio 6:1 for purpose of the comparison. All of the designs, incidentally, are in the domain of optics already manufactured or under consideration, and in fact are all less sensitive than some large optics currently being fabricated.

As evidenced by the table, sensitivities are increased by several orders of magnitude in some cases, highlights the care which must be given. This presentation is ended in this light, by a similar performance comparison of mirrors from actual or planned programs, involving the high aspect ratios being discussed here. This is given in Table 5, in which all values are ranked by sensitivity and normalized to the successful large optic of the European Southern Observatory (ESO), a 3.5 meter by 230mm meniscus with aspect ratio of approximately 15:1. Shaded programs are those completed or under completion at Itek Optical Systems.

16.0 CONCLUSION

Large optics with high aspect ratio are quite sensitive to the environment in which they are made and in which they must operate. In order to ensure their success, considerable consideration must be given to peculiar mirror properties, response to force errors, manufacturing approach, test methods, and the need for active control. Fabrication is not without challenge but can be thus approached with a degree of certainty in finishing to desired optical tolerances.

17.0 ACKNOWLEDGMENT

The author would like to thank the many present and past contributors at Itek Optical Systems who accomplished the design analysis and fabrication tasks herein described, as well as the other contributors referenced in this document.

18.0 REFERENCES

1. Yoder, Paul R., Jr., Opto-Mechanical Systems Design, Chs., 9-11, Marcel Dekker, Inc., New York, 1986

2. Schwesinger, G. "Optical Effect of Flexure in Vertically Mounted Precision Mirrors", J. Opt. Soc. Am. 44:417, 1954

3. Barnes, W.P., Jr., and Pepi, J.W., "Analytical Predictions for Lightweight Optics in a Gravitational and Thermal Environment", SPIE, San Francisco, CA, 1985

4. Pepi, J.W., "Test and Theoretical Comparisons for Bending and Springing of the Keck Segmented Ten Meter Telescope", The Hague, The Netherlands, International Congress on Optics, March 1990

5. Enard, D., "The European Southern Observatory Very Large Telescope", J. Optics (Paris), 22:2, pp. 33-50, 1991

6. Nelson, J.E., et al, "Stressed Mirror Polishing, Fabrication of an Off Axis Paraboloid", Appl. Opt., Vol 19, No. 14, 2341-2352, 1980

7. Pepi, J.W., "Delayed Elastic Effects in Zerodur at Room Temperature", Applied Optics, 31:1, pp. 115-119, January, 1992

TABLE 1
Nomenclature

F	=	random force error
E	=	material modulus of elasticity
α	=	coefficient of thermal expansion (CTE)
K	=	thermal conductivity
ΔT	=	axial thermal gradient
T	=	thermal soak
$\Delta\alpha$	=	material CTE inhomogeniety
ρ	=	density
σ_r	=	residual stresses
σ_c	=	coating stress
σ_a	=	material allowable stress
S	=	strength
y	=	performance error
D	=	facesheet diameter
t	=	facesheet solid thickness
k_n	=	geometrical/mount constant
R_s	=	mirror radius of curvature
$\Delta\varepsilon$	=	residual strain
Z	=	Zernnike polynomial

TABLE 2
Response to Soak and Gradient for a 300 x 5 Inch Thick ULE Curved Optic

	1°C Soak Change		1°C Axial Gradient	
	Error From Perfect CTE Value	Error From Radial CTE Variation	Error From Perfect CTE Value	Error From Radial CTE Variation
Uncorrected	0.0024	0.032	0.131	0.073
Focus	0.0000	0.019	0.037	0.046

- values in waves (surface) at 633nm

TABLE 3
Theoretical Warping (Normalized to 1.0 micron power)

Zernike Term	Edge Cut	Center Cut	Total
C(2,0)	+0.70	+0.30	+1.00
C(4,0)	-0.06	-0.13	-0.19
C(6,0)	0.00	+0.05	+0.05
C(8,0)	-0.01	-0.08	-0.09

- values correspond to 35psi stress in outer zone
- positive power is concave

TABLE 4
Sensitivity of High Aspect Ratio Optics Relative to Conventional

Size (meters)	Aspect Ratio	Grind/Cut/Coat	Metrology	Thermal Soak	Gravity
8.0	150	580	1760	190	37300
8.0	40	40	33	51	2620
4.0	25	16	16	16	256
2.0	15	6	7	5	24
1.0	6	1	1	1	1

TABLE 5
Senstivity Ratings for Large Optics

Size (Meters)		Aspect Ratio		Grind/Cut Coat		Metrology		Thermal Soak		Gravity	
VLT	8.2	LOS	235	LOS	240	LOS	3250	LOS	18	LOS	78
JNLT	8.1	ALOT	153	ALOT	100	ALOT	1350	VLT	7	ALOT	55
LOS	4.0	LAMP	118	LAMP	60	LAMP	810	ALOT	7	VLT	52
AEOS	3.6	VLT	47	VLT	9.5	VLT	12	JNLT	6	JNLT	38
ESO	3.5	JNLT	41	JNLT	7.0	JNLT	8	LAMP	4.5	LAMP	20
ALOT	2.6	KECK	25	KECK	2.8	KECK	8.6	AEOS	1.6	AEOS	2.6
LAMP	2.0	AEOS	24	AEOS	2.5	AEOS	3.8	ESO	1	ESO	1
KECK	1.9	ESO	15	ESO	1	ESO	1	KECK	.9	KECK	.8

• Listed by Degree of Difficulty Relative to ESO

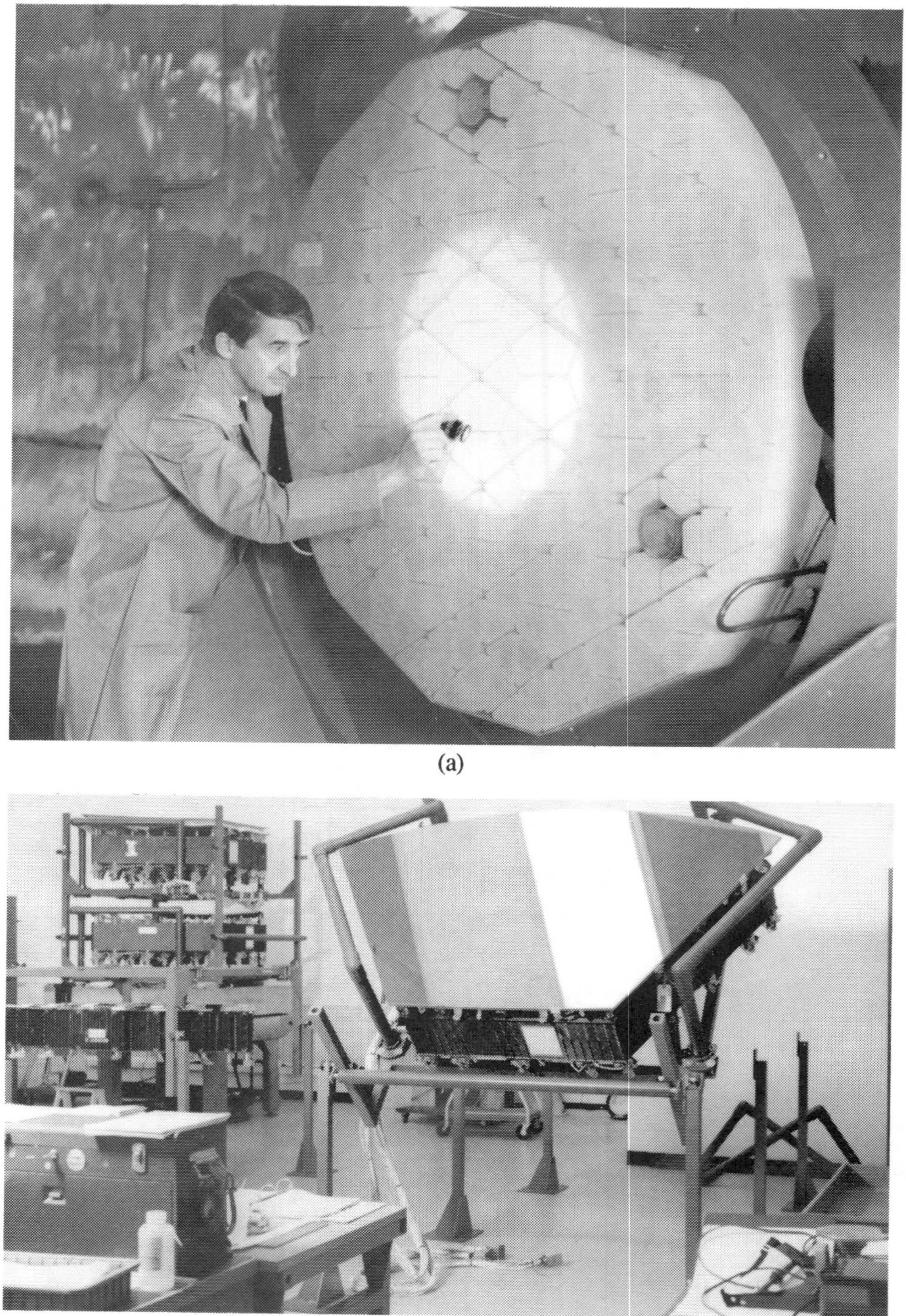

(a)

(b)

Fig. 1. Large-aspect-ratio optics.

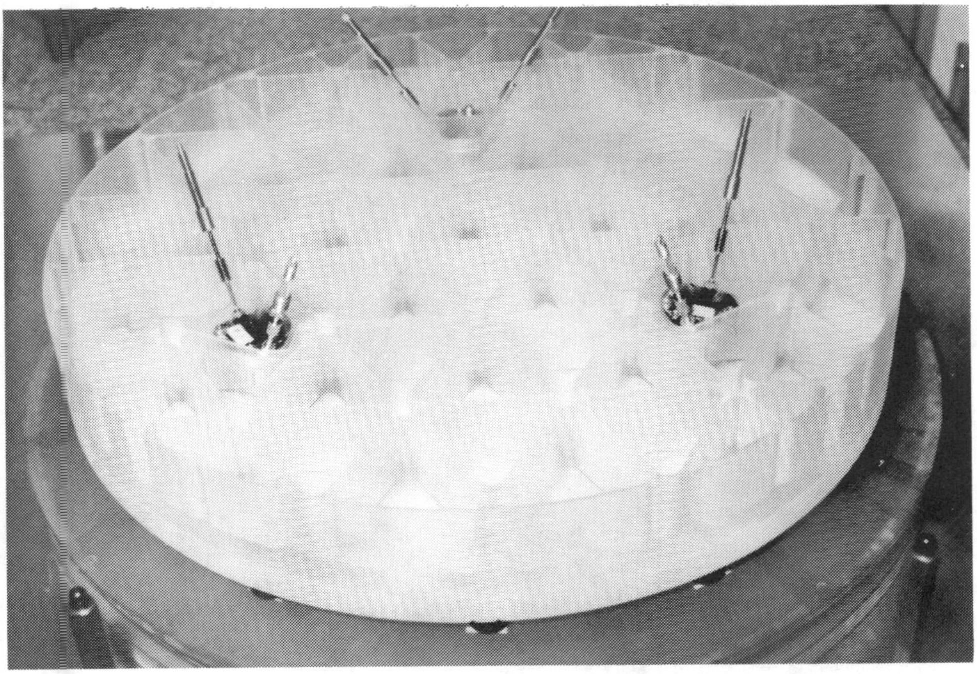

(a)

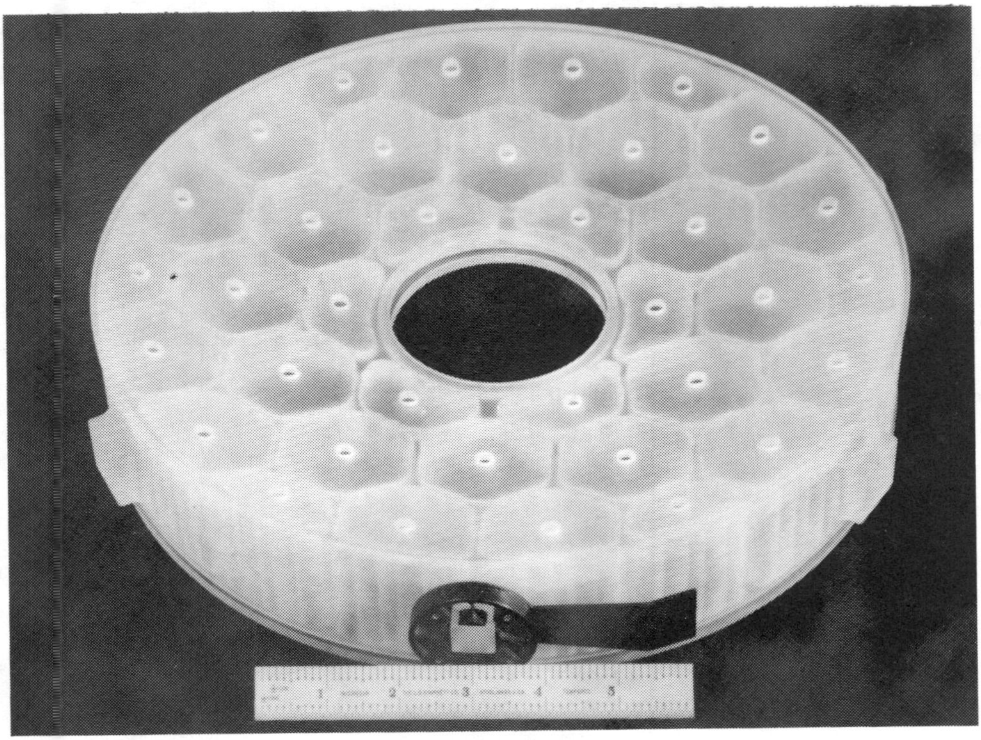

(b)

Fig. 2. Three-point near-kinematic mounts

Fig. 3. Multi-point whiffle tree mount.

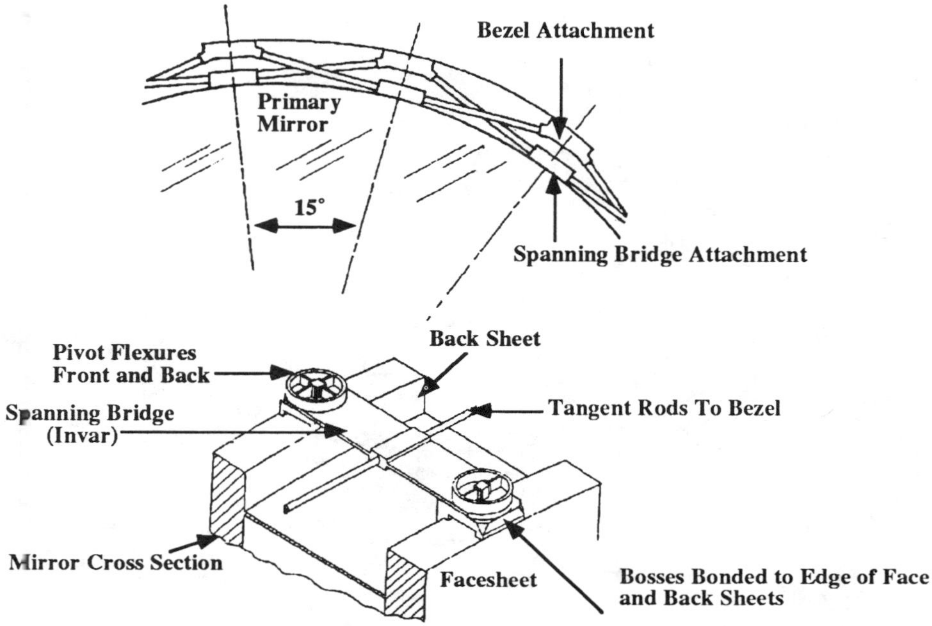

Figure 4
Tangent Rod Design

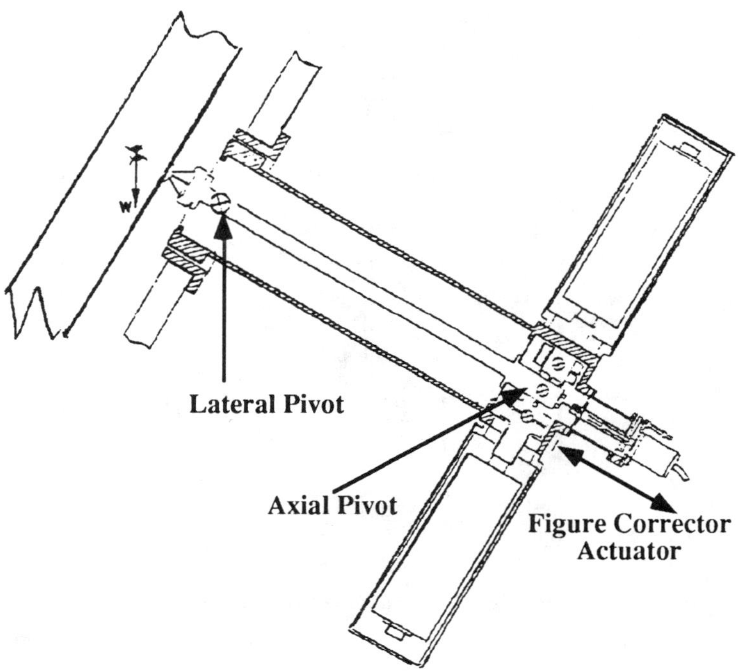

Figure 5
Dual Back Mount Concept

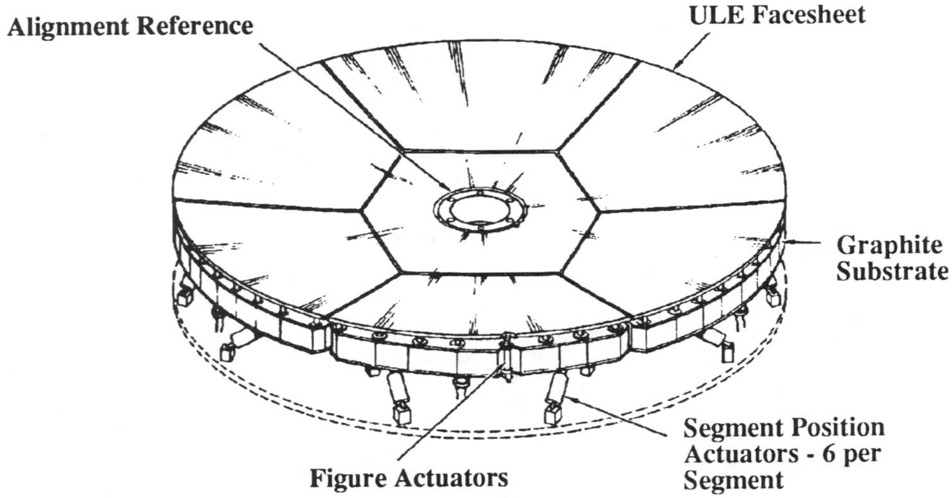

Figure 6
Lightweight Active Optic

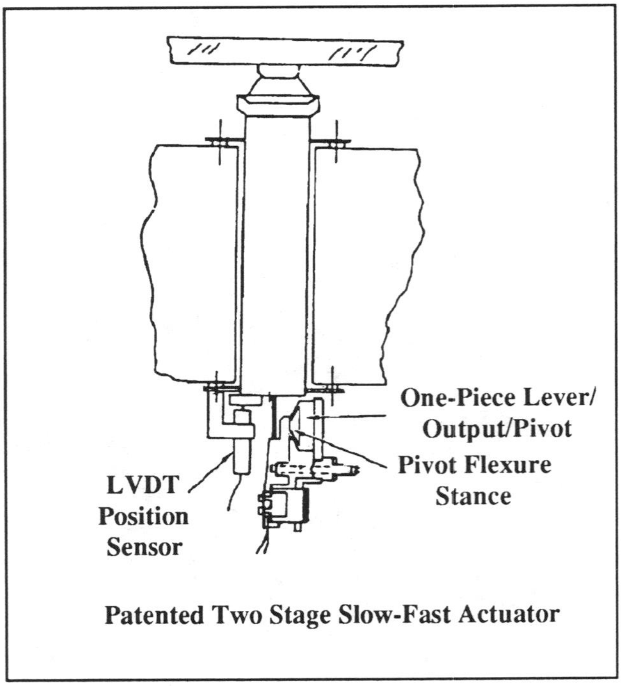

Figure 7
Surface Control Actuator Schematic

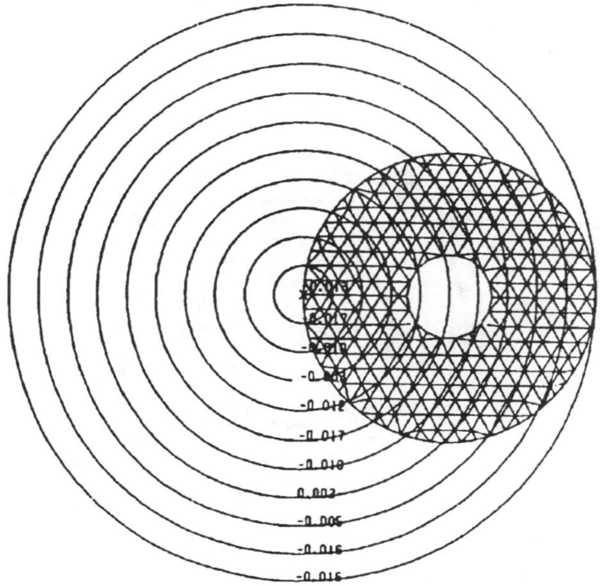

Figure 8
Prescribed ULE Variation

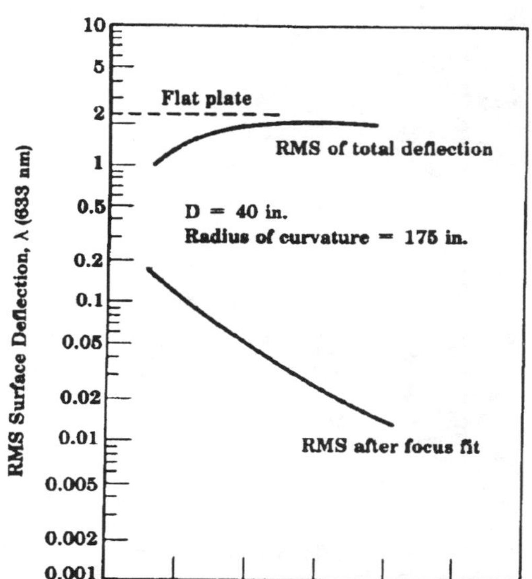

Figure 9
The Effect of Inhomogeneities on Performance - Solid Mirror, ULE-Prescribed Variation in Plan

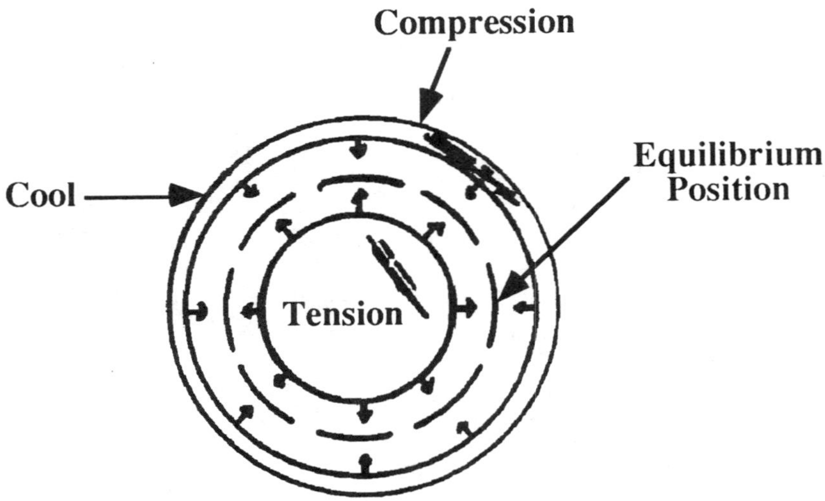

Figure 10
Internal Stress Due to Cooling

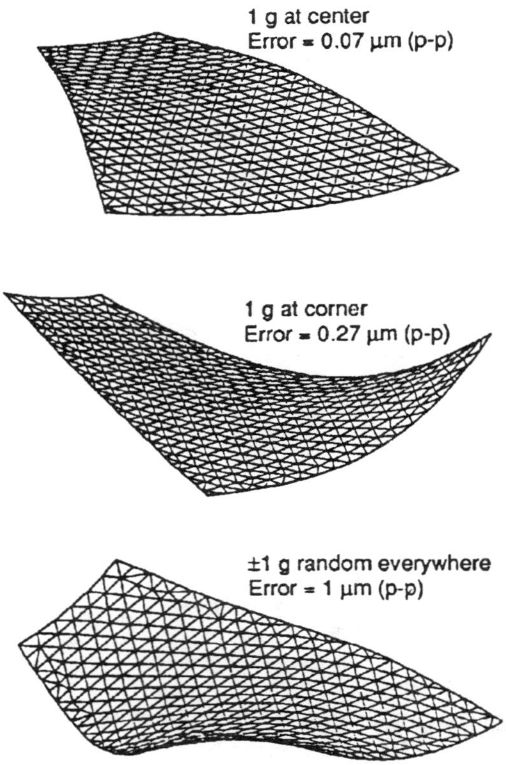

Figure 11
Fabrication/Metrology Mount
Sensitivity to Force Errors

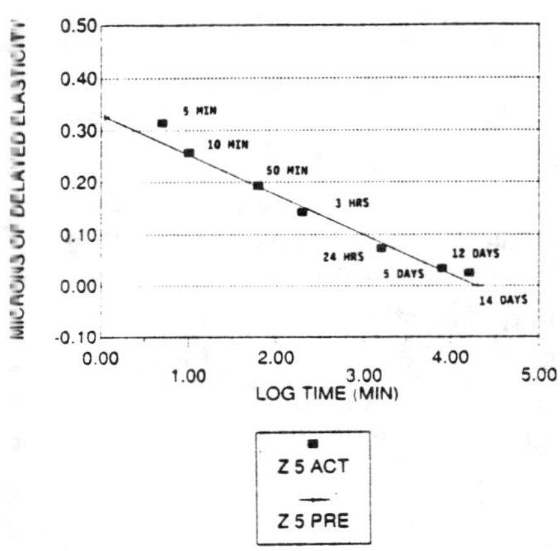

(a) Delayed elasticity after 2 weeks loading (30 kg) in Zerodur.

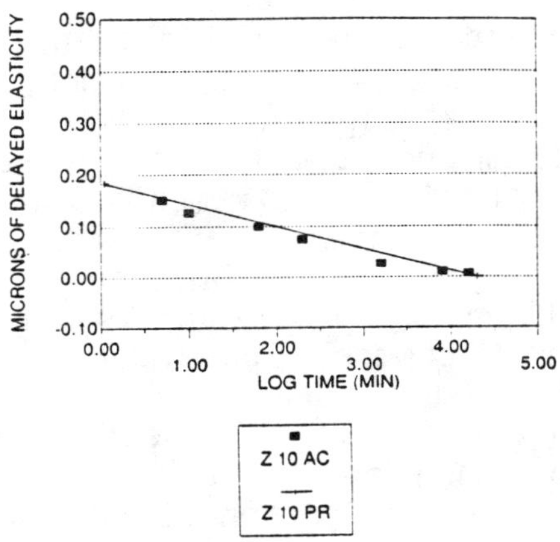

(b) Delayed elasticity after 2 weeks loading (30 kg) in Zerodur.

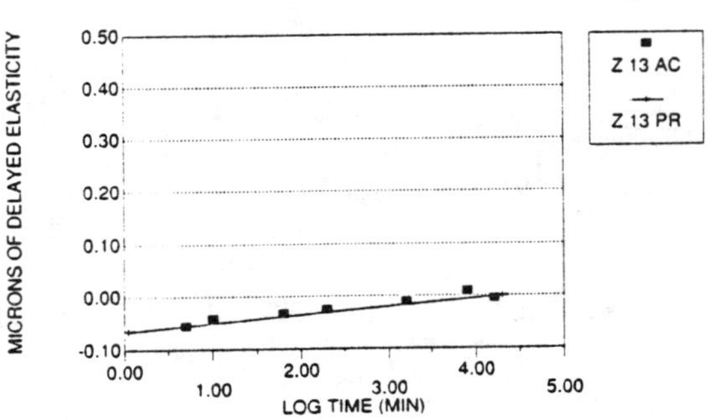

(c) Delayed elasticity after 2 weeks loading (30 kg) in Zerodur.

Figure 12
Delayed Elasticity over Time - Actuals (squares) vs. Predicted (line) in Zerodur for Z(5), Z(10), and Z(13)

OPTICAL WINDOWS

William P. Barnes

Precision Optics Corporation
22 East Broadway
Gardner, Massachusetts 01440-3338

ABSTRACT

In the context of this review, an optical window is a solid barrier with a principal function of transmitting some portion of the electromagnetic spectrum between the short wave atmospheric cutoff of 180 nanometer wavelength and the 12 micrometer thermal infrared. Uses range from protection of optical instruments against hostile natural environments to containment of exceptional man-made environments for study (e.g., combustion) or for production of special forms of energy (e.g., lasers). Flat, uniform thickness panels are most common. Spherical segments and conical elements have found use. Other curved elements and multiple flat glazings are often used to better conform to the shape of a vehicle or other enclosure.

Application of such barriers requires that we examine the effects, other than the desired unimpeded and undistorted transmission, which may be introduced.

1. INTRODUCTION

Perhaps "optical window" doesn't need the above definition. We are surrounded by windows. However, an unabridged dictionary has nine definitions, my (somewhat ancient) Britannica limits itself to the architectural fenestration definition, and Van Nostrand's Scientific Encyclopedia, Sixth Edition, completely ignores the word. Thus, I have taken advantage of this window of opportunity.

The insertion of a solid barrier for protection of an optical system, or isolation of an object or process, may introduce several potential penalties including:

a) reduced transmission resulting from the reflection, absorption, and scatter of the window material,

b) contrast reduction caused by lack of control of the reflected and scattered energy,

c) further increase in scatter or reduction of transmission by environmental effects such as erosion or radiation darkening,

d) addition of wavefront aberrations characteristic of the window shape, the window area in use, and positioning of the associated imaging system,
e) further aberrations induced in the window by the mechanical and thermal environments,
f) even absent the above, a possible significant restriction of the field of view, and,
g) the possibility of catastrophic failure.

My objective is to provide a review of some such penalties for several window configurations, and mounting methods. No claim is made for completeness. I attempt to provide some calculation procedures which may assist in judging whether a particular effect is important or to determine that finite element or other digital solutions are yielding reasonable results. The specification of units and numerical values is left, for the most part, to the other articles of this volume.

In critical applications the window behavior will require more precise study which then should be included as an integral part of the lens design analyses.

2. WINDOW CONFIGURATIONS

2.1. Flat Panes

The most common window is a single flat sheet of transmitting material. If perfectly flat and perfectly optically homogeneous, it adds no aberrations to subsequent imaging of an object at infinity, but it may limit the field of view. For close objects, a focal shift and other aberrations may be added to subsequent imagery. Accelerations, pressure differentials, other mechanical forces and thermal perturbations will affect both surface flatness and optical homogeneity.

Where the window replaces a shaped and otherwise opaque structure, or to increase the available field of view, multiple flat glazings may be used. These may approximate a curved surface without introducing the aberrations of a curved optical element. It is noted that small aberrations are sometimes tolerated for aesthetic or aerodynamic reasons- when is the last time you saw completely flat automobile windshield glazings?

2.2. Spherical Domes

A near hemisphere is an natural shape for a wide field scanning system such as the acquisition and tracking optics of a self guided missile. It also provides a substantial mechanical strength advantage in applications subjected to large pressure differentials, such as deep submergence vehicles. Some aspects of these particular applications are considered in more detail in subsequent sections.

2.3. Conformal Shapes

For a supersonic or hypersonic missile, considerations of shock wave aerodynamics may suggest a preference for a true cone rather than a spherical dome as the forward facing element. Cylindrical segments may be recommended for other locations in high speed vehicles. One possibility has been treated in detail by Gearhart[1].

3. FLAT WINDOWS

3.1. Reflection and Absorption

The presence of a window will affect both the amplitude and phase of the useful wavefronts it is designed to transmit. Useful energy may be removed by reflection from the window surfaces, by absorption in the bulk material and at times by mechanisms which can be described only as poorly understood "surface absorption", and by scattering, again as both a bulk and surface phenomenon. A penalty arises less from the loss of energy than from the ultimate disposition of the "lost" energy.

For small values of bulk absorption and at normal incidence, the fraction of incident energy reflected at each surface is given by the usual Fresnel relation for an index of refraction, n, as $[(n-1)/(n+1)]^2$. Moderate values of bulk absorption are well characterized by an exponential decay of intensity, I, equal to $\exp(-\beta z)$ in the thickness direction. Bulk scattering is often included (intentionally or otherwise) in the β coefficient. Ignoring multiple reflections, the single pass transmittance of the window, I/I_0, then may be calculated as:

$$I/I_0 = [4n/(n+1)^2]^2 \exp(-\beta h). \tag{1}$$

For non-normal incidence, the variation of reflectance with incident polarization becomes important at incidence angles of 30° or more. In materials with large values of refractive index and with surfaces not precisely parallel, multiple reflections may also affect subsequent imagery or instrumentation. Of course, anti-reflection coatings can substantially mitigate all the reflection effects.

3.2. Window Bending Effects

Consider a thin, nominally flat, window of an initially homogeneous and isotropic material. Transverse forces or temperature gradients through the thickness cause the window to bend. Within the limits of thin plate theory, the bending stresses vary linearly through the thickness and average to zero. For normally incident rays, the possible optical effects resulting from thickness change (via Poisson's ratio), and of stress induced refractive index change and birefringence are, at least theoretically, zero. Perceptible effects on optical path differences and on distortion in the imagery of sensors following the window may occur for large incidence angles or in wide field imagery when the stress pattern varies rapidly over the plan dimensions of the window. Such cases are respectable challenges for the best current ray-tracing programs, and will not be pursued here.

Bending alone, however, does introduce an optical path difference, even for small deflections, as a result of the the local slope of the deflected element. For circularly symmetric cases, the OPD was estimated by Sparks and Cottis[2], and applied to pressure bending by Sparks and by Klein[3]. The derivation did not include the effect of refraction at the first surface. If we include the refraction (for a surrounding index of 1.0), and generalize to the case for which the deflection, w, is a function of both plan coordinates, the optical path difference, OPD, may be shown to be:

$$\mathrm{OPD} = [(N-1)h/2N]*[(\delta w/\delta x)^2 + (\delta w/\delta y)^2]. \qquad (2)$$

N is the window index of refraction and h its thickness, with δ used here to denote partial differentiation. With this expression as our principal tool, we can explore several cases of bending by transverse loads and axial temperature gradients.

Let us note that windows of precisely circular plan form are quite uncommon. Pressure and vacuum vessel viewports are often circular because pipe flanges are circular. Some lasers use circular output windows, but many, such as the high energy device shown in Figure 1, use other shapes and multiple glazings to resist substantial pressure pulses. Reconnaissance applications, FLIR systems, passenger aircraft and cruise ships all use a variety of elliptical, square, rectangular or trapezoidal plan forms.

Figure 1. High energy laser output window array. (Courtesy Thermo Electron Technologies Corp.)

3.2.1. Twisted Window

As a first example, representing perhaps a worst case of window bending by a poorly matched mounting or by deflection of the frame of a vehicle, consider a square plate supported at two opposite corners and loaded symmetrically by forces, P, at the other corners. (A third, potentialy small and negligible, reaction is needed to keep the window from rotating, of course.) This case of pure bending is sketched in Figure 2, and is one subject treated in Chapter 2 of Timoshenko[4].

With my choice of coordinate axes parallel to the sides of the plate, and origin at the center, the deflection, w, is given by:

$$w = [6P(1+\mu)/Eh^3]*(xy), \tag{3}$$

where μ is Poisson's ratio, and E is Young's modulus.

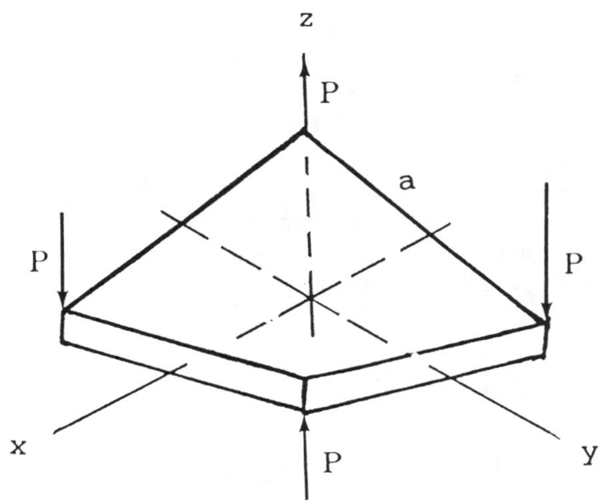

Figure 2. Corner loaded square plate.

Applying Equation (1) yields the following:

with $C_1 = [6P(1+\mu)/E]^2/h^5$

$$OPD = C_1 \, [(N-1)/2N]*[(x^2+y^2), \quad (4)$$

$$OPD_{p-v} = C_1 \, [(N-1)/4N]a^2 \quad (5)$$

$$OPD_{rms} = C_1 \, [0.0526(N-1)/N]a^2. \quad (6)$$

This case can be viewed as the application of equal and opposite pure bending moments (per unit width of plate) along axes through the corners of the plate, or as a pure twisting moment along the axes of Figure 2. The bending moments have constant and opposite values equal to P/2, thus the extreme bending stresses are $\pm 3P/h^2$. Alternatively, since the bending stresses are equal and opposite, the extreme shear stresses have the same values. Additionally, if the edges of the window are not otherwise constrained (i.e., are kinematically supported), the solution is applicable to windows of any plan form subjected to a given torsional deflection. The torsional deflection is the angle of twist per unit length in the direction of the torsional axis, and is equal to $w(a/2,a/2)/(a/2)^2$ for the square plate. In these terms, the extreme stresses developed as a function of the torsional deflection, Θ/L, are given by:

$$\tau_{xy} = \pm 2Eh(\Theta/L)/(1+\mu) \quad (7)$$

For final emphasis on this case, we note that the optical path difference introduced is (again, small deflections) a purely spherical change in the wavefront, the same effect as a purely spherical deflection.

Adopting a somewhat similar approach for the case of a window supporting a pressure differential, consider a square plate supported at the four corners subject to a uniform pressure on one surface.The solution for this case, sketched in Figure 3, is more complex. Some numerical results are included in Article 48 of reference 4.

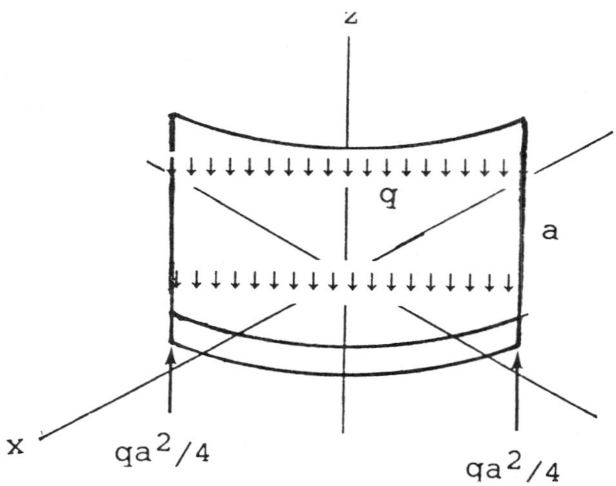

Figure 3. Uniformly loaded plate, corner supports.

A more explicit derivation, included as Appendix A, results in:

with $C_2 = (N-1)q^2/2NE^2h^5$,

$$OPD_{p-v} = C_2(4-3\mu)^2a^6/4.5, \quad (8)$$

$OPD_{rms} = C_2a^6*$	0.482	for μ = 1/6,(9)
	0.456	...1/5,
	0.418	...1/4,
	0.359	...1/3.

The maximum bending moment per unit width in this case occurs at the supports and is equal to $qa^2/4$, producing extreme bending stresses of $\pm 3qa^2/2h^2$.

The third case, which I consider useful for circular windows, is that of an edge support reaction which lies somewhere between a perfectly uniform simple support and the unlikely case (unless so designed) of three equally spaced concentrated forces. The supporting force at the edge is postulated to vary as $(1+\cos(3\Theta))$. For a plate of radius b, and supporting force of $(qb/2)(1+\cos(3\Theta))$, the deflection, w, may be shown to be:

$$w = 3q(1-\mu^2)/(16Eh^3)*\{2(3+\mu)b^2r^2/(1+\mu) - r^4$$
$$+\cos(3\Theta)[8(5-\mu)br^3/(9(1-\mu)(3+2\mu))$$
$$- 4r^5/(3(3+2\mu)b)]\} \tag{10}$$

The algebra is less daunting if one chooses a value for μ, and notes that the maximum value of the slope should occur at the edge. For $\mu = 0.3$, and $r = b$, the slope components are:

$$\delta w/\delta r = (qb^3/Eh^3)[1.05+0.460 \cos(3\Theta)] \text{ and,}$$
$$\delta w/r\delta\Theta = (qb^3/Eh^3)[-0.659 \sin(3\Theta)]) \tag{11}$$

The OPD is a maximum at $\Theta = 0$; for plate diameter D:

$$OPD_{max} = 0.0178((N-1)/N)(q^2D^6/E^2h^5). \tag{12}$$

Except for the additional N in the denominator, this is twice the result shown by Klein[3] for a uniform edge support.

As a fourth case, we consider an axial temperature gradient which is uniform over the plan dimensions of the window. The window deflection is a purely spherical change, and is independent of the plan form as in the first case above. For a steady state heat flux per unit area, F, through the window, a linear axial temperature dependence, is developed, and the OPD may be obtained by substituting:

$$C_3 = (\alpha F/k)^2h = (\alpha\Delta T)^2/h, \tag{13}$$

for C_1 in Equations 4, 5, and 6. Here α is the coefficient of thermal expansion, k the thermal conductivity, and ΔT the temperature differential across the window.

Often, for unsteady state cases, the thermal strain may be expressed as a power series in the thickness dimension:

$$\alpha\Delta T = \Sigma c_m(z/h)^m.$$

Then, noting that $\alpha\Delta T = \Sigma c_m - c_0$, the calculation of the curvature change may be rearranged as:

$$1/R = (1/h)\Sigma c_m[1-(m-1)(m-2)/((m+1)(m+2))]$$

and then expanded to:

$$1/R = (\alpha\Delta T/h)* \left[1- \frac{(c_3/10+c_4/5+2c_5/7+5c_6/14...)}{(c_1+c_2+c_3+c_4...)}\right] \quad (14)$$

Thus, unless the higher terms in the power series are quite large, the deflections and optical path differences may be closely approximated using the total temperature difference alone.

3.3. Lateral Temperature Gradients

It is normally necessary to both support and seal a window at the entire edge, thus providing a heat conduction path. Aerodynamic heating, the absorption of part of a high power transmitted beam, or other radiation to or from the window will develop lateral temperature gradients. The circularly symmetric lateral gradient case in which the temperature distribution replicates the intensity distribution in a high power laser is treated by Klein[5] in terms of the effect on the Strehl ratio of the focussed beam. The derivation may be put in the form of the following optical path differences for the average over polarization of the incident beam and for the birefringent OPD:

$$OPD_{unpol} = hT[dN/dT+\alpha(N-1)(1+\mu)+\alpha N^3E(q_p+q_s)/4],$$

$$OPD_{biref} = h[\alpha N^3E(q_p-q_s)/4](T'-T). \quad (15)$$

where T is measured as the temperature increase from the center of the window, and T′ is the area weighted average temperature increase between the center and the radial position ,r,or:

$$T' = (2/r^2)\int_0^r T\, r'dr'. \quad (16)$$

q_p and q_s are the piezooptic constants for, in this case, an isotropic material and stresses parallel and perpendicular (senkrecht) to the polarization of the incident beam. Klein treats the "short time approximation" case which neglects the thermal diffusion of the energy absorbed from a high power beam, but is important to assessment of the performance of high power, pulsed laser systems.

Environmental effects, such as convective heat transfer and radiation to or from the window may also be treated under the assumption of cylindrical symmetry. We assume convection to one window surface from a medium at temperature, T_1, with convection coefficient, H_1, and an absorbed uniform radiation flux, Q. The circular window of thermal conductivity, k, thickness ,h, and outer radius, b, is thermally coupled at the edge by a conductance, C, to a structure at a constant temperature T_s. The steady state solution is found as a Bessel function of the second kind (I_n) as[6]:

with $\Gamma = \sqrt{(H_1/kh)}$

$$T(r) = C[T_s-(H_1T_1+Q)/H_1][I_0(\Gamma r)]$$

$$\div [2\pi kh\Gamma b I_1(\Gamma b)+CI_0(\Gamma b)]$$

$$+ (H_1T_1+Q)/H_1, \qquad (17)$$

or simply, $T(r) = AI_0(\Gamma r)+B$.

The vitreous and polycrystalline dielectric window materials are relatively good thermal insulators. For these, this suggests that we model the edge conductance, C, as the outer 5% of the window radius. Thus:

$$C = k(2\pi bh)/0.05b = 20(2\pi kh).$$

Additionally redefining T as zero at the center, the temperature distribution is rewritten as:

$$T = (T_s-T_1-Q/H_1)[I_0(\Gamma r)-1]$$

$$\div [\Gamma b I_1(\Gamma b)+20I_0(\Gamma b)] \qquad (18)$$

The integration required to determine the T′ term of Equation 16, is probably best addressed using the polynomial approximations given, for example in Abramowitz and Stegun[7] for values of the argument up to 3.75. For larger values, the asymptotic expansion given there may be appended.

4. SPHERICAL SEGMENT WINDOWS

The decision to use a spherical segment as a window element is usually driven more by environmental, rather than optical, factors. Additionally, the optical effects of even the unperturbed window are of sufficient magnitude that they, and the perturbations, cannot be treated by the separable and fairly elementary treatments outlined above.

Analysis of the basic elastic, temperature, and thermal stress effects of the environment is also made more complex by the spherical boundary conditions. For example, in high speed missile domes of near hemispherical configuration, the pressure differential and rate of heat transfer will vary significantly over the surface of the dome.

One application treated in some optical detail in the past is that of underwater viewports. A (two volume and appendix) design report was prepared by the U.S.Navy Ocean Systems Center[8] and a portion reproduced for an SPIE symposium by Lones[9]. The presence of water (index 1.33) reduces the optical effect of the outer surface by a factor of 3 for the acrylic viewports considered. With flat windows, the effect is to make objects appear nearer, the "bent stick" effect. The spherical window is a much stronger negative lens than when air is present on both sides. Its effects on the apparent object distance, distortion and binocular viewing are reported in some detail. A complete underwater imaging system, the E. Leitz Canada "C201 ELCAN Water Contact Lens" is discussed by McNeil[10].

Klein[11] reviews the design criteria for shallow domes of CVD diamond as might be applied in supersonic or hypersonic missile guidance. (Flat windows for high power lasers are also treated.) Much of the current thrust in development of materials, processes, and application evaluation is reviewed in the SPIE symposia proceedings Window and Dome Technologies and Materials, Volume 1112 (1989), and Window and Dome Technologies and Materials II, Volume 1326 (1990).

5. SURVIVABILITY

5.1. Fracture

From a practical point of view, all of the window materials are brittle. Their failure mode, under constant load, may proceed from the slow growth of cracks (from flaws initially present) to a critical size at which the crack then grows rapidly. For glasses, the location of susceptible flaws is normally at the surface. For polycrystalline materials, such flaws may exist throughout the bulk. Single crystal materials may also be susceptible to cleavage along particular planes, with this form of damage being caused at times by only moderate thermal shock.

The relationship between crack size, crack orientation, and applied macroscopic stress has been the subject of considerable study. A large body of literature exists on this subject, and on the similar slow crack growth behavior of ductile materials under cyclic loading. Scanning of a recent issue of "Mechanics", a title citation publication of the American Academy of Mechanics, reveals the following journals devoted to this technology:

Engineering Fracture Mechanics,
Fatigue and Fracture of Engineering Materials and Structures,
International Journal of Fatigue, and
International Journal of Fracture.

In the same issue, related papers are included in the European Journal of Mechanics A Solids, and the International Journal of Engineering Science.

I can only provide a glimpse of this technology, abstracted from reference 12.

For glasses, the velocity of propagation of a crack has been shown to depend on crack dimension, L, and transverse tensile stress, σ, combined in a stress intensity factor, K_I, as:

$$K_I = \sigma\sqrt{(\pi L)}.$$

The value of this factor when catastrophic failure occurs is the critical stress intensity factor, K_{Ic}. Crack velocity and K_{Ic} data for several glasses were reported by Wiederhorn and Bolz[13], and a straight line correlation of the logarithm of crack velocity versus K_I reported. Representative values are those for fused silica (Corning 7940):

$$K_I(10^5 N/m^{3/2}) = 6.931+1.342\ln(v),$$

$$K_{Ic} = 7.3 \text{ to } 7.5\ 10^5 N/m^{3/2}.$$

For these correlations, the time to failure, t_f, at constant stress may be expressed as:

$$t_f = 2/(\pi\sigma^2)\int_{K_{Ii}}^{K_{Ic}} (K_I/v)dK_I.$$

The results of this calculation for fused silica as presented in reference 12 are reproduced in Figure 4.

A more recent communication[14], cites a study[15] of the effect of residual stress introduced with the initial flaw by indentation. The principal effect appears to be a reduction in the critical flaw size by a factor of four, or reduction in the value of K_{Ic} by a factor of two.

These and similar data must be handled with great care. In any practical situation, the presence, size, orientation and spatial distribution of initial flaws will be poorly defined. Additional flaws will be introduced in handling and exposure. Finally, the applied stress will be constant neither in space nor time.

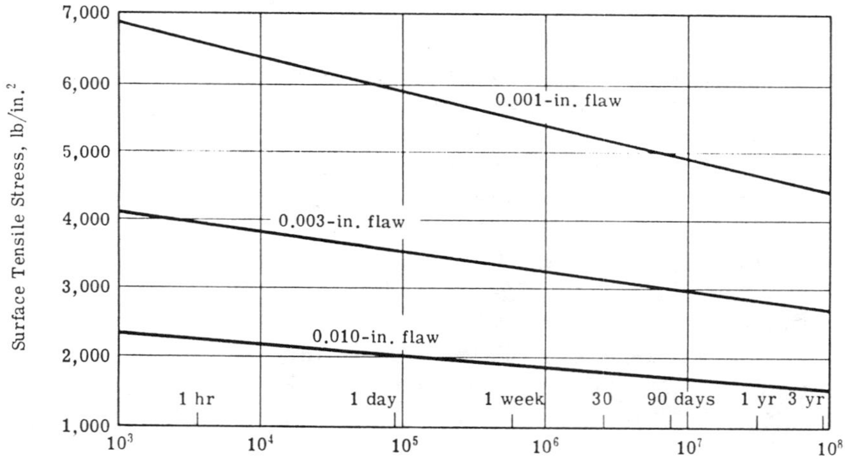

Figure 4. Time to failure at constant stress for fused silica.

5.2. Stress Analysis

The stress levels noted in Section 2 for cases of pure bending do not account for any edge or localized support reaction stresses. The same is true for the solutions for spherical domes in textbooks such as Timoshenko. That recourse to a careful finite element analysis should often be taken is illustrated for a shallow shell by Orringer and Tong[16]. For the design sketched in Figure 5, (Figure 19b of Ref. 16) constant thickness shell theory indicates that only compressive stresses will exist.

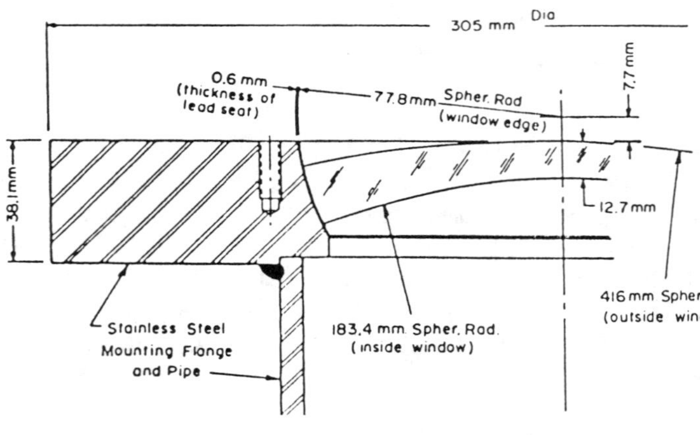

Figure 5. A shallow shell window.

A modeling which accounted for the thickness variation and the edge curvature yielded the stress contours of Figure 6 (partial of Figure 20 of Ref. 16) for a 300 psi pressure load. That is, a local tensile stress of about 1.3 times the applied pressure differential.

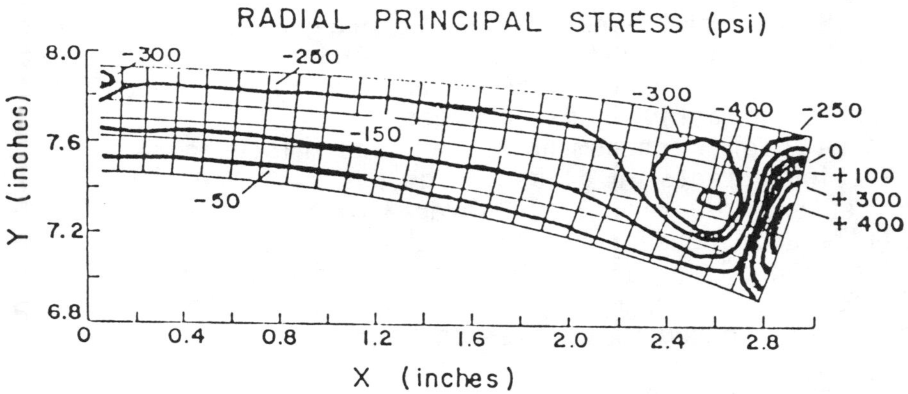

Figure 6. Finite element analysis for Fig. 5.

This window design was analyzed and reported initially by Bailey[17] for service as a radiation counter window to resist a 20 atmosphere pressure differential. The window was seated in lead, and subjected to a proof test and pre-pressurization of 100 atmospheres for 5 minutes and 32 atmospheres for 14 hours, then put in service without further disassembly. Bailey's additional comments are of interest:

> the edge was made thicker to "...keep the unit stress on the lead seat at a level where the creep rate is acceptable...", "...it is of paramount importance that the lead surface be relatively free of oxides at the time of application...we used HCl and methylene chloride...", "...the lead lining which seats the window under went considerable extrusion during this initial pressurization...the seat should not be disturbed...".

5.3. Radiation Effects

Ionizing radiation, including the shorter ultraviolet near the atmospheric cutoff, may cause color center formation and fluorescence. The potential impacts of reduced transmission or increased noise should be considered in those applications where exposure to natural or man-made radiations exceeds that of our everyday environment. Even long term exposure to normal solar radiation may cause darkening- for example, special UV grades of polymethyl methacrylate are available the resist this effect. Cerium doping of optical glasses has also been shown to be effective in improving resistance to radiation darkening. Color centers also may be bleached by subsequent heat treatment[18].

High dose rates of high energy electrons have been known to result in electron discharge damage (Lichtenberg figures[19]). High radiation doses may also induce density, and thus refractiveindex, changes[20].

A useful phenomenological and historical survey of radiation effects in glass is the paper by Sun and Kreidl[21]. The effect of cerium on several glasses is detailed by Kreidl and Hensler[22]. Pellicori et al.[23], provides a collection of transmission loss data for a number of window glasses, filter glasses, crystals, and transmitting fibers in the 400 to 900 nm wavelength region.

5.4. Impact Erosion

In our 1975 Handbook[12], we included a very brief review of the data then available. In part:

> "Data on impact erosion by solids (dust erosion) have been presented by Sheldon (Ref.[24] here). In these data there is no evidence of an incubation period or a threshold velocity. For brittle materials the volume of material removed (W) given by $W = k_1 r^a V^b$ is in reasonable agreement with experiment. Here r is the radius of the impacting particle and V its velocity. The other parameters, which may dependon the statistics of the flaw distribution in the material impacted, approach the values a=3, b=2.4, $k_1 = E^{0.8}/\sigma_0^2$ where E is Young's modulus and σ_0 is related to the failure strength of the material."

I have not, at this point, pursued later citations of the Sheldon[24] work.

The investigation of water droplet impact at high subsonic and supersonic speeds has continued apace, however. A substantial body of literature exists, and study of the current state of the technology might best begin with the papers of Session 5 of SPIE Proceedings Volume 1326, working backward if necessary. To set the framework, I quote from Adler and Boland[25]:

> "The mechanics of single waterdrop impact is fairly well understood (Adler, 1979b). The waterdrop contacts the surface and the contact area between

> the drop and the surface expands as the square root of the time. The loading cycle is relatively short: on the order of a tenth of a microsecond. A critical condition is reached where the pressure at the surface is catastrophically released. The magnitude of the pressure and the loading period generate the transient stresses in the target material which are directly responsible for the primary damage produced in the target. The unique feature of the waterdrop impact is the precipitous demise of the drop during the loading cycle."

This proceedings includes complementary papers on coatings for erosion protection and reflectance reduction. Continuing development of diamond-like coatings (and perhaps free standing diamond windows) offers future promise.

6. MATERIALS

In any discussion of materials, one is tempted to generate a nearly all inclusive listing of all conceivably useful materials and their important (and sometimes their obscure) properties. I have easily resisted that temptation because such an at once useful and compact tabulation has been accomplished by Wolfe[26]. A more detailed collection of optical property data is provided by Weber[27]. The AIP[28] and OSA[29] Handbooks are also valuable sources of data. Optical "constant" data for a range of wavelengths well beyond the scope I attempt to cover here may be found in the two volumes edited by Palik[30,31]. Weast[32], and other scientific and engineering handbooks are occasionally useful. Manufacturers' data is of particular value when you get into the nitty-gritty of hardware, and is indispensable for setting the final specifications and property certifications needed.

Some comments here on the materials suitable for certain wavelength ranges may be helpful.

6.1. 180 to 400 nanometer wavelengths

The alkali halides have limited value because of their solubility in water. The alkaline earth halides are much less soluble but subject to cleavage on (111) planes. Some recently developed fluoride based glasses are useful in the upper end of this wavelength range. Synthetic fused silica transmits well throughout, fused natural quartz exhibits reduced transmission below 250 to 220 nanometers.Both sapphire and magnesium fluoride are quite transparent but birefringent.

The principal practical candidate materials for windows (and lenses) are calcium fluoride and synthetic fused silica.

6.2. 400 to 2500 nanometer wavelengths

Almost everything labelled an optical material transmits in some part or all of this wavelength range. Ordinary soda-lime window glass cuts out some of the blue and red, and thus looks green when viewed edge-on. A number of optical glasses have been formulated for reduced sensitivity to optical path change with temperature (with zero stress change), i.e. reduced values of $[dN/dT+\alpha(N-1)]$ (see Section 3.3 above). It is worth noting that, although the α of fused silica is small, dN/dT is larger than, for example, BK-7 glass, and the optical path change with temperature cannot be assumed to be small.

Above 550 nanometers, zinc sulfide and zinc selenide "cut-on", and the transparent regions of silicon and germanium begin at about 1+ and 2+ micrometers respectively.

6.2. 2.5 to 3 micrometer wavelength

Materials synthesized or processed at high temperatures <u>in the strict absence of water</u> may transmit well in this region. Others, such as flame-fusion synthesized fused silica have strong O-H absorption bands in this region. Special processing, and the electrical fusion of natural quartz can produce transparent silicas, however.

6.3. 3 to 5 micrometer wavelengths

The transparency of the alkaline earth halides and sapphire extends to this region. Calcium aluminate and germanate glasses are also useful here, as is arsenic trisulfide (rather soft, however). Most of the materials used in the 8 to 12 micrometer region are also transparent here, though some may scatter more than desired. In general, oxide glasses are not recommended at these and longer wavelengths.

6.4. 3 to 12 micrometer wavelengths

The transparency of silicon declines significantly beyond 5.5+ micrometers, and zinc sulfide has a moderate absorption band near 6 micrometers. The alkali halides are transparent throughout and beyond this region (cesium iodide to about 50 micrometers). The calcogenide glasses (GeSeSb and GeSeAs), thallium bromide/iodide (soft), zinc sulfide, zinc selenide, and germanium may all find application.

Here it is of interest to note the development of a zinc sulfide/zinc selenide laminated window material[33] which takes advantage of the better erosion resistance of zinc sulfide and the better transparency to 12-14 micrometers od zinc selenide. Practical application has been made in the U.S.Air Force LANTIRN (Low Altitude Navigation and Targeting InfraRed system for Night) program. Klein[3] treats the flexural stress distribution which develops assuming zero stress as the zinc sulfide is deposited on a thicker zinc selenide substrate at about 650° Celsius. He also derives an expression for the window transmission in terms of the reflectance of the three interfaces and the material absorption coefficients and thicknesses.

Two additional effects of the laminated construction may be of concern in some applications. First, interfacial shear and normal tensile stresses are developed at window edges by the differential expansion. These may be estimated from a closed form solution published by Suhir[34]. Pionke and Wempner[35] point out that neither the Suhir solution nor finite element calculations can account for the likelihood of a stress singularity right at the edge. They also note, however that this singularity can be avoided by bevelling the individual layers away from the interface at the edges (suggested by Professor S.Lukasiewicz, University of Calgary). I'm not sure this discussion is ended.

Secondly, the refractive indices of the materials are different. Thus, if one first prepares a polished flat substrate of zinc selenide, deposits zinc sulfide, cools, and then refinishes the now curved outer surfaces to flat, one is left with a more or less spherical interface and a window with some optical power.

ACKNOWLEDGMENTS

I have attempted to acknowledge some of those who have contributed to my education by citing what I hope are the more useful recent references and other sources of data. I'm sure there remain many others whose particular contributions are locked in my subconscious.

7. APPENDIX A

Uniformly loaded square plate,
4 corners supported.

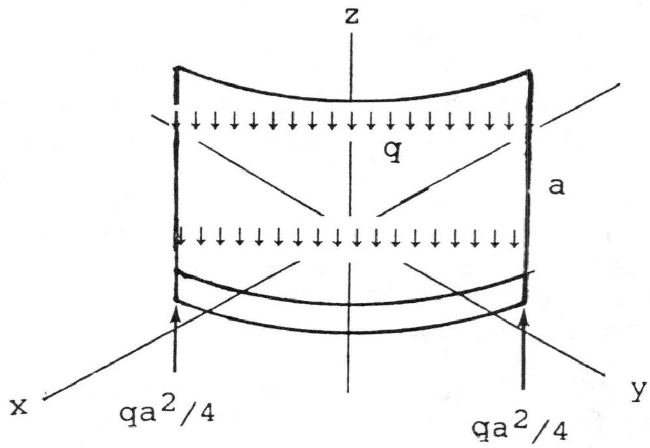

Figure A-1

For Cartesian axes as shown in Figure A-1, the bending moments per unit width ,(M), may be expressed as:

$$M_x = qa^2/4 - (q/12)(a - x\surd 2)^2,$$

$$= -D((\delta^2 w/\delta x^2) + \mu(\delta^2 w/\delta y^2)), \text{and} \qquad \text{(A-1)}$$

$$M_y = qa^2/4 - (q/12)(a - y\surd 2)^2,$$

$$= -D(\mu(\delta^2 w/\delta x^2) + (\delta^2 w/\delta y^2)). \qquad \text{(A-2)}$$

for plate dimensions a, transverse force per unit area q, deflection w, and material properties given by Poisson's ratio μ and flexural rigidity D. Solving for the derivatives and expressing D in the usual terms of Young's modulus, E, and plate thickness, h, yields, for example:

$$(-Eh^3/q)(\delta^2 w/\delta x^2) = 3(1-\mu)a^2 - (a - x\surd 2)^2 + \mu(a - y\surd 2)^2, \qquad \text{(A-3)}$$

and a similar expression for the y derivative.Integrating, and noting from symmetry that $\delta w/\delta x$ is zero at x = 0 for all values of y, and $\delta w/\delta y$ is zero at y = 0 for all values of x, one obtains:

$$(-Eh^3/q)(\delta w/\delta x) = 3(1-\mu)a^2x + [(a-x\sqrt{2})^3-a^3)/3\sqrt{2}] + \mu x(a-y\sqrt{2})^2, \text{and} \quad (A-4)$$

$$(-Eh^3/q)(\delta w/\delta y) = 3(1-\mu)a^2y + [(a-y\sqrt{2})^3-a^3)/3\sqrt{2}] + \mu y(a-x\sqrt{2})^2. \quad (A-5)$$

More algebra then provides the following maximum value (at the corners of the plate) for the optical path difference factor:

$$(\delta w/\delta x)^2+(\delta w/\delta y)^2 = (2a^6(4-3\mu)^2/9)(q/Eh^3)^2. \quad (A-6)$$

Even more algebra results in an rms value for this slope factor given by:

$$(qa^3/Eh^3)^2 [\sqrt{(0.2457\mu^4 - 0.8056\mu^3 + 1.3519\mu^2 - 1.1592\mu + 0.3918)}] \quad (A-7)$$

Some numerical values for the total optical path difference coefficients are tabulated in the body of the paper.

8. REFERENCES

1. Scott A. Gearhart, "Image degradations of an aerodynamically shaped optical window", Johns Hopkins APL Tech. Digest, 12, 1, 81-85, 1991.
2. M.Sparks and M.Cottis, "Pressure-induced optical distortion in laser windows", J.Appl.Phys., 44, 2, 787-794, Feb.1973.
3. Claude A. Klein, "ZnS, ZnSe, and ZnS/ZnSe windows: their impact on FLIR system performance", Opt. Eng. 25, 519-531, April 1986.
4. S.Timoshenko, and S.Woinowsky-Krieger, <u>Theory of Plates and Shells</u>, 2nd Edition, McGraw-Hill, 1959.
5. Claude A. Klein, "Optical distortion coefficients of high power laser windows", Opt. Eng. 29, 343-350, April 1990.
6. W.P.Barnes, "Some effects of the aerospace thermal environment on high-acuity optical systems", APPLIED OPTICS, 5, 5, 701-711, May 1966.
7. M.Abramowitz and I.E.Stegun, Eds., <u>Handbook of Mathematical Functions</u>, AMS • 55, National Bureau of Standards, Tenth Printing, December 1972.
8. J.D.Stachiw, <u>Acrylic Viewports for Ocean Engineering Applications</u>, NTIS ADA-070535,-070536,&-070537.

9. Joe J. Lones, "Optical aspects of underwater windows with an overview on mechanical installation", Ocean Optics VI, SPIE Vol.208, 84-123, 1979.

10. Gomer T. McNeil, "Metrical fundamentals of underwater lens system", Opt.Eng. 16, 2, 129-139, Mar./Apr. 1977.

11. Claude A.Klein, "Diamond windows for the infrared: fact and fallacy", Infrared Optical Design and Fabrication, Rudolf Hartmann and Warren J.Smith, Eds.,SPIE Volume CR 38, 218-257, 1991.

12. W.P.Barnes, Ed., Reconnaissance and Surveillance Window Design Handbook, AFAL-TR-75-200, October 1975.

13. S.M.Wiederhorn and L.H.Bolz, "Stress corrosion and static fatigue of glass", J.Amer.Ceram.Soc., 53, 543, 1970.

14. R. Nagle, Litton Itek Optical Systems, "Window life test review", Private communication, July 1991.

15. B.L.Symonds, R.F.Cook, and B.R.Lawn, "Dynamic fatigue of brittle materials containing indentation line flaws", J.Matl.Sci.,18, 1306-14, 1983.

16. O.Orringer and P.Tong, "Uses and abuses of the finite element method",SPIE Volume 450, 1983.

17. B.M.Bailey, "Quartz window design for gas-pressurized PMT installation," J.Nuclear Inst.& Methods, 125, 3, 353-355, Apr. 15, 1975.

18. M.J.Treadaway, B.C.Passenheim, and B.D.Kitterer, "Radiation coloring and bleaching of glass", IEEE Transactions on Nuclear Science, NS-23,6,1820-25, December, 1976.

19. G.C.Lichtenberg, Novi. Comment. Gott., 8, 168, 1777.

20. W.Primak, "Dependence of the compaction of vitreous silica on the ionization dose", Nuclear Science and Engineering, 65, 141-196, 1978.

21. Kuan-Han Sun and Norbert J. Kreidl, "Coloration of glass by radiation", in three parts, The Glass Industry, 33,10-12, October-November-December, 1952.

22. N.J.Kreidl and J.R.Hensler, "Gamma radiation insensitive optical glasses", J.Opt. Soc. Am., 47, 1, 73-75, January, 1957.

23. S.F.Pellicori, E.E.Russell, and L.A.Watts, "Radiation induced transmission loss in optical materials", APPLIED OPTICS, 18, 15, 2618-21, 1 August,1979.

24. G.L.Sheldon, "Similarities and differences in the erosion behavior of materials", Trans. ASME, J.Basic Engineering, 92, Ser.D(3), 619-626, Sept.1970.

25. William F.Adler and Peter L.Boland, "Multiparticle supersonic impact test program", Window and Dome Technologies II, SPIE Volume 1326, 268-279, 1990.

26. William L.Wolfe, "Optical materials for the infrared", Infrared Optical Design and Fabrication, SPIE Volume CR38, 55-68, 1991.

27. Marvin J.Weber, Ed., Handbook of Laser Science and Technology, Volume IV, Optical Materials: Part 2, CRC Press, Boca Raton, 1986.

28. Dwight E.Gray, Coord. Ed., American Institute of Physics Handbook, McGraw-Hill, 3rd Ed.,1972.

29. Walter G.Driscoll and William Vaughan, Eds., Handbook of Optics, McGraw-Hill, 1978.

30. Edward D.Palik, Ed., Handbook of Optical Constants of Solids, Academic Press, 1985.

31. Edward D.Palik, Ed., Handbook of Optical Constants of Solids II, Academic Press, 1991.

32. Robert C.Weast, Ed., CRC Handbook of Chemistry and Physics, CRC Press, Boca Raton, 61st Ed., 1980.

33. R.Donadio, J.Connolly and J.Pappis, "Sandwich type FLIR windows", AFML-TR-79-4113 (1979).

34. E.Suhir, "Interfacial stresses in bimetal thermostats", J. Appl. Mech.,56, 3, 595-600, 1989.

35. Christopher D.Pionke and Gerald Wempner, "The various approximations of the bimetallic thermostatic strip", J. Appl. Mech., 58, 4, 1015-1020, 1991.

The adjustment mechanisms: types and their applications in optical systems

Anees Ahmad
Center for Applied Optics
The University of Alabama in Huntsville
Huntsville, Alabama 35899

ABSTRACT

The various optical elements in a sophisticated optical system must be precisely aligned to each other to obtain an aberration-free image. In optical systems with very tight alignment tolerance requirements, the optics and their mounts are usually manufactured to rather loose tolerances, and adjustment mechanisms are then used to align the optics relative to each other at assembly. Another class of adjustment mechanisms is employed to move one or more optical elements of a system in real time to compensate for the image degradation due to environmental effects. This paper discusses the three basic types of adjustment mechanisms namely: linear, rotary and tilt mechanisms. Each mechanism consists of a number of parts such as the actuator, locking and preloading components. The selection criteria for these components of the adjustment mechanisms are presented in detail. Finally, some design guidelines for applications of the adjustment mechanisms in complex optical systems are presented.

1. INTRODUCTION

Certain optical systems, such as those for submicron lithography, can have 10-15 mirrors and lenses that must be axially positioned relative to each other and centered on a common optical axis within tolerances of a few microns. In order to achieve these positioning accuracies, it is impractical and extremely cost prohibitive to manufacture the optics and its mounts to micron-level machining accuracies. For such optical systems, it is more practical and economical to fabricate the optics and the mounting hardware to rather loose tolerances, and provide adjustment mechanisms to align the optical elements relative to each other at the time of assembly. This class of adjustment mechanisms are designed for infrequent use and generally have manual actuators. Once the optical system has been aligned, these adjustments are locked in place to retain the alignment.

Another class of adjustment mechanisms are employed to move or tilt an optical element in real time to compensate for the degradation of the image quality due to environmental effects. These mechanisms have motorized actuators and position readout sensors operating in a closed loop control system. Such mechanisms are generally used to correct focus and magnification errors in the optical systems due to thermal effects or any other environmental degradations.

2. TYPES OF ADJUSTMENT MECHANISMS

The three basic type of adjustment mechanisms are linear, tilt and rotary adjustments. A rigid body in space has six degrees of freedom, which are the three translations and the three rotations about x, y and z axes. An optical element in a system may need one or more of these translation or rotary adjustments for alignment purposes. In order to avoid cross-coupling between different adjustments, the preferable approach is to stack single axis adjustments on top of each other to achieve a multiple type of adjustment.

A typical adjustment mechanism consists of five basic components. These are:

1. An *interface* between the moving optical element and the fixed structure.

2. An *actuator* to adjust the moving element relative to the fixed structure.

3. A *coupling* between the actuator and the moving element.

4. A *preloading device* to eliminate backlash in the mechanism.

5. A l*ocking mechanism* to retain the adjusted position.

For each of these five components, a number of choices are available to the designer depending on the type of adjustment mechanism. The most commonly used components for linear, rotary, and tilt mechanisms are shown in table 1, table 2 and table 3 respectively.

3. SELECTION CRITERIA FOR COMPONENTS OF LINEAR ADJUSTMENT MECHANISMS

It is clear from the tables that a designer has a wide choice in selection of the components for a particular type of mechanism. The selection of a particular type of component is dictated by the performance requirements such as the frequency, range and resolution of adjustment, and other factors such as the size, cost and the load capacity of the mechanism. In the following sections, a number of options for different parts of an adjustment mechanism are discussed, and general guidelines are presented to help the designer in selecting suitable components for a particular application.

3.1 *The interfaces*

The interface between the moving optical element and the fixed mounting structure is generally determined by such design factors as the travel range and frequency of adjustment, shock, load capacity, cost and size

Table 1. Choice of components for linear mechanisms

Interface	*Actuator*	*Pre-load*	*Locking*	*Coupling*
Flexure	Coarse screw	Compression spring	Set screw	Ball /cone
Kinematic	Fine screw	Extension spring	Jack screw	Ball/flat
Ball bearing	Micrometer	Flat spring	Lock nut	Ball/socket
Roller bearing	Differential micrometer	Belleville washer	V clamp	Threads
Air Bearing	DC motor	Curved washer	Collar clamp	Flexible coupling
Dovetail Slide	Stepper motor		Epoxy	Lead screw
Flat slide	Piezoelectric		Control system	

Table 2. Choice of components for rotary mechanisms

Interface	*Actuator*	*Pre-load*	*Locking*	*Coupling*
Cross flexure	Coarse screw	Compression spring	Set screw	Ball /cone
Ball bearing	Fine screw	Extension spring	Lock nut	Ball/flat
Spherical bearing	Micrometer	Flat spring	V clamp	Ball/socket
Journal bearing	Differential micrometer	Belleville washer	Collar clamp	Flexible coupling
Roller bearing	DC motor	Curved washer	Epoxy	Worm/gear
Air bearing	Stepper motor	Torsion spring	Control system	Rack/pinion
	Piezoelectric			Belt/pulley

Table 3. Choice of components for tilt mechanisms

Interface	*Actuator*	*Pre-load*	*Locking*	*Coupling*
Cross flexure	Coarse screw	Compression spring	Set screw	Ball /cone
Kinematic	Fine screw	Extension spring	Jack screw	Ball/flat
Spherical bearing	Micrometer	Flat spring	Lock nut	Ball/socket
Journal bearing	Differential micrometer	Belleville washer	V clamp	
Blade flexure	DC motor	Curved washer	Collar clamp	
	Stepper motor		Epoxy	
	Piezoelectric		Control system	

requirements. If a long travel range is required and the mechanism is going to be adjusted frequently, a bearing interface must be used between the moving element and the fixed structure. Ball and roller slides, shown in figures 1 and 2, have low friction and are suitable for long travel ranges. Ball slides are less expensive than roller slides, but also have lower load capacity and accuracy (straightness of travel) as compared to roller slides. A dovetail slide, shown in figure 3, has high stiffness and load capacity and has a low cost. The main disadvantages of a dovetail slide are stiction and high friction. These slides are generally used in prototypes for laboratory type setups, where the adjustments have to be simple and economical.

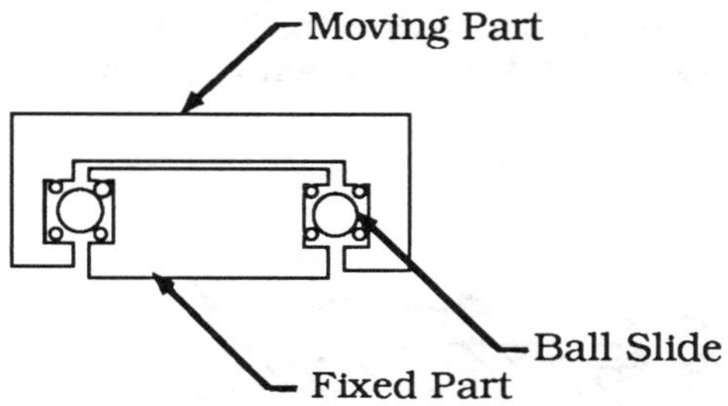

Figure 1. A typical ball slide interface for linear mechanisms

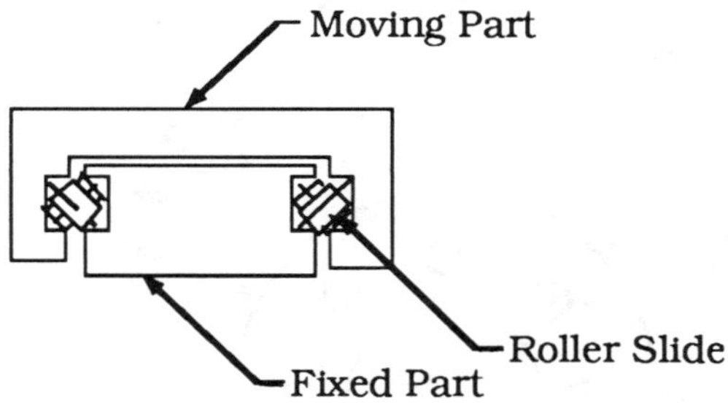

Figure 2. *A typical roller slide interface for linear mechanisms*

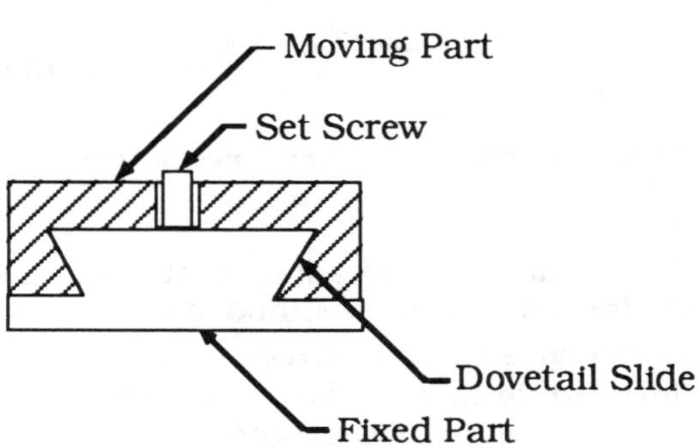

Figure 3. A typical dovetail slide interface with set screw locking

Hydrostatic bearings, which include both gas and oil bearings, are virtually free of friction and wear and have negligible cross-axis run out. In optical systems, oil bearings are not commonly used because these are messy and present the risk of contaminating the optics. The optical system that require adjustment mechanisms with a long travel range and high accuracy and load capacity use gas bearings such as shown in figure 4. The pressurized gas is generally very clean and dehumidified air but some applications use dry nitrogen or helium also. The main disadvantages of gas bearings are their cost and complexity. A remote and elaborate pumping and filteration system is required to supply clean and dry air. The design and fabrication of gas bearings is complex and expensive. The number and size of air jets, size of the air relief pocket and surface area of the bearing, and supply pressure of the air must all be taken into account when designing an air bearing for an application.

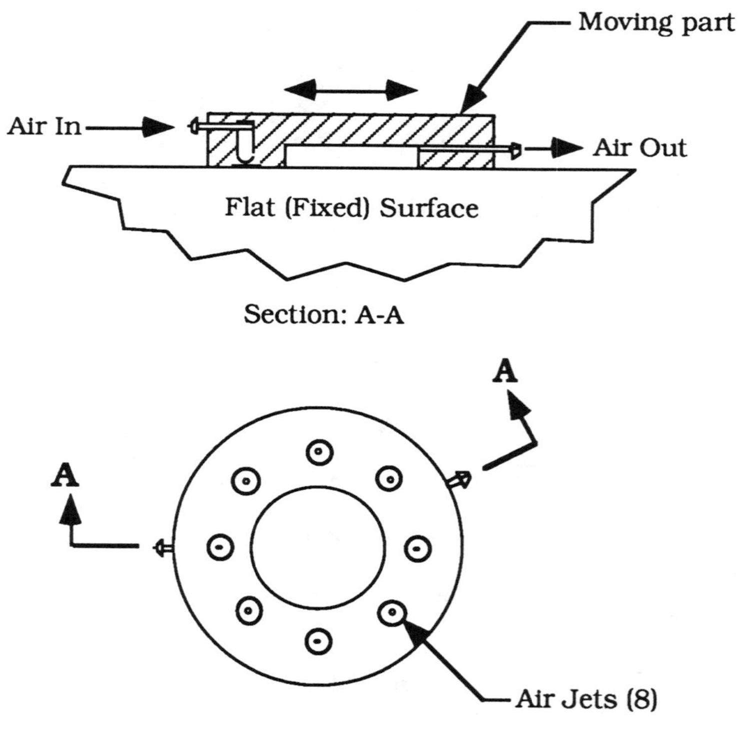

Figure 4. An air bearing interface for linear mechanisms

Flexures are suitable for backlash free adjustments over short travel ranges. These have low friction and hysteresis, and do not require any type of lubrication. Flexures can have several shapes such as a flat strip, circular or universal. A flat flexure interface is shown in figure 5. The design and fabrication of flexures is quite complex and is discussed in detail in by Billing, Weinstein, Paros and Weisbord[1]. For an application, a flexure must be designed to have a specific stiffness which is determined by its length, width , thickness, shape and material. When flexures are bent to move a component, a reaction force is induced in the component attached to the flexure. In applications where flexures are directly bonded to optical components, the reaction force from the flexure can produce localized surface distortion. Therefore, the flexures must be designed for a proper stiffness to keep this distortion within acceptable limits. The flexure interface is generally suitable for those precision mechanisms which require high resolution over short travels.

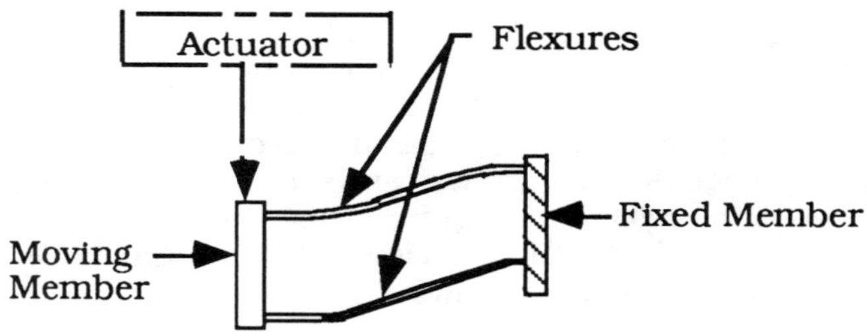

Figure 5. A flat flexure interface for linear mechanisms

For low cost and light loads, a sliding contact between the the moving and fixed part, shown in figure 6, may be the best alternative. This type of interface has high friction and there is no control over the linearity of the adjustment i.e. a linear translation may be accompanied by a slight in plane rotation. This type of sliding interface is suitable for infrequent adjustments.

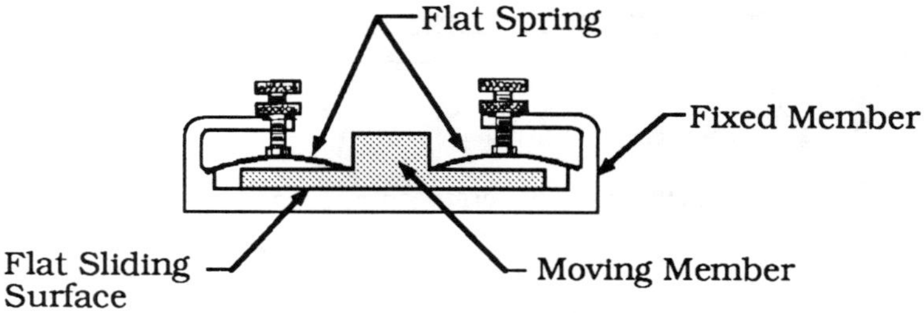

Figure 6. A linear mechanism with a flat sliding interface

3.2 *The actuators*

The choice of a suitable actuator depends on travel speed, range, resolution and frequency of adjustment, and cost, size and weight requirements for the adjustment mechanism. For example, the motorized actuators are generally used for making frequent adjustments in real time. These include DC, linear and stepper motors and piezoelectric devices. The main advantages of such actuators are long travel range, high resolution and velocity, and position readout capability. These actuators can come with built in position encoders and can be used in a closed loop control system. Therefore, the position of an optical component can be monitored, and the drifts due to environmental effects can be corrected in real time. The principle disadvantages of motorized actuators are their high cost and weight and large size.

A piezoelectric actuator is used for short travel range requiring high resolution. A high voltage is needed to produce a movement of the order of a few microns. The piezo actuators have high load capacity. Typical disadvantages are hysteresis, creep and nonlinearity of travel versus the applied voltage.

If an application does not require frequent adjustments, it is more cost effective to use screws or micrometers. If a position readout is not required, the screws, as shown in figure 7, are more economical and compact as compared to micrometers. The screws can be coarse, fine or of differential type depending on the resolution requirements. The differential screws and micrometers are used for small travels, and are quite bulky and more expensive as compared to regular screws.

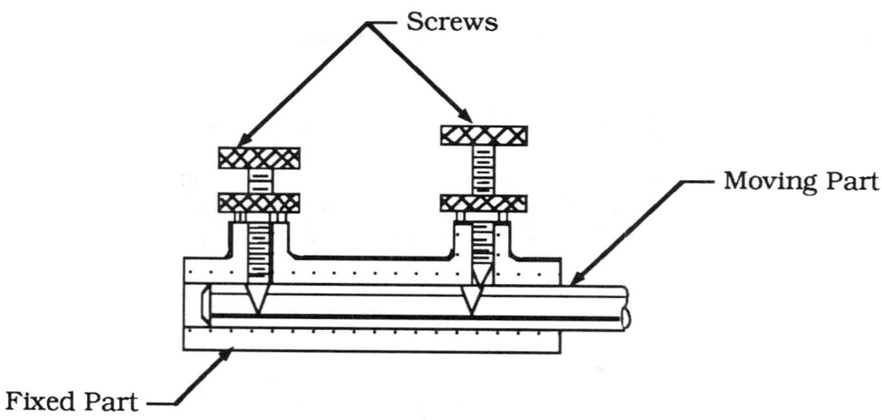

Figure 7. A linear mechanism with screw actuators

3.3 *The coupling methods*

A number of options are available to couple the actuator to the moving component of an adjustment mechanism. The choice of a suitable method depends on such design factors as frequency of adjustment, shock requirements, weight, cost and size of the mechanism. The tip of the actuator is generally rounded and polished to a high finish to minimize wear due to friction. The tip of the actuator can bear against a flat surface, a cone or a spherical socket. The flat surface, shown in figure 8, results in a point contact with the tip of an actuator and therefore, produces high contact stresses. The ball/cone interface, shown in figure 9, results in a line contact, and produces much lower contact stresses. For infrequent adjustments of light weight components, a ball tip acting directly against a flat surface, is the most simple and economical choice. The ball cone interface is more expensive to machine and larger in size, and is generally used for adjusting heavy components which must be adjusted frequently.

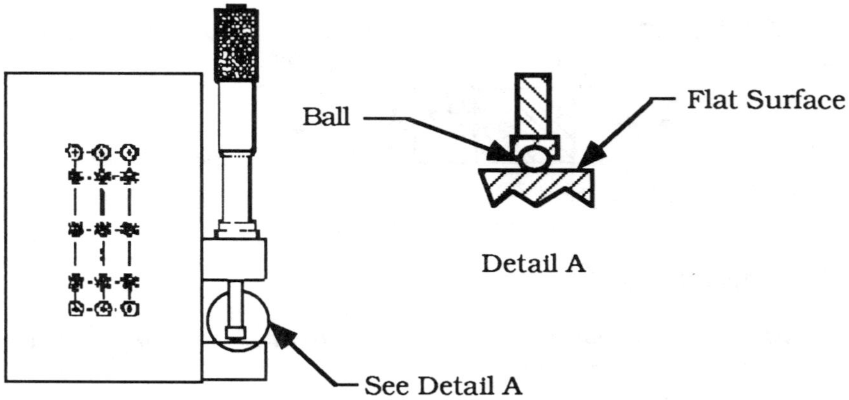

Figure 8. A translation stage with a micrometer and a ball/flat coupling

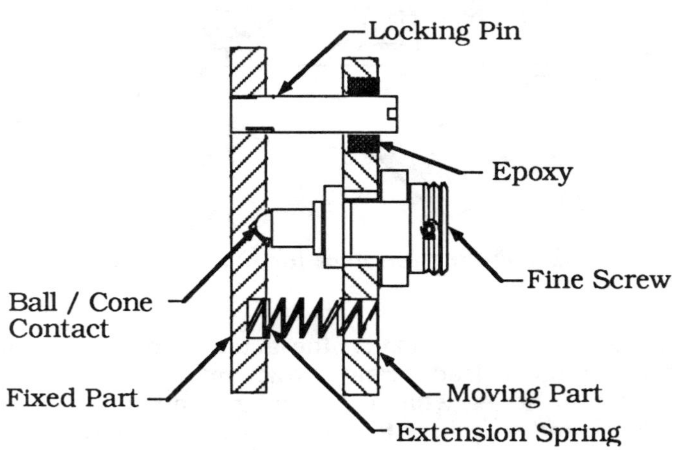

Figure 9. A linear mechanism with a ball/cone interface, epoxy locking and extension spring preloading

The moving component can also be physically attached to the actuator through a threaded attachment or through a flexible type of coupling. The threaded coupling, shown in figure 10, is not commonly used because a slight misalignment of the line of travel relative to the actuator axis induces lateral loads in the actuator and can cause rapid wear and damage to the actuator. The flexible coupling, shown in figure 11, does not have this problem because slight misalignments are compensated by the flexibility of the coupling. The flexible couplings come in many types such as bellows, spring, Oldham, jaw and Schmidt type. All these flexible couplings are relatively bulky and heavy, and are expensive and should be used in large mechanisms in which heavy components need to be adjusted frequently.

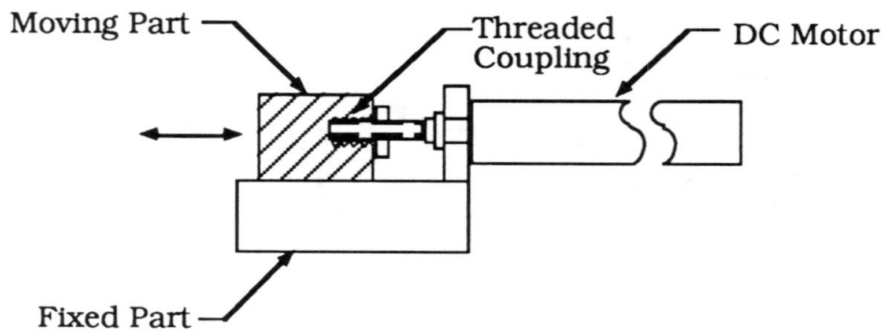

Figure 10. A linear mechanism with a threaded coupling between the actuator and the moving component

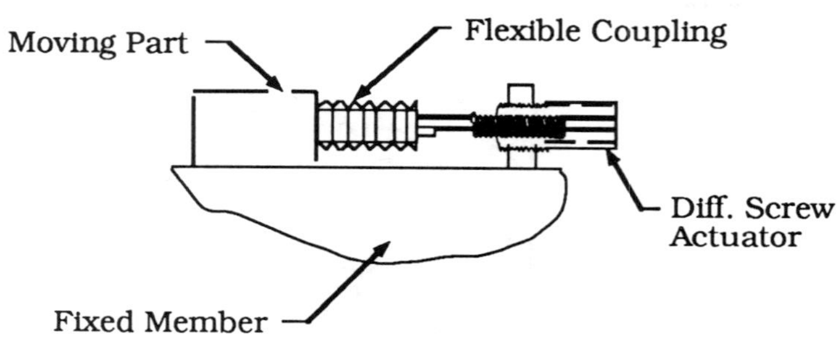

Figure 11. A flexible coupling interface for linear mechanisms

For large travels and heavy loads, the actuator may be coupled to the moving part through a lead screw shown in figure 12. The actuator is coupled to the main screw while the moving component is attached to the nut. The backlash and wobble of the nut can be minimized by preloading the nut. The friction can be reduced by using ball or roller bearings between the screw and nut. The lead screws are large in size and expensive but can produce very linear travel over a long range.

3.4 *The preloading methods*

Most of the mechanisms employ some form of preloading arrangement to ensure a positive movement, free of backlash, when the adjustment is made. The selection of a suitable type of preloading method depends on such design factors as the range and frequency of adjustment, load capacity, cost and size of the adjustment mechanism. In adjustments with a long travel range, a compression or an extension spring must be used for preloading the moving part against the stationary part. Typical spring preloading methods are shown in figure 9 and 13, although there can be many variations of these methods. Normally the springs can be placed around the adjustment screw or can be centered between two adjacent adjustment screws to preload both screws with one spring.

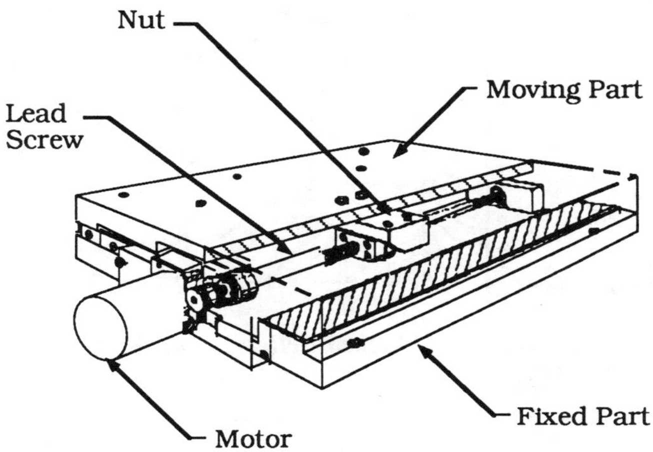

Figure 12. A translation stage with a lead screw coupling

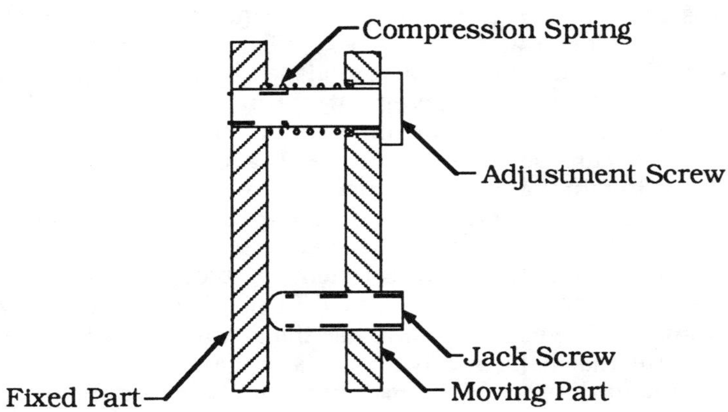

Figure 13. A mechanism with a compression spring for preloading and a jack screw for locking

Belleville and curved washers are used in compression mode to obtain high preloads over very small travels. These washers have very high stiffness and therefore, produce high loads as a result of relatively small compressions. These washers are compact in size and are very economical and suitable for high shock applications. The travel range of mechanisms using Belleville washers can be increased by stacking the washers as shown in figure 14.

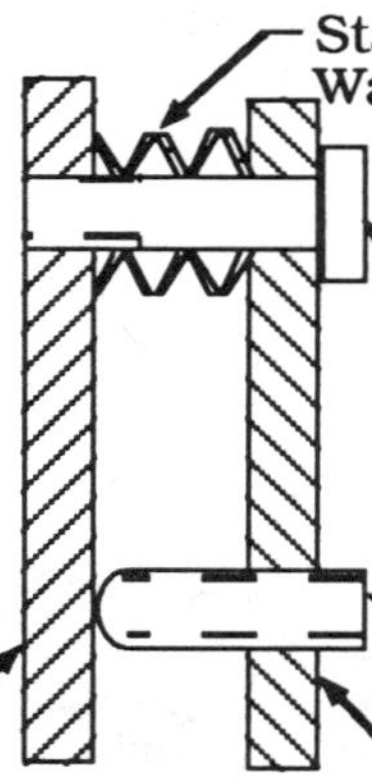

Figure 14. A mechanism with stacked Belleville washers for preloading and a jack screw for locking

3.5 *The locking methods*

After an element has been adjusted, it must be locked in place to retain the adjusted position. The design factors affecting the choice of a locking method are the frequency of adjustment, shock, vibration, weight, size and cost requirements. The locking can be accomplished by locking the actuator itself to prevent its accidental movement, tampering or its drift under environmental vibrations. For micrometers or screw type actuators, simple caps or covers can be used to prevent accidental movements. The movement due to vibrations can be prevented by using set screws to lock the rotation of the actuator screws. The second option is to positively lock the moving element relative to the fixed structure, and several methods are available for doing this. If the optical element is going to be adjusted and aligned only at assembly, it can be locked in place by using epoxy bonding shown in figure 9. The epoxy locking has the advantage of being a very low cost method but the main disadvantage is that the mechanism can not be readjusted without disassembly, which is generally very difficult and poses a great risk of damaging the parts.

Jack screws or locknuts, shown in figure 14 and 15, are economical ways of locking the mechanisms that do not require a great precision. The reason for lower accuracy is that a large force can be exerted on the adjusted element when a locknut or jack screw is tightened. The advantages of these methods are that these are compact in size and can be disassembled very easily without posing any risk of damage to the mechanism parts.

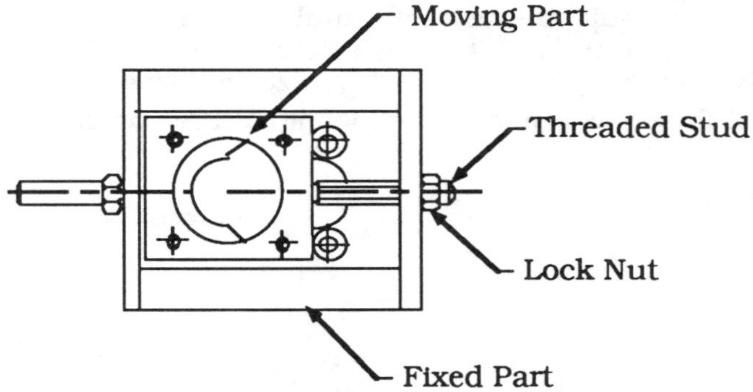

Figure 15. A linear stage with locknuts to retain the adjusted position

The vee and collar clamps are not commonly used because these are bulky and expensive to machine. Moreover, the bearing surface on the moving element must be parallel to the fixed structure to achieve effective locking. For motorized actuators, a closed loop feedback control system can be employed to retain the adjusted position of the moving element. As mentioned earlier, this method is expensive and is used only in those mechanisms which require real time adjustment.

4. SELECTION CRITERIA FOR COMPONENTS OF ROTARY ADJUSTMENT MECHANISMS

A number of choices are available for the components of a rotary mechanism depending on the performance requirements such as frequency, range and resolution of angular travel, shock, load capacity, cost and size of the mechanism. A number of components that are used in linear mechanisms can also be used for rotary mechanisms either in the same configuration or in their rotary version. These choices along with some design guidelines are discussed below.

4.1 *The interfaces*

The choice of a suitable interface between the rotating part and the fixed structure depends on such design criteria as the range and frequency of adjustment, shock, load capacity, cost and size requirements. As in the case of linear mechanisms, rotary versions of the bearings discussed in section 3.1, are more commonly used to interface the rotating part to the fixed part. Spherical and journal bearings, shown in figures 16 and 17 are used for light duty cycles because of friction and wear problems. These bearings are compact in size and are inexpensive as compared to rotary ball, roller and air bearings, which are used for heavy loads that are adjusted frequently.

For small angular adjustments, a flexural pivot (Bendix type) offer the advantages of friction and backlash free angular adjustment. These commercially available pivots are small in size and have low hysteresis. The application of these pivots in a mirror mount has been described by Rundle[2] in detail.

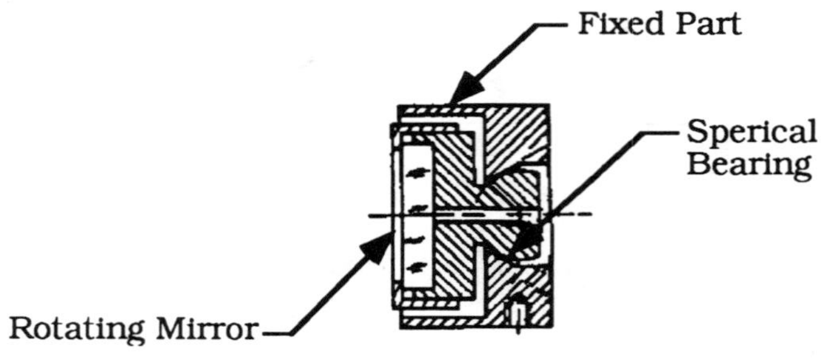

Figure 16. A mirror mount with a spherical bearing interface

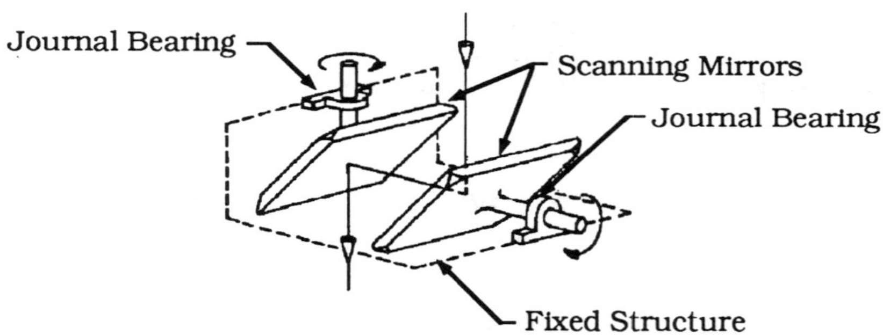

Figure 17. A rotary mechanism with journal bearings

4.2 *The actuators*

The actuators used in rotary mechanisms are very similar to those used in linear mechanisms discussed in section 3.2. Screws are used in low cost applications for small angular adjustments as shown in figure 18. Micrometers and differential micrometers are used when more precise adjustment with a readout is required. Commercially available rotary stages use both types af micrometers extensively. Once again the motorized actuators are used in applications requiring large and frequent angular travels or for real time adjustments..

4.3 The *coupling methods*

Some of the coupling methods for linear mechanisms discussed in section 3.3 can also be used in rotary mechanisms. The round tip of an actuator can bear against a flat, cone or a spherical socket. The advantages and disadvantages of these arrangements have already been discussed in section 3.3. These interfaces can only be used for very small angular adjustments since the tip of the actuator is sliding relative to the rotating part and loses a good contact with the rotating part for larger angles. A typical mechanism using a ball/flat type of coupling is shown in figure 18.

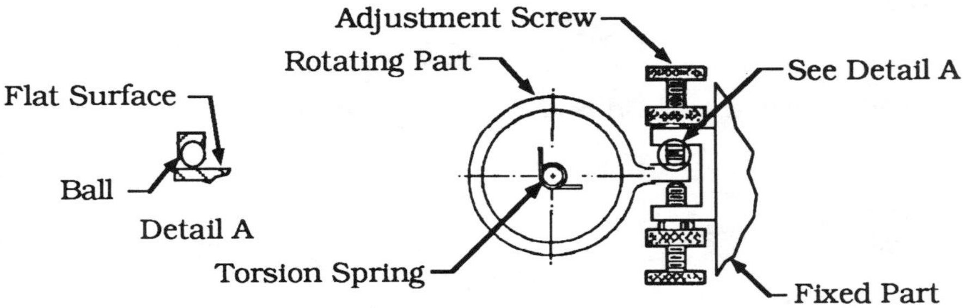

Figure 18. A rotary mechanism with screw actuators, a ball/flat interface and a torsion spring for preloading

For large and frequent angular adjustments, an actuator can be coupled to the rotating part through a flexible type of coupling or a universal type of joint as shown in figures 19 and 20. These types of couplings can tolerate relatively large misalignment between the axes of rotation of the actuator and the moving part. The principle disadvantages are large size, weight and cost.

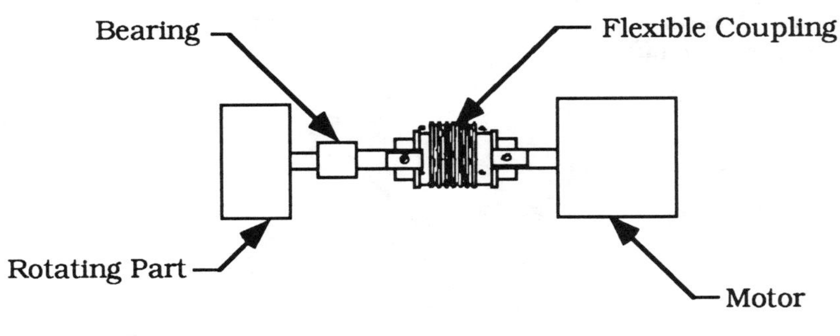

Figure 19. A flexible coupling is used to interface the motor with the rotating part

Sometimes, linear actuators are used in rotary mechanisms because of their lower cost and compact size. In such cases the linear motion of the actuator is converted to a rotary movement through a worm and gear or a rack and pinion arrangements shown in figures 21 and 22. In these mechanisms, the part to be adjusted is attached to the rotating gears. These mechanisms are expensive because of their complexity.

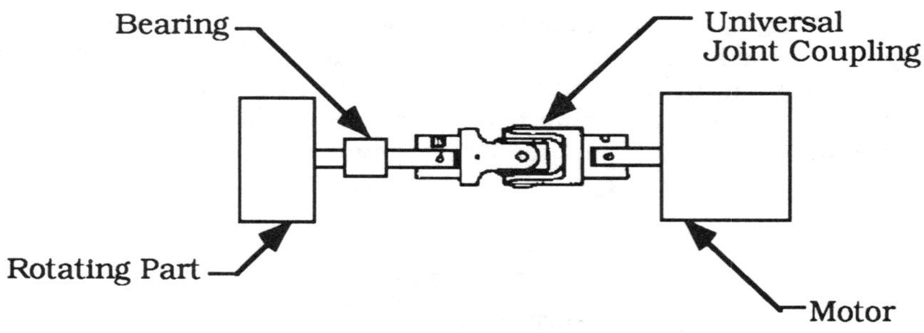

Figure 20. A universal joint coupling for rotary mechanisms

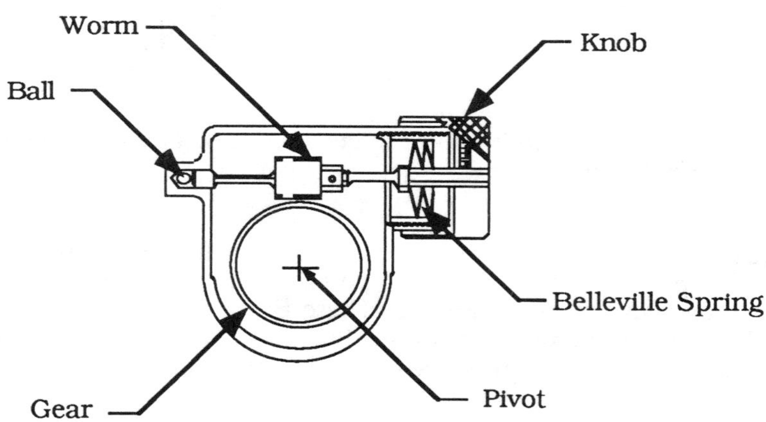

Figure 21. A rotary mechanism with a worm/gear interface and Belleville washers for preloading

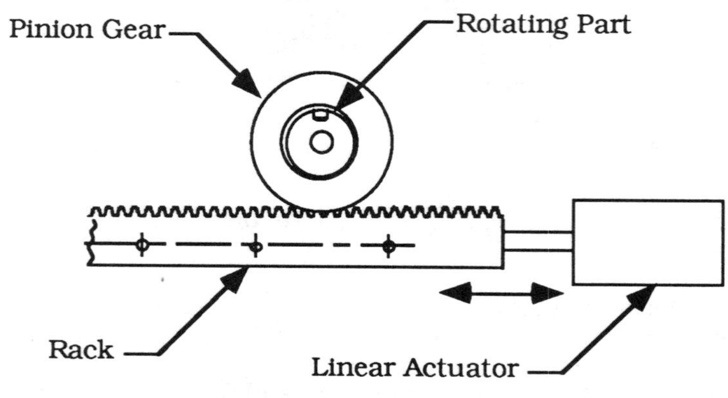

Figure 22. A rack and pinion interface for rotary mechanisms

4.4 *The preloading methods*

The selection of a proper preloading method to obtain a backlash free rotary adjustment is based on the same design factors that are discussed in section 3.4 for the linear mechanisms. The tension or compression springs, shown in figure 23, are extensively used for preloading purposes because of their low cost, and also because these can be used over a fairly large angular range. The application of Belleville washers for preloading is shown in figure 21. As mentioned earlier, these washers are used for high preloads over small adjustment ranges. For larger angular rotations, a torsion spring , shown in figure 18, can be used for preloading the rotating part. One end of the spring is attached to the rotating part, while the other end is attached to the fixed structure.

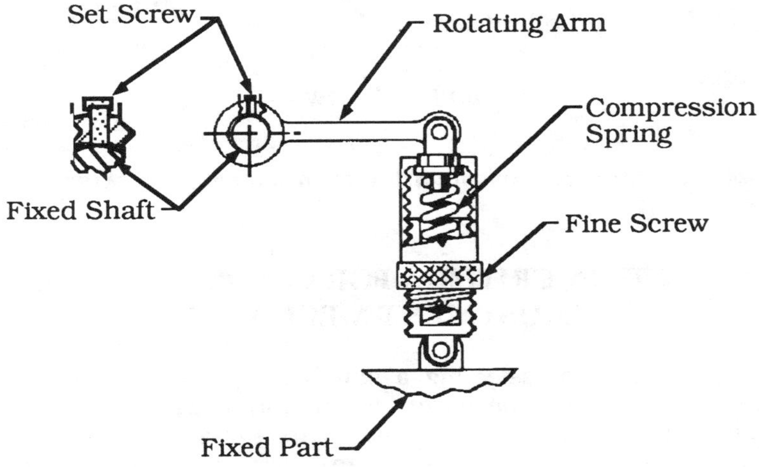

Figure 23. A rotary mechanism with a compression spring for preloading and a set screw for locking

4.5 *The locking methods*

The design factors affecting the choice of a suitable locking method are similar to those for the linear mechanisms as already discussed in section 3.5.

These include set screws, clamps, epoxy, and locknut. The mechanism shown in figure 23 employs a set screw to lock the rotating arm against the fixed shaft after the required adjustment has been made.The mechanism shown in figure 24 uses a collar clamp for locking purposes. The clamp is rigidly attached to the rotating part. After the adjustment has been made, the screw in clamp is tightened to achieve the clamping action to lock it against the shaft which is part of the fixed structure. In the same fashion, a locknut or epoxy bonding may be used for locking purposes. The relative advantages and disadvantages of all these locking methods have already been discussed in section 3.5.

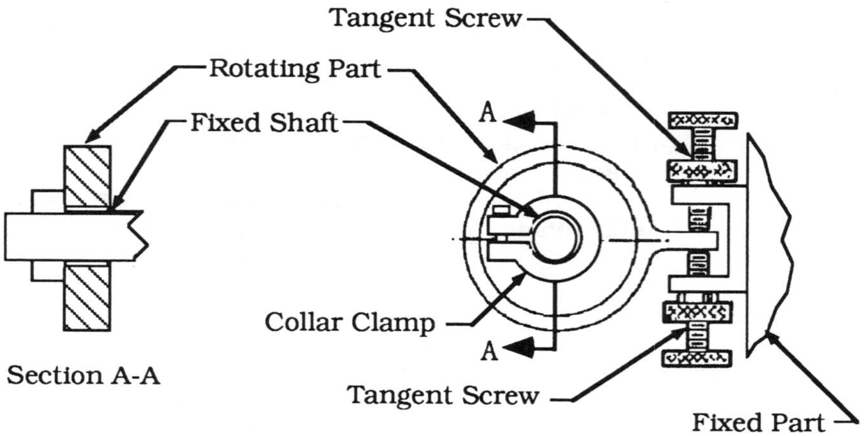

Figure 24. A rotary mechanism with a collar clamp for locking

5. SELECTION CRITERIA FOR COMPONENTS OF TILT ADJUSTMENT MECHANISMS

The design factors for selecting a suitable type of components for tilt mechanisms are the same as for linear and rotary mechanisms. If an adjustment mechanism is designed with three mutually orthogonal adjustment points, it can be used to perform linear as well as tilt adjustments. When the three actuators are moved equally, an axial movement results. However, if only one of the actuator is moved, a tilt adjustment about an axis defined by the two others is achieved. Therefore, a number of the same components that are used in linear mechanisms, can also be used in tilt mechanisms. The application of these parts in tilt mechanisms is discussed briefly below.

5.1 *The interfaces*

The tilting component can be attached to the fixed structure through rotary bearings (journal, ball, roller, air), flexures (Bendix, flat blade) or a traditional kinematic interface. The examples of mechanisms using two of these interfaces are shown in figures 25 and 26. The tradeoffs for these type of interfaces have already been discussed under linear and rotary mechanisms.

5.2 *The actuators*

The actuators used in linear and rotary mechanism can also be used in tilt mechanisms. The screws and micrometer adjustments are used in manual mechanisms, and are extensively employed in commercial tilt stages and mirror mounts. One such stage using a micrometer is shown in figure 27. Once again, the motorized actuators are large in size and expensive, and are used when real time and frequent adjustments are required.

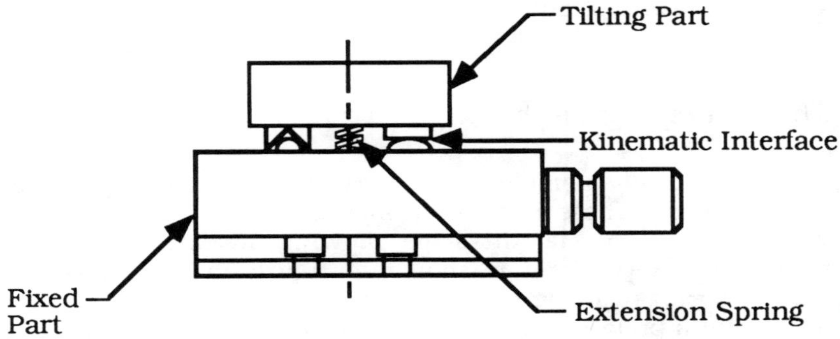

Figure 25. A tilt stage with a kinematic intaerface and an extension spring for preloading

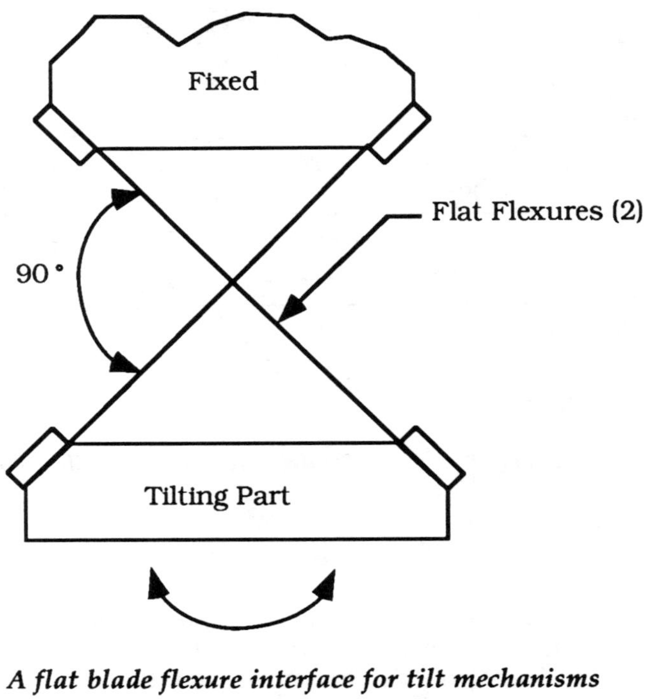

Figure 26. A flat blade flexure interface for tilt mechanisms

5.3 *The coupling methods*

As in linear and rotary mechanisms, the rounded tip of the actuator can bear against a flat surface, cone or a spherical socket in tilt mechanisms The ball/flat contact is illustrated in figure 27. The flat surface can be replaced by a cone or a spherical seat to increase the contact area and to minimize the contact stresses in those tilt mechanisms , which are adjusted frequently and require a higher load capacity.

5.4 *The preloading methods*

The preloading methods employed in linear mechanisms can also be used in tilt mechanisms. These include springs and washers as discussed in section 3.4. A commercial tilt stage showing an extension spring for preloading the tilt platform is shown in figure 25 The extension or compression springs can be centered between two adjacent actuators to achieve a uniform preloading. For smaller tilts and higher preloads, stacked Belleville or curved washers can be used in place of springs. The advantages and disadvantages of these preloading methods have already been discussed under linear adjustment mechanisms.

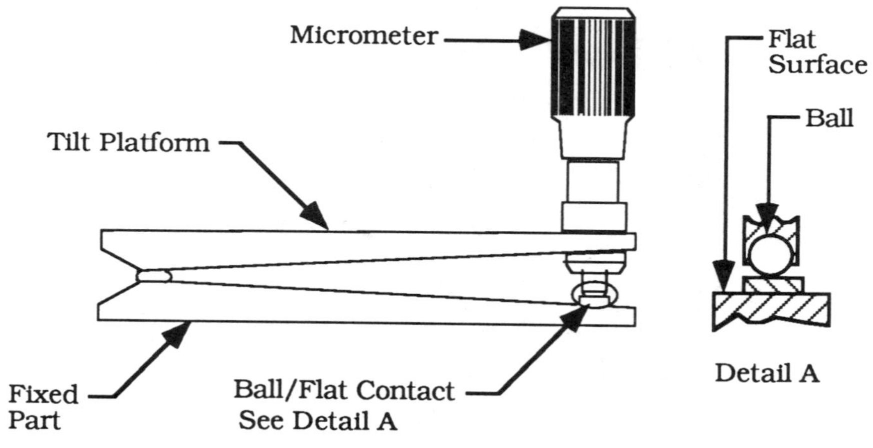

Figure 27. A tilt stage with a micrometer actuator and a ball/flat contact

5.5 *The locking methods*

The locking methods for tilt mechanisms are similar to those employed in the linear adjustment mechanisms. These include set screws, jack screws, locknuts, clamps and epoxy. The set screws and locknuts are used for temporary locking in coarse adjustment tilt mechanisms. The epoxy locking is economical and simple in design but is used those mechanisms which are adjusted at initial assembly and alignment only. As in the linear and rotary mechanisms, a closed loop control system can be used to retain the desired tilt adjustment in real time position control applications.

6. DESIGN GUIDELINES FOR USING ADJUSTMENT MECHANISMS IN OPTICAL SYSTEMS

An adjustable optical mount is inherently more complex, expensive and less stable than a comparable fixed type of mount. Therefore, a careful trade-off design study must be conducted to weigh the benefits of an adjustable optical mount against the potential stability and cost drawbacks. A good summary of how and when adjustment mechanisms must be incorporated in optical systems has been presented by Vukobratovich[3].

The decision to use adjustable optical mounts is dictated by the sensitivity analysis performed by the optical designer. This analysis defines the assembly tolerances for the optical system to achieve the image resolution and quality requirements. For systems with high image resolution requirements, usually the alignment tolerances of the optical elements exceed the practical fabrication tolerances for the optical elements and the associated mechanical hardware. This problem is further compounded by the stack up of machining tolerances and inspection uncertainties when the optical system has a large number of mirrors and lenses. In such applications, it is economical and time saving to manufacture the optical and mechanical parts to rather loose tolerances and provide adjustments for a few sensitive optical elements. These elements are then adjusted to compensate for the errors and uncertainties in the positions of other optical elements. Therefore, a rigorous optical sensitivity analysis must be performed to identify the optical elements that have the strongest effect on the image quality of the system.

Once the type of the adjustments required for an optical element have been determined, the next step is to design an adjustable mount to provide the required adjustment range and resolution. The glass optical elements are normally assembled in a metal frame, and an adjustment mechanism is provided between the frame and the fixed mounting structure. This way, a direct contact between the actuator and the glass element is avoided, and the element is shielded from the high contact stresses at the tip of the actuator. Providing an adjustment range, which is longer than that determined by the optical analysis, can be very risky. Such mounts can be assembled far off from their required nominal position, and a lot of time may be wasted during alignment to obtain any image at all. In such cases, it is better to design the mount such that it can be shimmed at assembly in case the range of adjustment is found to be inadequate. After the optical system has been aligned, the adjustable mounts must be positively locked to prevent misalignment due to accidental tweaking or drifts due to vibrations. The adjustments that are used at initial assembly only can be designed to have lockable mounts with removable actuators. This approach not only saves on the cost of actuators, but also eliminates the risk of accidental misadjustment later on.

The adjustable mounts should be designed to tilt or rotate about the principal points to avoid cross talk between axial and tilt adjustments. The cross coupling of adjustments can be very frustrating because several iterations of tilt and axial adjustments may be required to achieve the alignment. The adjustable mirror mounts must be designed to tilt the mirror about its vertex to avoid an unwanted image shift. The adjustment points in a tilt mechanism should be positioned in a mutually orthogonal pattern relative to the axis of the optical element. Such mounts are easy to adjust and produce predictable movements.

The adjustment mechanisms must be designed to have a large mechanical advantage such that a large axial movement or rotation of the actuator will incrementally move the optical element. This design feature can also save valuable alignment time because there is a less likelihood of accidentally overshooting the optimum position. If feasible, the mechanisms must be designed to have a coarse as well as a fine adjustment. The coarse alignment can be quickly achieved by using the coarse adjustment, while the fine part of the adjustment is only used to optimize the quality of the image.

A number of translation, rotary and tilt stages, mirror, lens and gimbal mounts are available commercially. These mounts are economical, precise and quite rugged, and are very suitable for prototypes and laboratory setups. Their main disadvantage is their bulky size and weight, which makes their use impractical in the systems with several optical elements that are packaged together tightly. If weight and size are not a problem in an application, it is far more economical and time saving to use commercial mounts rather than designing and fabricating the custom mounts.

While the adjustment mechanisms offer several advantages, the disadvantages of the adjustable mounts must not be overlooked, and provisions must be made in their design to minimize the negative effects on the optical system. First of all, the number of adjustable optical elements in a system must be kept to an absolute minimum. This not only save on the fabrication cost but also maximizes the long term stability of the system. The adjustable mounts are less rigid and more unstable than fixed mounts, and therefore, experience drift with time. The adjustments often induce non linear and unpredictable effects in optical systems. The adjustable mounts are mechanically weak and are more susceptible to drifts due to shock loads, vibrations and temperature variations.

7. CONCLUSIONS

The design choices for various components that make up the linear, rotary and tilt adjustment mechanisms have been presented. The advantages and disadvantages of these choices have been presented to help the designer in choosing the most suitable parts for a particular applications. The design guidelines for incorporating the adjustment mechanisms in optical systems have been presented. The adjustment mechanisms play an important role in the integration and alignment of sophisticated optical systems which have tight positioning tolerances. By incorporating these mechanisms in the optomechanical design of the systems, it becomes feasible and economical to produce and assemble such optical systems to very high alignment accuracies, which in turn greatly enhances the image quality produced by these systems. For this purpose, the application and designs of two adjustable mirror mounts for a submicron lithography system have been discussed by Ahmad[4].

8. REFERENCES

1. D. C. O'Shea, "**Selected Papers on Optomechanical Design**", SPIE Milestone Series, Vol. 770, pp. 321-355, 1989.

2. W. J. Rundle, "**Design and performance of an optical mount using cross-flexure pivots**", SPIE Proceedings Vol 1167, pp. 306-312, 1989.

3. D. Vukobratovich, "**Introduction to Optomechanical Design for Mechanical Engineers**", SPIE Short Course Notes U8, pp. 115-141, 1989.

4. A. Ahmad, "**High Resonance Adjustable Mirror Mounts**", SPIE Proceedings Vol. 1167, pp. 313-317, 1989.

The Influence of International Standards

on Opto-mechanical Design

Robert E. Parks
Optical Sciences Center
University of Arizona
Tucson, AZ 85711

ABSTRACT

In the last 10 to 15 years, a considerable body of international standards literature has been published on both mechanical and optical design. We discuss the influence of these internationally developed standards on the design and fabrication of optical systems. We conclude that while there are large benefits to be gained from using these international standards, there will have to be a substantial educational effort at all levels from project scientist to worker on the shop floor to take advantage of the benefits. Many sources to help in this education process are outlined.

INTRODUCTION

While this paper is about the influence of internationally developed standards on opto-mechanical design, we start by indicating that there is no direct influence. The standards developed by the International Standards Organization (ISO) are voluntary standards and can be used or not used as any manufacturer sees fit. The influence of these standards will, rather, be felt indirectly through such forces as having to bid on components specified according to the ISO standards, not having market access to certain countries unless ISO standards are used and met or because some customer mandates the use of international standards in the design and fabrication of some product.

Because there is no direct connection between the ISO standards and the need to use these standards, it is the thesis of this paper that an extensive educational process must take place before the benefits of the ISO standards can be truly recognized. This paper is an attempt to begin that process and point out the various sources of further information about the international standardization effort.

We begin by giving a little background about international standards work in general. As part of this background, some of the theses upon which international standardization is based are discussed.

Then we move on to technical areas of direct interest to the opto-mechanical designer and outline some of the standards work done in these fields. The actual work in optics has been very recent and much of it is still in the draft form. However, these drafts are readily available for study and use in areas where there is presently no other suitable method of specifying needed features on optical components.

Once we have outlined much of the available ISO literature, we discuss other sources of opto-mechanical standards literature and comment on that literature. We then address the problem of introducing the features of the new standards into designs and drawings in sensible way. It seems to us that simply decreeing from some day forward that all drawings will be done using an certain format or that all parts will be inspected to a particular standard is counter productive. It is suggested that there is a method of gradual introduction that allows all involved to get acquainted with these new standards ideas in a logical and systematic fashion.

Finally, it is suggested that the move toward using the ISO standards begins at the top. The opticians and inspectors can only make and inspect what is on the drawings. It is the project leaders and designers that start the process of using these new tools. Before using the tools, however, these same people must be familiar with the standards and the advantages they offer.

Because the ISO standards in the opto-mechanical area are substantially more thorough than the existing national standards, it is like giving a craftsman more and better tools to work with. With the better tools, the work itself will become better. In this time of increased competition, we all need all the new tools we can get to keep up.

BACKGROUND

Efforts toward international standardization began early in the 20th century at a meeting in the United States. It was the rapid spread of electrification that served as the stimulus for the International Electrical Congress of St. Louis to pass a resolution "that steps should be taken to secure the cooperation of technical societies of the world by the appointment of a representative commission to consider the question of standardization of the Nomenclature and Ratings of Electrical Apparatus and Machinery".[1]

The 'commission' that resulted from this resolution was the International Electrotechnical Commission (IEC) founded in 1906 with headquarters in Geneva, Switzerland. While this international standards body is concerned with electrical and electronic matters, the objective of the Commission so well states the purpose and methods of international standardization that we quote it. "The object of the Commission is to promote international co-operation on all questions of standardization and related matters in the fields of electrical and electronic engineering and thus promote international understanding. This object *inter alia* is achieved by issuing publications, including recommendations in the form of international standards, which the National Committees are expected to use for their work on national standards, in so far as national conditions will permit."[2]

There are several key features in this statement of objectives aside from the principal one of promoting international understanding. First of all, the IEC is a *voluntary* standards organization. The use of its standards are not mandatory and the extent of any use is up to the individual national committees. Of course, there is nothing stopping a governmental body (national or otherwise) from adopting an international standard into its own regulatory code. At that point, the standard becomes mandatory in the region where the governmental body has jurisdiction.

Another feature is that membership in the IEC (and as we will see below is similarly in ISO) is by "National Committee". The national committee in the United States is the American National Standards Institute (ANSI), a private standards coordinating and approval organization that has been appointed by the US Government to fulfill this role. The US is the only country whose National Committee is a private rather than a government organization. This creates more than a few problems in international standardization because ANSI cannot speak for the US Government whereas all the other national committees can.

The next big push in international standardization came after World War II when the International Standards Organization was established as part of the United Nations Charter. ISO is also headquartered in Geneva and physically housed in the same building as the IEC. ISO has the responsibility for coordinating and approving standards in all fields of technology except those under the narrower IEC charter. Of course these days almost everything uses electronics and electricity in one way or another so the dividing line is fuzzy. The IEC and ISO have been working hard at "harmonizing" their rules and procedures so that the approval criteria and cycle for new standards is the same. There are also some "Joint Committees" between the two groups in the areas of computer and image formation technology.

ISO was founded in 1947 at a meeting of the national committees of 25 countries. The National Committee for the US is ANSI. Other national committees that may be familiar are the British Standards Institute (BSI), the Deutsches Institut fur Normung (DIN) and the Association Francaise de Normalisation (AFNOR). A complete list of all the national committees belonging to ISO along with addresses and phone numbers are listed in the ISO

Memento[3], an index of members and technical committees.

In terms of how it functions, ISO has over 190 Technical Committees (TC's) representing fields of technology from Tractors and farm machinery to Micrographics and optical data storage. One of the ISO national committees heads each of these TC's and serves as the secretariat or administrative body for the TC's it heads. Further, each TC is broken down into several Subcommittees (SC's) that again have national committees that serve as secretariats. Within the SC's are Working groups (WG's) where the actual standards writing gets done. Some of the TC's most closely allied with opto-mechanics are listed in Table 1.

STANDARDS WRITING PROCESS

Now that the organization of ISO has been explained, the process of writing standards will be discussed. We will describe not only the mechanism behind the standards writing but also some of the guiding principles. At first glance, some of the procedures look somewhat cumbersome but there are well thought out reasons for doing things according to the procedure. On the other hand, the system has a great deal of flexibility. This permits reasonable people to accomplish the difficult task of coming to consensus without the burden of too rigorous a set of rules to contend with. Of course, too, the secretariat has a great deal of influence on how some matters are dealt with even though they are supposed to be completely neutral.

Before any standards work is undertaken, there has to be a consensus of the national committees that there is the need for standardization in a particular area. If the technical area is outside the scope of any existing TC's, a new TC will be formed and given a scope that defines its work. In the case of the TC we are most interested in, TC172 - Optics and optical instruments, it was formed in 1979 because many of the other TC's were writing application oriented standards that dealt with optical principles. However, since each TC consid-

Designation	Technical Committee Name	Secretariat
ISO/TC10	Technical (mechanical) drawings	DIN
ISO/TC12	Quantities, units, symbols, etc.	SIS
ISO/TC36	Cinematography	ANSI
ISO/TC42	Photography	ANSI
ISO/TC57	Metrology, properties of surfaces	GOST
ISO/TC112	Vacuum technology	GOST
ISO/TC130	Graphic technology	AFNOR
ISO/TC171	Micrographics	AFNOR
ISO/TC172	Optics and optical instruments	DIN
ISO/TC180	Solar energy	SAA
ISO/TC187	Colour notations	SIS

Table 1
Various ISO Technical Committees Related to Optics

ered the optical aspects from its own applications viewpoint, there was a lack of consistency between the work of the particular TC's. It was this concern over consistency that led to the consensus to form TC172.

Almost always, the initiative is taken by one national committee to undertake new work. In the case of TC172, it was DIN, the German Institute for Standardization. Because DIN had a long record of standards writing for optics it was easy for DIN to write a scope for the technical work and then solicit other national committees to join it in establishing the TC. Implicit in this effort was that DIN would volunteer to be the secretariat for the TC and that the first chairman for the group would be a representative of the German optics community.

At the inaugural meeting of TC172 in Pforzheim, Germany, the various national delegates were asked to volunteer for being the secretariats of the 9 SC's originally proposed. Because there was little interest in the technical scope of some of the SC's, some were not initially activated. Within the activated SC's, Working Groups were established and appropriate scopes agreed upon. Only at that time was the structure in place to begin writing standards. Table 2 shows the structure of ISO/TC172.

Once the structure was in place and the scope of the technical work was agreed upon for each SC and WG in which there was interest, the procedure is then to place those applicable national standards on the table and use them as the basis for an international standard. In the case of optics, the Germans had a long history of work in this area while few other countries had done much. Thus the basis for much of the ISO work in optics is based on DIN standards.

TYPES OF STANDARDS

The type of standards writing is also considered important. Basically there are three types of standards, those dealing with vocabulary or nomenclature, those specifying test methods for products

Subcommittee		Working Groups		Secretariat
SC1	Fundemental standards			DIN
		WG1	General optical test methods	DIN
		WG2	Indications in optical drawings	SNV
		WG3	Environmental test methods	DIN
SC3	Optical materials and components			AFNOR
		WG1	Raw optical glass	DIN
		WG2	Coatings	ANSI
SC4	Telescopes			GOST
SC5	Microscopes			DIN
		WG3	Terms and definitions	AFNOR
		WG4	Interfacing imaging optics	ON
		WG5	Stereo microscopes	ANSI
SC6	Geodetic instruments			SNV
		WG1	Terminology	DIN
		WG2	Tests	SNV
SC7	Ophthalmic, endoscopic, metrological instruments and test methods			DIN
		WG1	Endoscopes	JISC
		WG2	Tonometers	DIN
		WG3	Other ophthalmic instruments	DIN
		WG4	Tests for contact lenses	ANSI
SC8	Ophthalmic optics			AFNOR
		WG1	Contact lenses	BSI
		WG2	Spectacle frames	DIN
		WG3	Corrective lenses	AFNOR
SC9	Electro-optical systems			DIN

Table 2
Organizational structure of ISO/TC172 - Optics and optical instruments

and those setting performance criterion for products within the technical field in question. It is necessary to work on the nomenclature first so that all parties agree what it is they are talking about. This work also helps clarify what are the important aspects of the work to follow.

Next, a series of standard test methods or procedures must be worked out. These are generic test methods for some specific performance aspect of the products in that technical field. For optics, these might be tests for resolution or OTF, or they might be environmental tests that simulate the conditions that real instruments would see under actual use. Depending on the use, the test conditions could be made more or less severe.

Finally, standards involving the actual performance of instruments or products are written. These might involve such things as what uniformity in resolution must an instrument have to be labeled as having a flat field or what uniformity in illumination must a copy lens have over the specified working field of view. It is clear now that the first two types of standards must have already been in place before the performance standard can be written. Obviously too, it takes a real knowledge of what is desirable and feasible in terms of real instrument performance to write a performance standard.

This talk about the types of standards also helps illustrate a point or define terminology that is sometimes used regarding standards. Some standards are referred to as being horizontal while others are vertical. A horizontal standard is one that applies to all aspects of a given technological field. In the case of optics, a standard that defines how to do a drawing of an optical component would be a horizontal standard because it applies to all optical instruments. On the other hand, a standard dealing with the thread size for microscope objectives is a vertical standard because it only applies to microscopes.

When it comes to writing performance standards there can be conflicts over what SC's have the responsibility for writing performance standards.

For example, environmental tests for optical instruments are obviously horizontal standards but are the people who wrote these broad based standards the best informed people to apply them to, say, microscopes. The microscope people would say no, and, in fact, at the last Plenary meeting of ISO/TC172, the vertical standards writing SC's were given the last say over performance standards although the horizontal and vertical groups are supposed to work together.

REQUIREMENTS OF STANDARDS WRITING

While this gives a feel for the mechanisms for writing standards, we have said little about the content of the standards being written and what are the rules for the content. First, all ISO standards must use SI units and the definitions that ISO has given for physical quantities. The SI units are given in ISO 1000-1981 - *SI units and recommendations for the use of their multiples and of certain other units.* This ISO standard and 14 other standards dealing with the definitions of quantities, units and their definitions are contained in *ISO Standards Handbook 2.* This Handbook and all ISO standards materials are available from ANSI[3].

Next, all ISO standards are written in English, French and Russian, the official languages of ISO. As a matter of actual practice, more and more, English is becoming the only official language of ISO. This is not official policy but the precedent for it started in the early days of ISO. At that time, it was agreed that Russian was an official language but that the Russians would have the total responsibility for translations both ways from Russian to English and back because they were the only ones truly qualified. As time went on, the French realized that they were holding up progress when the proceedings of the working group meetings had to be translated into French. Beside, most of the French attending the meetings spoke good English.

As a result of this, much of the actual standards work is written only in English (although the minutes of the meetings are done in both English and French). It is then up to AFNOR, the French National Committee to translate the final standard into French so it can be published by ISO in Geneva. If the French do not submit a suitable translation by the time the standard is to go to press, it is published in English only. This will probably be a more common experience.

Partly as the result of the above statements, but much more because there are so many other languages than just English, French and Russian, ISO standards make use of symbols wherever possible. The idea here is that the people actually making the product may not speak a word of English, but if they are familiar with the meaning of the symbols on an optical drawing, for instance, there will be no problem with its interpretation. Also it is up to the National Committee of any non-English speaking countries to translate the ISO standards into their own tongues.

Another aspect of ISO standards writing that is vigorously enforced is the making use of already written ISO standards when working on a new standard. Thus when ISO/TC172/SC1/WG2 was working on ISO 10110 - Indications in optical drawings[4], a standard on how to represent an optical element on a mechanical drawing, all the previously adopted mechanical drawing practice was used *in toto*. All the ISO mechanical drawing standards published through 1988 are bound together in two volumes. These are *ISO Standards Handbook 12, Technical drawings* and *ISO Standards Handbook 33, Applied metrology - Limits, fits and surface properties*[3].

Other things that are considered in the actual standard are the use of series of numbers in logarithmic progressions, for example. These progressions appear when preparing tables of, say, surface roughness. If one wants to cover a wide range of possible roughness values in a table, it makes more sense to divide the range of numbers this way rather than use a linear division. Also, the standards make use of tables of fits for shafts and holes such that if one knows the type of fit desired, it is simply necessary to look up the tolerance value in a table such as contained in ISO 286 - *ISO system of limits and fits*. This is the origin of the tolerances seen on European drawings where there is a dimension followed by, say, h8. This effectively means plus 0, minus something depending on the diameter of the part for shafts and external features. There is a corresponding H8 for the hole tolerance.

Somewhat along the same lines, many standards have default tolerances built into them. These are the tolerances on features not indicated on the drawing *per se* but covered by the default. The drawing should indicate that the default convention is being used. The default tolerances have the effect of making sure that non-toleranced features in fact have some reasonable but not tight tolerances.

For example, ISO 10110, Part 11 has default tolerances for 10 different features of lenses. These are all loose enough that 2 elements each made to the default tolerances would not work well together in a single lens cell, but at least the lenses would fit into laboratory type lens mounts so they could be adjusted relative to one another.

TECHNICAL ASPECTS OF OPTO-MECHANICAL STANDARDS

Now that we have covered many of the details of how standards get to be written and what some of the ground rules are for writing them, we turn to some of the technical details of interest to the designer and fabricator. Here we will not go into any great detail, but rather indicate the type of standards material available for study and use. Since they were written first and make up the foundation of opto-mechanical design, we will cover the ISO mechanical standards first.

As we already mentioned, the *ISO Standards Handbooks 12 and 33* contain all the mechanical drawing standards published through 1988. The ISO standard most similar to the US ANSI Y14.5M - 1982, *Dimensioning and Tolerancing* is ISO 1101:1983[5]. The forward to ANSI Y14.5M lists the few differences from ISO 1101 that existed in 1982. Those differences are being reduced steadily as the US and ISO work to "harmonize" these 2 standards. In addition to ISO 1101, however, there are 10 other ISO standards, all relating to various aspects of dimensioning, tolerancing and inspection.

Perhaps the most useful of these is ISO/TR 5460-1985 - *Geometrical tolerancing - Verification principles and methods - Guidelines*[5]. The "TR" in the title designation indicates that this is a technical report rather than a standard although it could be adopted as a standard if there were the consensus to do so. In any case, what makes this useful is that it is a companion to ISO 1101. For every feature designation in ISO 1101, ISO/TR 5460 shows in schematic form, kinematically correct methods for verifying that each feature meets the specified criteria.

In other words, for every kind of feature control call out, ISO/TR 5460 shows several examples of how to fixture the part to a surface plate and how to set up an indicator to verify that the feature is in tolerance. The effect of this technical report is to make clear in a functional way what each feature control call out means in terms of the instruments used to inspect the feature. Thus the TR 5460 can almost be used as a textbook to illustrate the meaning of the feature callouts in ANSI Y14.5M or ISO 1101. An example from ISO/TR 5460 is shown in Fig. 1 for verifying flatness.

Other ISO standards contained in *ISO Standards Handbook 33* that go far beyond US documents include ones on surface roughness. In the US we have 2 standards on roughness, ANSI B46.1 - 1978, *Surface Texture* and ANSI Y14.36, *Surface Texture Symbols*. In the *ISO Standards Handbook 33*, there are 4 standards dealing with definitions related to surface roughness, one on how to specify roughness and one on how to use a stylus instrument for measuring roughness. In addition, there are 2 standards covering 2 different methods of measuring roundness, either using a rotary table and stylus or using a V-block and indicator.

Principle 1 — Verifying flatness deviations by comparing with a flat element

Symbol	Tolerance zone and application example	Verification method	Comments
▱	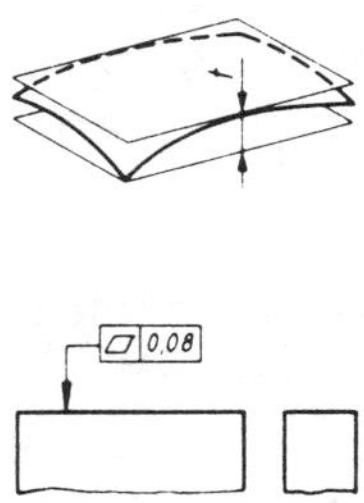	**Method 8.1.3** 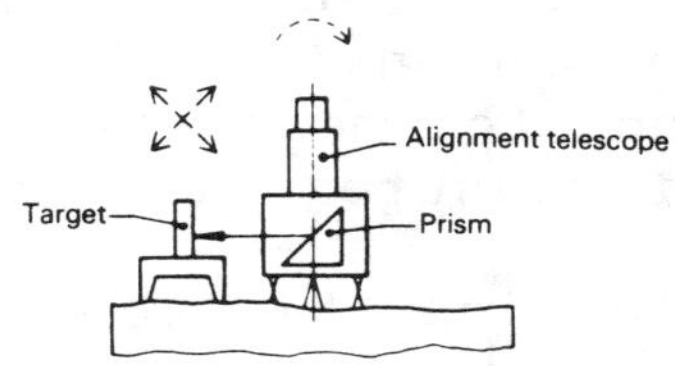 Place the alignment telescope on the object. Align the rotating axis perpendicular to the superimposed surface of the object. The flatness deviation is the maximum difference from the calculated superimposed surface.	This method is suitable for large surfaces. The alignment of the rotating axis can be corrected mathematically.
	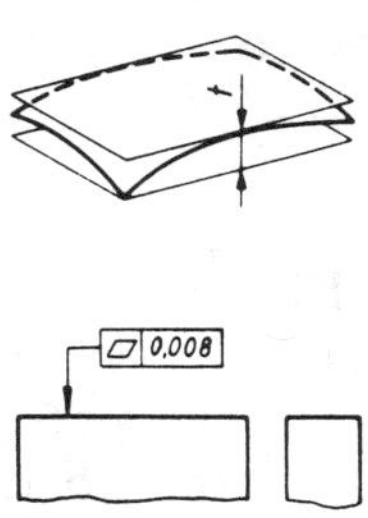	**Method 8.1.4** 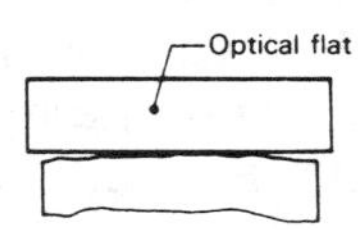 Place the optical flat on the object and observe it in monochromatic light. The flatness deviation is the number of interference lines counted, multiplied by $\lambda/2$ of the light used. $\left(\frac{\lambda}{2} \approx 0{,}3\ \mu m\right)$	This method demands a highly reflective surface. This method is practical only for small objects with flatness deviations up to 20 µm, depending on the size of the optical flat. The optical flat should be adjusted to the object in such a way that the deviation is minimized.

Fig. 1 An example of two of the methods for verifying flatness deviations as shown in ISO/TR 5460. (Example taken directly from ISO/TR 5460.)

The last part of *ISO Standards Handbook 33* contains 14 standards and recommendations on using the measuring instruments of precision engineering and inspection such as dial indicators, vernier calipers, external micrometers, roughness comparison standards and gauge blocks. The standard on gauge blocks, ISO 3650-1978, is ten pages long and gives information on 4 grades of gauge block accuracies, methods for measurement of their length and conditions of measurement.

Before leaving this topic, we should mention 2 US publications that bear on ANSI Y14.5M. These are a book called *Geo-metrics II*[6] by Lowell W. Foster and a text/workbook called *Design Dimensioning and Tolerancing*[7] by Bruce A. Wilson. Mr. Foster was vice chairman of the ANSI Y14.5 committee when the 1982 version of the national standard was published and remains active on the committee. He is also the head of the US delegation to ISO/TC10, the Technical Committee that wrote ISO 1101, among others. In his preface, Mr. Foster says "*Geo-metrics II* is dedicated to the promoting of standardization of engineering drawing techniques in support of ANSI Y14.5. This text is also dedicated to furthering worldwide efforts through the ISO and its standards and programs such as ISO/TC10/SC5 - Dimensioning and tolerancing".

The other book is written more like a textbook than Mr. Foster's and is a little more basic. If one did not know anything about ANSI Y14.5, Wilson's book is a good place to start. He considers ANSI Y14.5 the authoritative document and says that his book simply expands on the explanations given there. We think he is being rather modest.

INTERNATIONAL OPTICAL STANDARDS

We move on now to describe the tangible effects of the work of ISO/TC172 as it affects opto-mechanical design. At the moment there are 12 published ISO standards concerning optics and about 150 drafts of standards. Since we cannot possibly cover all these even by just mentioning the titles, we will give a general idea of what is available and how to get a listing of the individual titles.

All of the published and draft ISO standards concerning optics *per se* are listed on the SPIE electronic mail bulletin board, OPTOLINK. OPTOLINK maybe reached by dialing (206) 733-2998 by computer modem. If this is the first time calling OPTOLINK, you must give yourself a password. Once into the system, the main menu will let you select "Member services" and the second of these is "Standards". The listing of standards may be viewed screen by screen or downloaded to your computer so that you may look over the listing at your leisure. As the listing will tell you, the published ISO standards are available through ANSI[3] and the drafts from NAFM[4].

FUNDAMENTAL STANDARDS

ISO/TC172/SC1 - Fundamental standards, has been one of the most active SC's. There are 3 WG's in this horizontal standards SC working on Optical testing, WG1, Indications in optical drawings, WG2 and Environmental testing, WG3. Much of the work of WG1 has been standards on OTF definition and methods of testing. There are also some standards on the measurement of distortion and veiling glare.

All of WG2's time has been spent working on a 13 part optical drawing standard, ISO/DIS 10110. When adopted, this will be the defining document on how to do a drawing representing an optical element or subassembly. The standard includes how to indicate the mechanical features of an element that are unique to optics, how to specify the material properties, figure, centering and surface imperfections. There are also sections on indicating surface roughness or texture including a power spectral density method, coatings, default tolerances and laser damage threshold. Fig. 2 shows a lens element drawn according to the principles of ISO 10110.

ISO 10110 is a parallel standard to ANSI/ASME Y14.18M-1986, *Engineering drawings and related documentation practices - Optical parts*, and the surface imperfection part of ISO 10110 is parallel to ANSI PH3.617-1980(R1985), *Optical elements and*

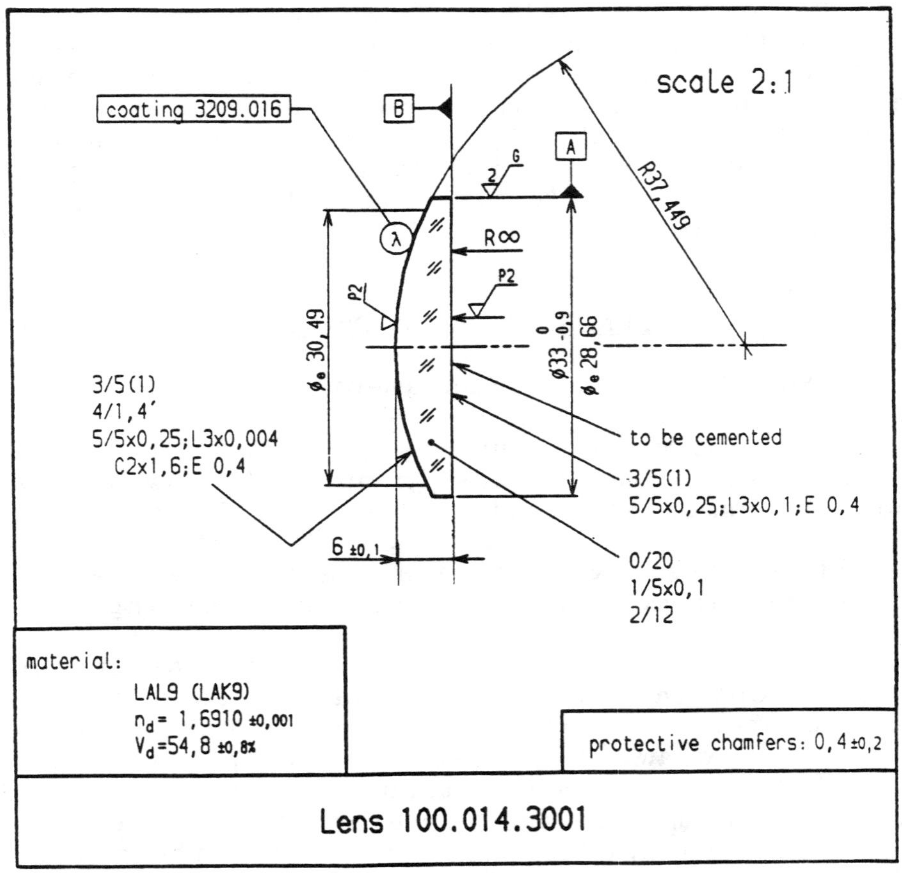

Fig. 2 An example of a simple lens drawing done according to ISO/DIS 10110, Part 1. Notice that the specifications concerning strain birefringence, bubbles, inhomogeneity, striae, figure, centering, surface and coating imperfections (or beauty) and surface roughness are all given in symbolic notation so that notes concerning these specifications do not have to be translated from English.

assemblies, definitions, methods of testing and specifications for appearance of imperfections[3].

In WG3, they have been working on the specification of 20 or so tests of environmental testing conditions applicable to optical instruments. They are now beginning to write vertical requirements standards for various types of optical instruments. These standards specify the severity of the environmental tests depending on the type of service the instrument is likely to see in the field.

OPTICAL MATERIALS

There are 2 active WG's in SC3, one dealing with clear optical glass, that is, not filter glass, and on dealing with optical thin film coatings. WG1 has written several standards about the tests that specify the durability of optical glass under various climatic conditions but otherwise has not done too much work.

On the other hand, WG2 has been quite active and has written a 4 part standard, ISO/DIS 9211 on optical coatings. The 4 parts cover a definition of terms, the specification of the optical properties of thin film coatings, the environmental durability of coatings and a set of tests to demonstrate this durability. This document is far more encompassing than all the US MIL coating specs put together.

MICROSCOPES

Subcommittee 5 has been working on standards relating to microscopes and associated hardware such as slides and cover glasses. There are 5 published standards and 14 drafts from this group.

MEDICAL OPTICS

The title of this SC is somewhat misleading since most of the work concerns ophthalmological instruments and many aspects of the measurement of contact lenses. It also includes chemical and biological test methods relating to contact lenses. Anyone interested in design criteria relating to ophthalmological instruments is sure to find a wealth of information here. There 9 draft documents on these instruments. There are an equal number of drafts relating to the testing of various optical and biological properties of contact lenses.

OPHTHALMIC OPTICS

In this SC there are 3 WG's working on contact lenses, spectacle frames and corrective lenses. There are a total of about 15 drafts rather evenly distributed over the 3 WG's.

ELECTRO-OPTICAL SYSTEMS AND LASERS

This is potentially the most influential SC in TC172. There are 7 WG's in this SC, 6 of them working on laser related standards and one just getting started working on E/O systems. The activity in the laser area has been due to European Community pressure. One of the first things the EC is standardizing is tools and machinery used in production. Since lasers are used for materials processing, there is extreme pressure to work out a suitable set of international standards before the Europeans write their own standards. Since DIN is the Secretariat for both the European Standards Organization (CEN) and for ISO, the Secretariat has encouraged the ISO delegates to work hard and fast so these standards only have to be written once.

The various laser WG's are

WG1 - Terms, test methods and test instruments for lasers,

WG2 - Interfaces and system specifications

WG3 - Safety

WG4 - Laser systems for medical applications

WG5 - Laser systems for general applications

WG6 - Optical components and their test methods.

The work in these groups has progressed in 2

years from no committee to draft international standards in most cases. One has to look to OPTO-LINK just to keep up with the pace of the work in this area.

The work in WG7 - Electro-optical systems other than lasers, is headed by the Japanese Industrial Standards Committee (JISC), the Japanese national committee. They are interested in working on standards related to liquid crystal displays, integrated optics, infrared imaging and interferometry, both white light and laser sources. They will be starting with definitions and terminology and working into interfaces and quality requirements.

This covers the specifics of what is being worked on at present in international standards for optics and optical instruments. As is readily apparent, there is activity in almost all areas of optics and as standards relating to quality requirements are adopted, it will have a great impact on everyone in the optics field. While it will take a time to adjust to the new standards, the benefits in the long term are bound to be positive.

ADOPTION OF INTERNATIONAL STANDARDS

There are 2 issues concerning the adoption of new international standards, one is the legalistic issue and the other is the practical one. For thoroughness we will treat the first issue briefly but spend most of our time on the second, really important issue.

There are really 2 stages of adoption of an international standard. The first stage is a vote by all the national committees that participated in the writing of the standard as to whether or not to make it an international standard. The committees are saying at this point, yes, we agree that this is the standard we all wrote together and that our countries have no overwhelming objection to it. This vote does not have to be unanimous; "If 75% are cast in favor of the DIS, it is accepted for publication as an International Standard"[8].

At this point, the national committee in each country can decide whether or not to adopt the standard as a national standard. In the US, the ANSI/NAPM OP Committee would be the body that would actually vote on this issue after an announcement had been made in the ANSI *Standards Action*[3] and a public meeting has been held. Even prior to such a vote, the US military could, for example, adopt ISO 10110 as the new MIL spec for optical drawings. This action would make ISO 10110 a "mandatory" standard as far as the military were concerned.

Before we leave this discussion of the procedures followed during the adoption of international standards, we should point out 2 other very useful documents. The first is the 3 volume set of *ISO Directives*[3] that describes in detail the entire ISO standards writing process. The first volume, *Procedures for the technical work*, describes the make up and management of the working groups doing the standards writing work. Included are many pages of sample forms for the secretariats to use in assigning work and distributing the results of that work.

The second volume, *Methodology for the development of International Standards*, goes into the technical details of how an ISO standard should be written. There is information on units, symbols, and the mention of "standard reference materials". The third volume, *Drafting and presentation of International Standards*, deals with the physical layout of the standard and addresses such items as paragraph numbering, topic ordering and references to other ISO documents. This set of documents is necessary for anyone with the responsibility of a WG leader or higher in the ISO standards writing administration.

The other document needed by US people working in the international standards area is the ANSI publication, *Criteria for the Development and Coordination of the U.S. Positions in the International Standardization Activities of the ISO and IEC*[3]. This publication outlines how domestic standards writing and administrative groups in the US interface through ANSI with the ISO. It gives the rules for determining if there is US interest in certain technical areas, how to make sure that interest is broadly represented and how to present that interest in the ISO forum.

On the much more germane question of how do standards get adopted in a practical sense, we make 2 observations; companies that want work bid on what they are asked to make, and if a new optical drawing standard were put in place by *fiat* on some arbitrary day, there would be chaos. Taking a harder look at these points, if a drawing comes into a shop done in a new or peculiar format, the shop owner has 2 choices; he can return the drawing without bidding or he can learn what the new format means and then bid accordingly.

However, the shop owner will never make a part to a new standard unless he gets a drawing done in the new standard. For this reason, if the new standard is used, it will first be used by the people doing the design work. Adoption of a new optical drawing standard will not be a grass roots process, it must come from the top down.

Regarding the second point, nothing as complicated as the ISO 10110 drawing standard can be adopted overnight. It will take weeks of study to get proficient with each of the 13 parts. It will be similar for other new ISO opto-mechanical standards. It seems that the sensible approach to any sort of change over from one system to another is to use the old system for all those features that can be adequately communicated with the old system.

For features that cannot be indicated using the old system, use the new system if it has a feature that describes what is desired. Note on the drawing by an asterisk (or some other appropriate symbol) the feature(s) indicated by the new standard and list the standard in the title block along with the old.

When a shop gets such a drawing, they do not immediately have to learn a whole new system. They have one or 2 new items to learn, a much simpler process and one less likely to illicit a "No Bid".

Presumably the shop manager will get a copy of the new standard, just as he would if a new MIL spec were listed, and see what is being asked for. If it is within his capabilities, he is now in an advantageous position over those shops that did not want to take the time to learn something new.

The next time he gets a drawing with these same new call outs, he is ready to go to work. If a new feature is added next time, it is still not a problem. Of course, as the new standard is used more and more widely, the learning process goes on. Before long, the shop will be familiar with the whole new standard because by the time you have become familiar with half of it, it is a little hard to ignore the rest of it.

To make the learning process easier, the Optical Society of America is going to publish a "Handbook"[9] on ISO 10110 that takes each of the 13 parts and covers them in detail. The Handbook will interpret the standard and give the rationale for doing things as they were done in the standard. It will give examples of the application of the standard and contrast it with other standards the worker may already be familiar with. The Handbook will be to ISO 10110 as a book on using DOS is to the DOS Reference Manual. One has all the definitions and formats, the other tells how to use it.

CONCLUSION

We have shown that there are many international mechanical and optical standards either published or being reviewed as drafts. These standards and draft standards are readily available to use as guidelines for new design work. It was indicated that these new standards will not be used if they are not brought bear at the highest levels of design. Management must encourage the use of these international standards so that newly designed products will have access to all world markets.

It was also indicated that the adoption of the new international standards should not take place all at once but in a gradual move away from the present way of doing business. The new standards should be used where they can convey unambiguous meaning as to what is desired and where it is impossible to convey this meaning with the older standards or methods of doing things. As companies and their workers gain experience with the new standards, the switch to using all international standards will be that much easier.

REFERENCES

[1]*Statutes and Rules of Procedure*, International Electrotechnical Commission (IEC), Geneva, Switzerland, 1986, p. 5.

[2]*Op cit.*, p. 5.

[3]Available from ANSI, 11 West 42nd St., NY, NY 11036.

[4]Available from the National Association of Photographic Manufacturers (NAPM), 550 Mamaroneck Ave., Harrison, NY 10528.

[5]Contained in *ISO Standards Handbook 33*.

[6]Addison-Wesley, Reading, MA, 1986.

[7]Goodheart-Willcox, South Holland, IL, 1992.

[8]*ISO Memento*, 1990, Geneva, p. 5.

[9]Available from the Optical Society of America, 2010 Massachusetts Ave. NW, Washington, DC 20036.

Design of the Optomechanical Interface

Chair
Roger A. Paquin
Advanced Materials Consultant

Advanced considerations of the lens-to-mount interface

Paul R. Yoder, Jr.

Consultant in Optical Engineering
1220 Foxboro Drive, Norwalk, Connecticut 06851

ABSTRACT

The most frequently used lens-to-mount interfaces involve direct contact of shaped shoulders, spacers and/or retainers onto the polished lens surfaces or onto ground bevels. In some cases, the lens rim may contact a machined surface of the mount. Elastomeric suspension of the lens in the mount is a possible alternative design. Surface-contact mount types reviewed here include "sharp corner", tangent and toroidal interface versions using retaining rings to hold the lenses against shoulders or spacers. Means for conservatively estimating stress build-up with-in the lens due to axial preload at assembly and at extreme temperatures are suggested. Examples showing the influences of different material characteristics (such as coefficients of thermal expansion and Young's modulus) and of both positive and negative changes of temperature from that existing at assembly are discussed.

1. INTRODUCTION

A reliable lens-to-mount interface design should constrain the lens throughout the survival temperature range and under all anticipated shock and vibration loadings without inducing potentially damaging amounts of stress into the optical or mechanical materials. Under operating conditions, changes in surface locations and orientations should conform to the allowable decentration, despace and tilt budgets while the stresses should result in acceptable levels of surface deformations and induced birefringence. As discussed elsewhere during this conference,[1] costs of the parts and of the labor needed to assemble and align the optics are also important factors to be considered during the design process. These are frequently influenced by the optomechanical interface design.

In this paper, we describe some of the more successful types of optomechanical interfaces for rotationally symmetric optics such as lenses and windows and we review approximate analytical methods used to assess the suitability of these alternative designs. These methods are then illustrated with specific examples to show the influences of key design and material parameters. While the treatment of forces and stresses given here is somewhat simplistic, it can be used to compare alternate designs or to assess the need for experimental verification or for more rigorous analyses using, for example, finite element analysis methods.

Although we refer here to the optical parts as though they are always lenses made of glass and to mechanical parts (housings, cells, spacers, retainers, etc) as if they are always made of metal, it should be understood that many of these mounting considerations also apply to windows or filters or to rotationally-symmetric mirrors and to other materials such as plastics or crystals.

A force acting in the axial direction exerted by the mount onto the component is called axial preload. Frequently, a modest axial preload is created at assembly since this tends to prevent motion of the component relative to the mount under external forces such as acceleration. Forces exerted in the radial direction at the rim of the component are sometimes referred to as "hoop" forces. Both types of compressive forces tend to introduce stress into the glass and into the mount so should be kept within acceptable limits.

A glass lens can usually withstand compressive stress as large as $3.45X10^{8}$ N/m^{2} without failure so this value is generally accepted as a "rule-of-thumb" survival tolerance.[2] In some designs, lens surfaces can be placed in tension as a result of applied forces. A commonly accepted tolerance for tensile stress is $6.9X10^{6}$ N/m^{2}. Under operating environmental conditions, distortions of optical surfaces due to mounting stresses can degrade performance. No simple means for predicting refracting or reflecting surface deformations due to external forces are currently available. Techniques such as finite element analysis are usually employed to evaluate such deformations. Systems using polarized light may be sensitive to stress-induced birefringence within the refracting material. A "rule-of-thumb" tolerance for stress in such cases is $3.45X10^{6}$ N/m^{2}.

2. TYPICAL INTERFACE CONFIGURATIONS

The most frequently used lens-to-mount designs involve axial contact between annular areas on the mount and on the polished lens surfaces or on bevels ground into those surfaces. Radial contact between the rim of the lens and the inside diameter (ID) of the mount also may occur. The stresses within the materials depend upon the geometric shapes of the contacting surfaces, the forces driving the surfaces together and the elastic properties of the materials used.

Use of a threaded retaining ring to secure a lens against a reference surface such as a shoulder or a spacer within a cell or housing offers advantages over other types of lens mounts in that this type mount can easily be assembled and disassembled without damage, it automatically compensates for lens edge thickness variations, it can provide a reasonably predictable axial preload and it is compatible with sealing with an injected sealant or with an O-ring.

Figures 1 through 4 illustrate concepts for interfaces between a cell shoulder, a lens and a threaded retainer. The simplest provides line contact at a "sharp corner" around the periphery of each lens surface. As shown in Fig. 1, this corner can be a 90° intersection between a cylindrical hole and a plane or an obtuse angle such as a 135° intersection between a conical hole and a plane. Acute corner angles should be avoided. A "sharp corner" interface can be used with either convex or concave lens surfaces. Accuracy of the actual intersection angle is not essential; errors of ±2° usually are tolerable.

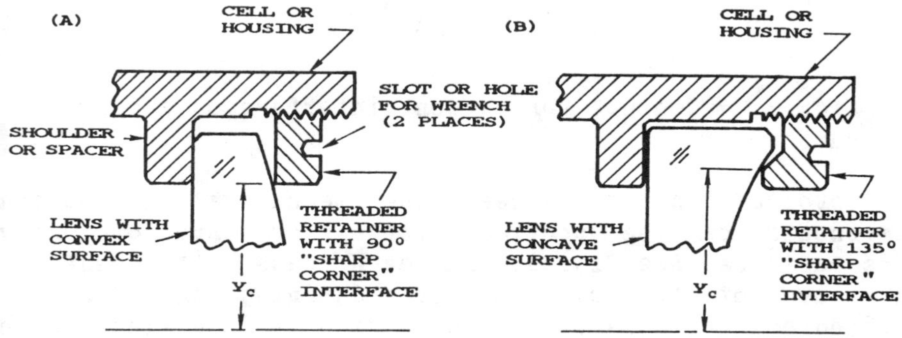

Fig. 1 - Concepts for "sharp corner" interfaces on (A) convex and (B) concave surfaces.

Good machining practice calls for the "sharp corner" to be burnished slightly to minimize burrs or nicks. Typically, a radius of the order of 0.05 mm then results.[3] A tendency for better (i.e., smoother) sharp edges to result from machining an obtuse angle intersection was reported by Hopkins.[4] The radial height of line contact, y_c, usually is just slightly greater than the radius of the lens' clear aperture.

Tangential contact at a height y_c, is illustrated in Fig. 2. It can be used only with convex lens surfaces. The half-angle, φ, of the right-circular cone in such an interface is given by the following equation:

$$\varphi = 90° - \text{arc sin } (y_c/R) \qquad (1)$$

where R is the surface radius of curvature and y_c is typically one-half the arithmetic average of the lens' clear aperture and its OD. The typical tolerance on φ is $\pm 2°$.

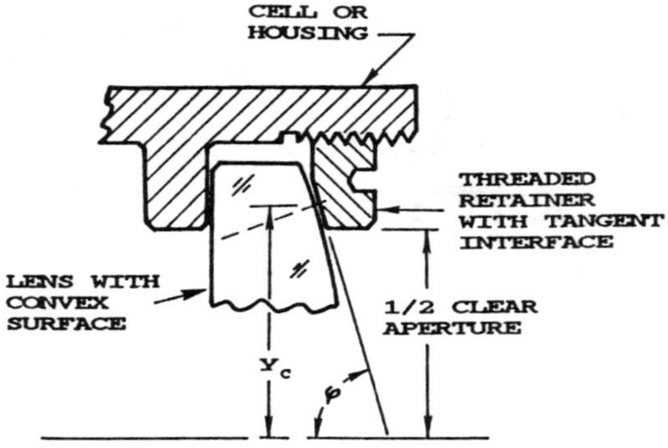

Fig. 2 - Concept for tangent interface on a convex surface.

Toroidal contact between the mechanical and optical components can be provided for either convex or concave lens surfaces. See Fig. 3. In each case, the center of curvature of the circular arc defining the "donut" is located off the axis by the dimension y_T and lies on the local normal to the lens surface passing through a point at height y_c defined as for tangent contact. Preferred values for the sectional radius of the toroids are 10R for a convex surface and 0.5R for a concave surface, where R is the radius of the lens surface.[5]

If the mechanical interface with the lens is lapped to the same spherical radius as the lens surface within a few

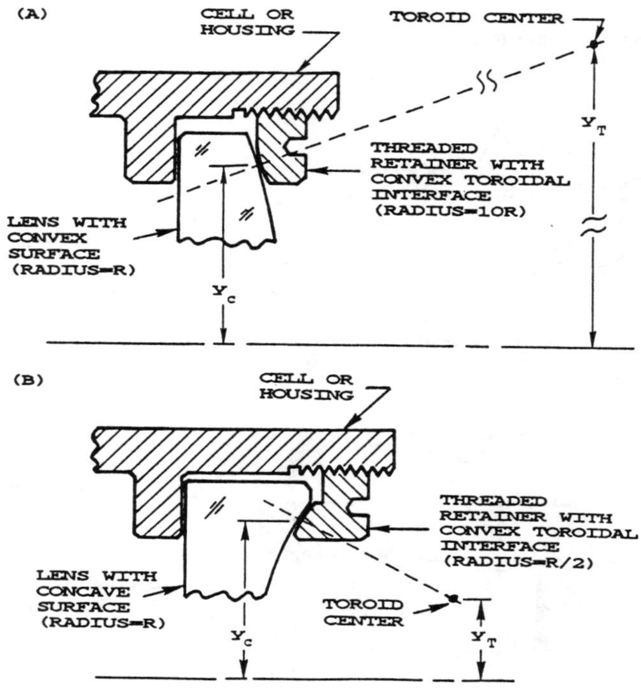

Fig. 3 - Concepts for toroidal interfaces on (A) convex and (B) concave surfaces.

fringes of visible light, a spherical interface mount can be produced. See Fig. 4. If the radii of the mechanical and optical surfaces do not match adequately, the interface will degenerate into line contact at the inner or outer edge of the annular mechanical area. Manufacture of the spherical interface mount is costly due to this need for radius matching. Hence, it is generally used only in special designs.

In each of the above described lens mounts, contact of the metal occurs on polished glass surfaces. These surfaces are usually accurately made and well aligned to each other due to the inherent nature of the optical grinding and edging processes. In most lenses, contact occurs at a zone where the optical surface is inclined with respect to the axially-directed force imposed by the retainer. Radial components of that force tend to center the lens with respect to the axis. The magnitude of this centering tendency depends upon the radius of curvature of the lens surface with sharply curved surfaces centering more easily.

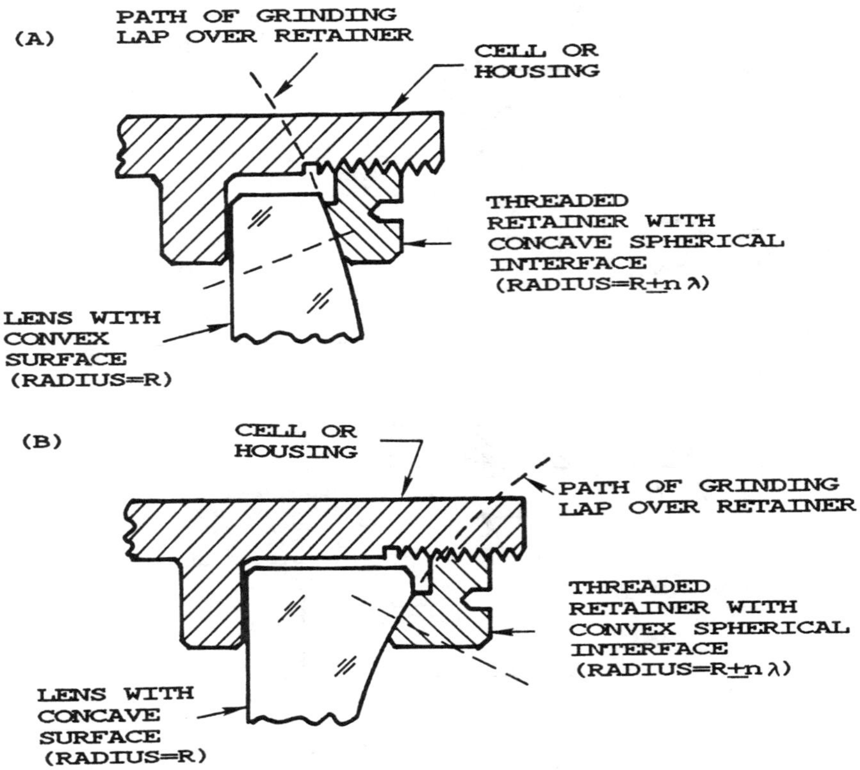

Fig. 4 - Concepts for spherical interfaces on (A) convex and (B) concave surfaces.

Contact against flat bevels ground into the edges of the lens' spherical surfaces can be provided as shown in Fig. 5. Such interfaces have the advantage of distributing axial preload due to the clamping action of the retaining ring over a large area on the lens thereby reducing stress build-up within the glass. They have the disadvantages of not being able to help self-center the lens and of referencing against secondary surfaces of potentially reduced angular accuracy as compared to the polished lens surfaces. While accurate flat bevels can be produced, they add significantly to the cost of the lens.

Resilient materials such as room temperature vulcanizing (RTV) elastomers are frequently used to seal optical components into their mounts with or without hard mechanical constraint. The resilient layer can provide some degree of protection against shock and vibration as well as some thermal isolation. Epoxy or urethane compounds are sometimes used for the same purpose. These materials are generally somewhat less resilient than RTV compounds, but they are superior in regard to structural bond strength.

Figure 6 illustrates a common technique for mounting a lens in an annular ring of elastomer. Registry of one optical surface against a machined surface of the mount helps align the optic. Centration can be retained by temporary shimming or external fixturing

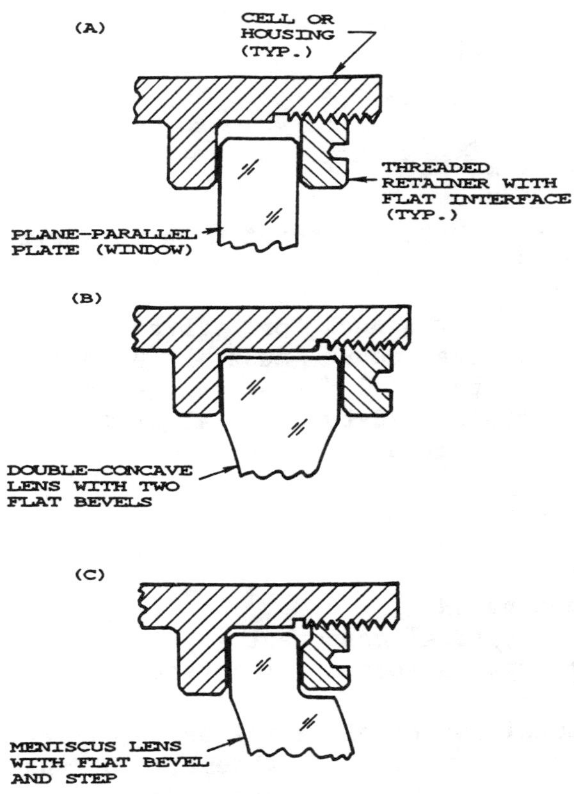

Fig. 5 - Concepts for flat bevel interfaces on (A) plano, (B) convex and (C) concave surfaces.

during the curing process. If the elastomer layer's radial thickness (t_E in the figure) is in accordance with an equation given in Section 4.5, the design becomes relatively insensitive to temperature changes.

3. AXIAL PRELOAD AND AXIAL STRESS EFFECTS

3.1 Effects at Assembly Temperature

If the optic is expected to experience acceleration

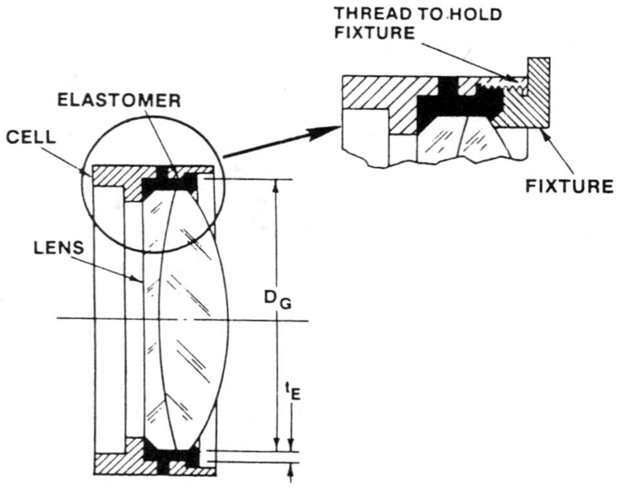

Fig. 6 - Concept for mounting a lens with an annular ring of elastomer. From Yoder.[11]

forces due to gravity, shock or vibration and its axial location within the mount must be held constant during such exposure by the axial force exerted by the retainer, it should be constrained by an axial preload, P_1, as given by the following equation:

$$P_1 = 9.807WA \qquad (2)$$

where:
P_1 is in newtons (N)
W = weight of optical component in kg
A = maximum imposed acceleration factor.

An additional amount of axial preload, P_2, might also be imposed upon the lens at assembly to ensure that the mount does not lose axial contact with the surface if it expands more than the glass as the temperature rises. This preload is discussed in Section 3.3.

If the glass and metal were inelastic, the contact would be along a line reaching around the perimeter of the lens surface at a radius y_c from the axis. Actually, both materials are somewhat elastic and deform slightly under pressure so contact occurs over a rectangular area of width Δy and length $2\pi y_c$. The magnitudes of Δy and of the contact area both vary by the square root of the preload. Large stresses can be created in the glass at the interface between the lens and the mount if the axial force is exerted upon a small area.

If we know the total preload, $P_T = P_1+P_2$, exerted by the retainer against the lens at the zonal contact radius, y_c, during assembly at a temperature of (typically) 20°C, the resulting peak axial stress, S_A, developed within the contact area can be estimated using the following equation from Rourk[6] as explained by Delgado and Hallinan[3] and by Yoder[5]:

$$S_A = 0.798(K_1 p/K_2)^{0.5} \tag{3}$$

where:

S_A is in N/m²

p = linear preload $= P_T/(2\pi y_c)$ in N/mm applied at the line contact near the rim of the lens

y_c = contact height in mm

$K_1 = (D_1+D_2)/D_1D_2$ for a sharp corner contacting a convex surface

$= (D_1-D_2)/D_1D_2$ for a sharp corner contacting a concave surface

$= 1/D_1$ for a conical surface contacting a convex surface (tangent interface)

$= 1.1/D_1$ for a toroid surface of sectional radius $10D_1$ contacting a convex surface

$= 1/D_1$ for a toroid surface of sectional radius $0.5D_1$ contacting a concave surface

D_1 = twice the lens radius of curvature in mm

D_2 = twice the sharp corner radius (typically 0.1 mm)

$K_2 = ((1-\nu_G^2)/E_G)+((1-\nu_M^2)/E_M)$

ν_G = Poisson's ratio for the glass

ν_M = Poisson's ratio for the metal

E_G = Young's modulus for the glass in N/m²

E_M = Young's modulus for the metal in N/m².

Equations for the contact stresses introduced into a lens with flat bevel or spherical interfaces are not included here because those stresses would be relatively insignificant as compared to those with the interfaces listed. Typically, those stresses would equal the total preload divided by the area of contact. For "sharp corner" interfaces, D_2 is so small compared to the radius of most surfaces contacted, the values for K_1 (and hence contact stresses) are essentially the same for convex and concave surfaces. We noted earlier that tangential contact can occur only on convex surfaces. As shown by Yoder,[5] the stresses computed for both toroidal interface cases would be within five percent of those computed for the tangential case and much smaller than those for "sharp corner" interfaces.

Another equation from Rourk[6] can be used to estimate the annular width of the contact area between the elastically deformed surfaces. This is:

$$\Delta y = C(p)^{0.5} \quad (4)$$

where

$$C = 1.6(K_2/K_1)^{0.5}$$

and all other terms are as defined above. The contact area is $2\pi y_c$ times this width.

As shown by Yoder,[7] the axial preload divided by the contact area is an average contact stress which must be multiplied by 1.2768 to give the peak axial stress as computed by Eq. (3).

3.2 Effects at Reduced Temperatures

When the temperature decreases by ΔT from that existing at assembly, the preload exerted by the retaining ring increases by P_3 due to differential contraction of the cell and the lens if the thermal expansion coefficient of the metal α_M exceeds that of the glass, α_G. P_T will then equal $P_1+P_2+P_3$. The contact stress within the glass, S_A' at the lower temperature depends upon the magnitude of this increased preload and can be computed by Eq. (3). The thermally-induced preload depends, in part, upon the cross-sectional areas within the mount and deep within the lens to which that force is applied. With the able help of Alson Hatheway[8] and Ralph Richard[9], the following approach has been developed for estimating these areas and the resultant thermally-induced preload.

Figure 7 shows the geometry of a typical lens cell. The wall thickness around the lens, the inner radius of this wall and the radius to the center of the cross-sectional area are indicated. This area is

$$A_M = 2\pi t_c((D_M/2)+(t_c/2)). \quad (5)$$

Two cases for the geometry defining the affected area within the lens are shown in Fig. 8. The lens is assumed to be clamped between line contacts at radius y_c. The dashed lines form squares so, to reasonable approximation, the stress is limited to the shaded zone of annular width equal to the thickness t_E of the glass at y_c. For the case of Fig. 8(A), this zone lies within the lens and the area is given by Eq. (6).

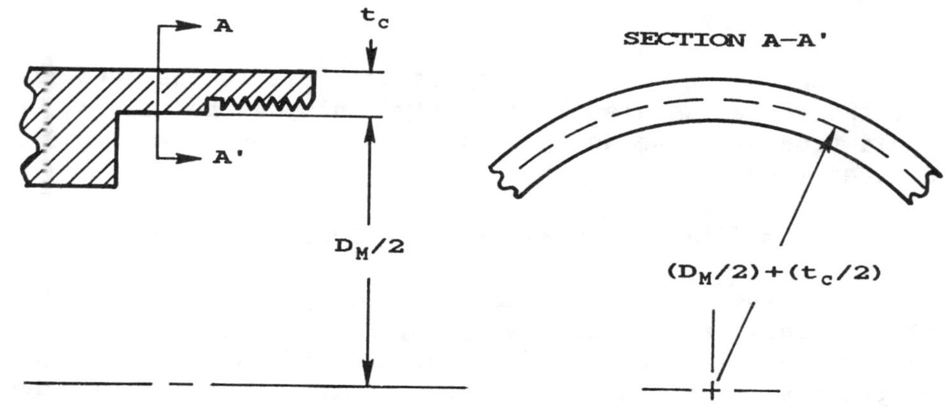

Fig. 7 - Geometric relationships determining the cross-sectional area of a lens cell wall.

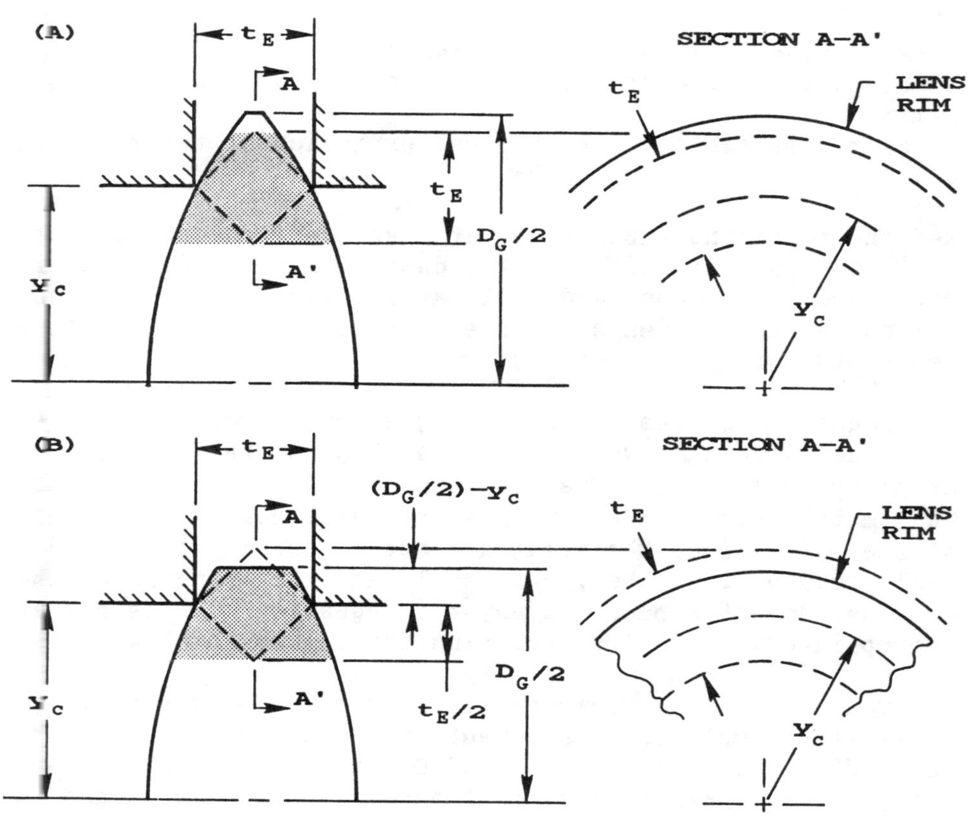

Fig. 8 - Geometric relationships determining the cross-sectional area of the stressed region within a clamped lens; (A) when this region lies within the lens rim and (B) when this region is truncated by the lens rim.

$$A_G = 2\pi y_c t_E . \tag{6}$$

If the line of contact is so close to the lens rim that the dashed square lies partly outside the glass, we should truncate the annular width as shown in Fig. 8(B). The area then is

$$A_G' = (\pi/4)(D_G - t_E + 2y_c)(D_G + t_E - 2y_c). \tag{7}$$

Since the force is not uniformly distributed over either glass area, it is appropriate to use an "effective" area equal to one-half the geometric area to determine the preload, P_3, produced by the temperature change from the following equation:

$$P_3 = K_3 \, \Delta T \tag{8}$$

where:
$K_3 = -E_G A_G E_M A_M (\alpha_M - \alpha_G) / (2)((E_G A_G/2) + E_M A_M)$
α_G = coefficient of thermal expansion of the glass in ppm/°C
α_M = coefficient of thermal expansion of the metal in ppm/°C

and the remaining parameters are as defined above. Note that K_3 is specific to the design since it depends, in part, upon the lens and cell areas. Also, we assume the expansion coefficients to be constant over the entire temperature range of interest.

Equation (8) resembles an equation given by Bayar[10] for the stress developed in a lens as the temperature drops by ΔT. His equation applies to columns of glass and metal with equal cross-sectional areas tied together at the ends. Since the areas are not generally equal in a real design and they, as well as the lengths of the columns, do not appear in Bayar's equation, it is not an appropriate vehicle for computing the thermal preload.

The usual low temperature specification for survival of military optical instrumentation is -62°C; the corresponding value of ΔT is -82°C It should be noted that any preload, $P_1 + P_2$, existing at assembly should be added to that estimated by Eq. (8) before computing the contact stress level at the reduced temperature. This is contrary to the approach suggested by Yoder[11] in which the stresses at assembly and at low temperature were added directly.

The bulk stresses at T_{min} in the glass and in the

metal are here defined as S_{BG} and S_{BM} respectively. They can each be estimated by dividing the total preload, P_T, by $A_G/2$ (or $A_G'/2$ with a truncated stressed region) and by A_M.

3.3 Effects at Elevated Temperature

As the temperature rises, the metal of the mount expands more than the glass component within. Any axial preload existing at assembly temperature T_a (=20°C) will then be reduced. If the temperature rises sufficiently, that preload will disappear and, if not otherwise constrained (as by a sealant), the optic is free to move within the mount due to external forces. Unless the changes in position and orientation of the lens within the small axial and radial gaps created by differential expansion are smaller than the allowable motions for the application, the design should prevent such freedom at the highest temperature that the instrument is to survive without damage. Let us define this as T_c. For military applications, T_c is typically 71°C. Instruments for laboratory, commercial or consumer use may have lower survival temperature requirements. In special applications, T_c may exceed 71°C.

The axial gap developed between the cell and the lens after preload has been released can be approximated as

$$\Delta Gap_A = (\alpha_M - \alpha_G)(t_E)(T_c - T_a) \qquad (9)$$

where all terms are as defined above.

One way of preventing release of preload is to provide a preload at assembly, P_2, such that the axial force on the lens is just reduced to zero at T_c. Defining the temperature reduction from T_c to that at assembly, T_a, as ΔT, we can use Eqs. (5) through (8) and the geometry of Figs. 7 and 8 to estimate P_2. By adding P_1 and P_2 we obtain the total axial preload, P_T, required to constrain the optic against both axial acceleration forces and to prevent complete release of assembly preload at the highest temperature. Note that whenever the temperature drops below T_a, we incur additional preload (P_3) as discussed earlier.

3.4 Retaining Ring Torque Required to Produce a Given Axial Preload

The torque, Q, that should be applied to the retaining ring to introduce a given axial preload, P, to the edge of the lens can be approximated from an emperically-derived formula given by Kowalski:[12]

$$Q = 0.2P_T D_R \tag{10}$$

where D_R is the pitch diameter (in cm) of the thread on the ring and P_T is in N. Q is then expressed in N-cm.

3.5 Bending of the Component due to Axial Forces

If the lines of contact between the mount and optical component occur at different radii from the axis, axial forces applied through those contacts can cause bending of the surfaces. See Fig. 9. One of the surfaces must necessarily be placed in tension; this is the condition in which glass is the weakest. The following equation due to

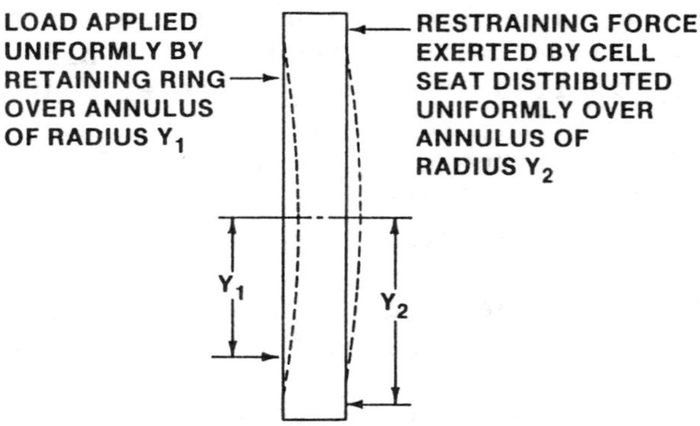

Fig. 9 - Bending forces exerted on a lens approximated by a plate of uniform thickness) when annular contacts are not at the same radial distance from the axis. From Yoder.[7]

Rourk[6] gives the tensile stress induced by bending:

$$S_T = \frac{3P_T}{2\pi m t_E^2}\left[0.5(m-1) + (m+1)\log\frac{y_2}{y_1} - (m-1)\frac{y_1^2}{y_2^2}\right] \tag{11}$$

where:

S_T = tensile stress in the component in N/m²
P_T = applied axial load in N
m = 1/Poisson's ratio
t_E = edge thickness of component in mm
y_1 = innermost contact radius in cm
y_2 = outermost contact radius in cm.

As mentioned earlier, the general tolerance for tensile stress is 6.9×10^6 N/m².

4. RADIAL EFFECTS

4.1 "Hoop" Stress at Reduced Temperature

In all designs considered above, radial clearance was assumed to exist between the optic and the mount. In some designs, this clearance is the minimum allowing assembly so, at some reduced temperature, the metal touches the rim of the optic and, at still lower temperatures, a so-called "hoop" stress develops. The magnitude of this bulk stress, S_R, for a given temperature drop, ΔT, can be estimated as[7]

$$S_R = K_4 K_5 \Delta T \tag{12}$$

where:

$K_4 = (\alpha_M - \alpha_G)/((1/E_G) + (D_G/2E_M t_C))$

$K_5 = 1-((2\Delta r)/(D_G \Delta T(\alpha_M - \alpha_G)))$

D_G = lens OD in mm

t_c = mount wall thickness outside the rim of the optic in mm

Δr = radial clearance in mm.

If Δr exceeds $D_G \Delta T(\alpha_M - \alpha_G)/2$, the lens will not be constrained by the cell ID and hoop stress will not develop within the temperature range ΔT due to rim contact.

4.2 Radial Stress Within the Cell Wall

As another consequence of differential contraction of the cell relative to the lens, bulk stress is built up within the metal in accordance with the equation:

$$S_M = S_R D_G / t_C \tag{13}$$

where all terms are as defined above. With this expression, we can determine if the cell is strong enough to withstand the force exerted upon the lens without exceeding its elastic limit. If the yield strength of the metal exceeds S_M, a safety factor exists.

4.3 Combined Axial/Radial Stress at Reduced Temperatures

Bayar[10] indicated that the combined effect of axial and radial bulk stresses within the lens can be estimated as the root sum square (rss) of the orthogonal stresses. This is most significant at the worst case reduced temperature. Hence:

$$S_{rss} = (S_{BG}^2+S_R^2)^{0.5} \tag{14}$$

where all terms are as defined above. This value should be compared to the allowable compressive stress for the glass (typically 3.45×10^8 N/m^2) to determine a safety factor for the design.

Similarly, the axial bulk stress within the metal can be combined with the radial stress in the cell as

$$S_{rss} = (S_{BM}^2+S_M^2)^{0.5}. \tag{15}$$

4.4 Stress-Induced Birefringence

The unit optical path difference (OPD/cm) for the two states of polarized light (perpendicular and parallel) due to a given amount of mechanical stress at a given temperature can be determined from:

$$OPD/cm = 10K_s S \tag{16}$$

where
OPD is in nm
K_s = stress optic coefficient of the glass in mm^2/N
S = bulk stress level in N/m^2 = S_{BG} computed with area A_G or A_G' as appropriate.

The OPD computed by Eq. (16) is that existing within the annular region shown shaded in Figs. 8(A) or 8(B) at the edge of the lens where the axial stress is concentrated. Towards the center of the aperture, the birefringence is significantly reduced. Unfortunately, no simple means to estimate its magnitude directly for a given lens configuration is currently available.

4.5 Growth of Radial Clearance at Increased Temperatures

The increase in radial clearance, ΔGap_R, between the optic and the mount due to a temperature increase of ΔT from that at assembly can be estimated by the equation:

$$\Delta Gap_R = K_6 \Delta T \tag{17}$$

where: $K_6 = (\alpha_M - \alpha_G)D_G/2$ and all other terms are as defined above.

4.6 Elastomeric Mounting

As mentioned in connection with Fig. 6, resilient

materials such as elastomers are frequently used to suspend optical components in their mounts. If the resilient layer has a particular thickness, the assembly will be athermal in the radial direction thereby resisting stress build-up due to differential expansion or contraction. This thickness is

$$t_E = (D_G/2)(\alpha_M - \alpha_G)/(\alpha_E - \alpha_M) \qquad (18)$$

where α_E is the thermal expansion coefficient of the elastomer in ppm/°C and all other terms are as defined above.

Valente and Richard[13] reported an analytical technique for estimating the decentration Δ of a lens mounted in a ring of elastomer when subjected to radial gravitational loading. Their method can be expanded to include more general radial acceleration forces resulting in the following equation:

$$\Delta = AWt_E/(\pi Rd((E_E/(1-\nu_E^2))+E_S)) \qquad (19)$$

where:

Δ is in mm
A = acceleration factor
W = weight of optical component in kg
t_E = thickness of elastomer layer in mm
R = optical component OD/2 in mm
d = optical component thickness in mm
E_E = Young's modulus of elastomer in N/m²
E_S = Shear modulus of elastomer in N/m²
ν_E = Poisson's ratio of elastomer.

The decentrations of modest sized optics corresponding to normal gravity loading are generally quite small, but will grow under shock and vibration loading. Fortunately, the resilient material will tend to restore the lens to its unstressed location and orientation when the acceleration loading dissipates.

5. MOUNTING EFFECTS UNDER OPERATING CONDITIONS

5.1 Distortions of Optical Surfaces

In the above discussions, axial forces were assumed above to be applied uniformly over a contact area on the lens. This may not be the case if the interfacing surfaces on the cell, spacer and/or retainer are not as true and smooth as the glass surfaces. In general, the glass will contact the three highest points on the metal first and the weaker component (usually the lens) will distort under high loads until sufficient area comes into contact to establish equilibrium.

Within limits, the distorting effects of nonuniformly applied axial forces can be reduced by making the mechanical component touching the glass somewhat resilient. Some techniques for doing this are illustrated schematically in Fig. 10. Note that such flexibility introduces a spring into the design that must be taken into account in the computation of preloads and resultant axial stresses. In fact, omission of resiliency of mechanical parts such as threaded retainers is one of the reasons the above analysis of axial stress is so conservative.

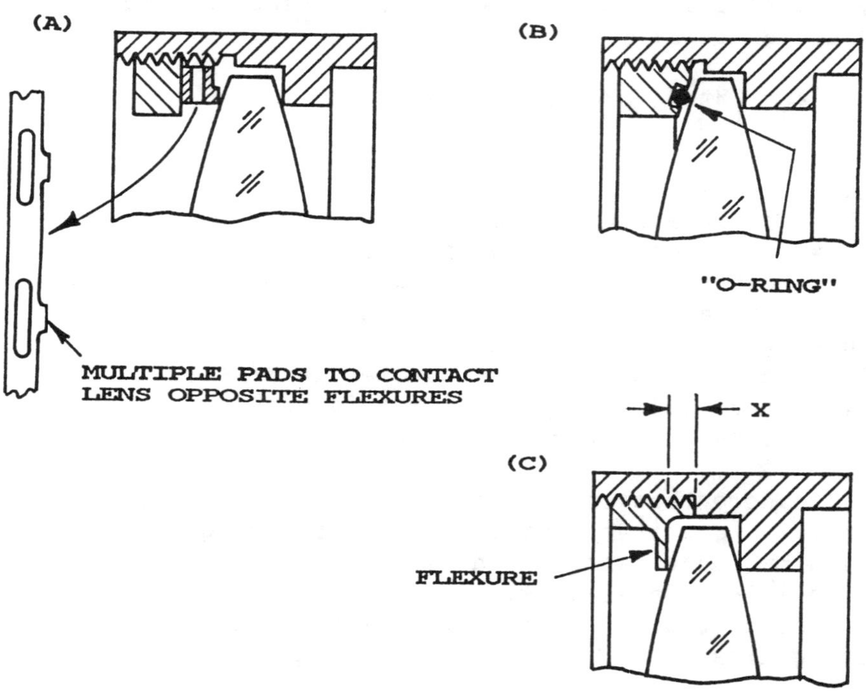

Fig. 10 - Concepts for resilient mounting interfaces to more uniformly distribute preload forces around the perimeter of a lens. From Yoder.[7]

Analytical means for predicting the surface distortions for a given optomechanical interface design, such as finite element analysis, exist. They require detailed knowledge of the shapes of the contacting surfaces and the stiffnesses of all the optical and mechanical parts.

5.2 Stresses Under Operating Conditions

Any of the above equations for stress induced into an optical component by forces exerted by the mount can be applied to operating conditions as well as worst case conditions. In general, the stress levels will be reduced, but they still may be significant in terms of their effects upon system performance. Surface deformations due to mount-induced forces may be the worst problem, but consideration of such is beyond the scope of this paper.

6. ILLUSTRATIVE EXAMPLES

The lens mount design shown in Fig. 11 is a good example to consider here. The SK15 meniscus lens is held against a shoulder machined into an aluminum cell by a threaded retaining ring with tangent interface. The rim of the lens is ground spherical to minimize the posssibility of damage when inserted into the ID of the cell machined at assembly to clear the glass by only 0.020 mm on the diameter. Table 1 lists the design parameters of importance here.

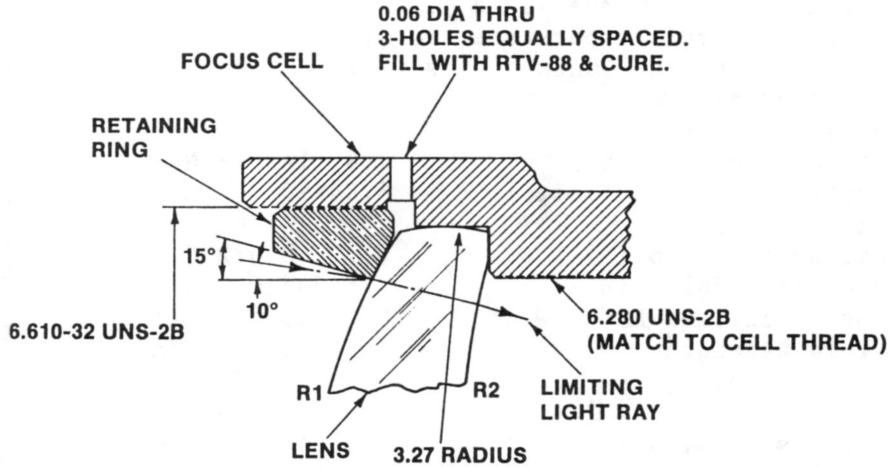

Fig. 11 - Lens mount example analyzed in the text. From Yoder.[14]

Two variations of the design have been analyzed. In the first case (see Table 2), the retainer torque is determined to give enough preload at assembly to hold the lens against 10 G acceleration as well as to clamp it at all temperatures up to 71°C. The gap that otherwise would develop between the cell and lens at that elevated temperature is about 0.006 mm. The preload increases at a rate K_3 = 1410 N/°C so, when the temperature drops to the minimum survival value of -40°C, the preload reaches 1.57×10^5 N. The resulting axial contact stress at T_{min} is 1.38×10^8 N/m². A safety factor for the glass of 2.5 then exists.

Table 1 - Design Parameters for Example

Parameter	Value
Lens Outside Diameter (D_G)	165.9890 mm
Lens Axial Thickness (t_a)	18.9992 mm
Lens Edge Thickness @ y_c (t_E)	6.5786 mm
Surface Radius (R)	220.2180 mm
Lens Weight (W)	0.9612 kg
Cell Inside Diameter (D_M)	166.0093 mm
Cell Wall Thickness (t_c)	4.9784 mm
Interface Type	Tangent
Glass/Metal Contact Height (y_c)	80.8673 mm
Radial Clearance (Δr)	0.0101 mm
Retainer Thread Pitch Diam. (D_t)	167.8940 mm
Max. Acceleration Factor (A)	10 G
Assembly Temperature (T_a)	20 °C
Maximum Operate Temperature (T_{maxop})	52 °C
Minimum Operate Temperature (T_{minop})	-30 °C
Maximum Survival Temperature (T_{max})	71 °C
Minimum Survival Temperature (T_{min})	-40 °C
Elevated Temp. for Contact Loss (T_c)	71 °C
Lens Material	SK15
Poisson's Ratio (ν_G)	0.275
Young's Modulus (E_G)	8.41×10^{10} N/m²
Coeff. Thermal Exp. (α_G)	6.84×10^{-6} ppm/°C
Cell Material	Al 6061
Poisson's Ratio (ν_M)	0.332
Young's Modulus (E_M)	6.83×10^{10} N/m²
Coeff. Thermal Exp. (α_M)	2.36×10^{-5} ppm/°C
Yield Strength	2.55×10^8 N/m²
Cell Material	Ti6Al4V
Poisson's Ratio (ν_M)	0.340
Young's Modulus (E_M)	1.14×10^{11} N/m²
Coeff. Thermal Exp. (α_M)	8.82×10^{-6} ppm/°C
Yield Strength	6.89×10^8 N/m²

Using the geometry of Fig. 8(A) and data from Table 1, we compute the cross-sectional area of the stressed region within the lens and then determine the total preload and axial bulk stress within the glass for the minimum operating temperature of -30°C. Birefringence in that region of the lens is then found to be 1620 nm/cm (see Table 2).

Table 2 - Evaluation of Mounting Effects:
Case 1, Aluminum Cell and High Preload

Parameter	Value
Preload for Acceleration (P_1)	$9.43X10^1$ N
Preload for Contact @ T_c=71°C (P_2)	$7.22X10^4$ N
Total Preload @ T_a (P_1+P_2)	$7.23X10^4$ N
Retainer Torque Required (Q)	$2.43X10^5$ N-cm
Preload Increase per °C (K_3)	$-1.41X10^3$ N/°C
$\Delta T_1 = T_{min}-T_a$	-60 °C
Preload Increase @ T_{min} (P_3)	$8.46X10^4$ N
Total Preload @ T_{min} ($P_1+P_2+P_3$)	$1.57X10^5$ N
Axial Contact Stress @ T_{min} (S_A')	$1.38X10^8$ N/m²
Safety Factor for Glass	2.5
Lens Cross Section Area (A_G)	$3.26X10^3$ mm²
$\Delta T_2 = T_{minop}-T_a$	-49 °C
Preload Increase @ T_{minop}	$6.92X10^4$ N
Total Preload @ T_{minop}	$1.41X10^5$ N
Lens Axial Bulk Stress @ T_{minop}	$8.64X10^1$ N/m²
Birefringence @ T_{minop}	$1.62X10^3$ nm/cm
Cell Cross Section Area (A_M)	$2.67X10^3$ mm²
Cell Axial Bulk Stress @ T_{min}	$5.88X10^7$ N/m²
Cell Radial Stress @ T_{min}	$1.15X10^8$ N/m²
Cell rss Stress @ T_{min}	$1.29X10^8$ N/m²
Safety Factor for Metal @ T_{min}	2.7

At minimum survival temperature, the cell develops a bulk stress of $5.88X10^7$ n/m². Combining this on a rss basis with the radial stress generated at that temperature, we obtain the composite stress of $1.29X10^8$ N/m²; this is one-half the yield strength of the metal and therefore probably reasonable.

The main problem with this example is its great sensitivity to temperature changes. An extremely large axial preload at assembly is needed to hold the lens in place at the high survival temperature. The torque required to achieve this preload is prohibitively large. It is apparent that this design could use some refinement. One would expect the thermal problem to be reduced if the coefficient of thermal expansion of the cell were made to match that of the glass better.

In the second case evaluated, we change the cell to titanium and apply a torque to the retainer of 560 N-cm which is equivalent to 50 lb-in; a moderately heavy, but feasible torque for this size lens assembly. Surface distortions due to mounting forces should be minimal. The mounting effects then are as listed in Table 3.

Table 3 - Evaluation of Mounting Effects:
Case 2, Titanium Cell With Reduced Preload

Retainer Torque Applied @ T_a (Q)	$5.65\text{X}10^2$ N-cm *
Resulting Preload @ T_a (P_T)	$1.68\text{X}10^2$ N
Preload Req'd for Accel. (P_1)	$9.43\text{X}10^1$ N
Preload Available to hold Contact @ T_c (P_2)	$7.40\text{X}10^1$ N
Preload Increase per °C (K_3)	$-2.05\text{X}10^2$ N/°C
$\Delta T_1 = P_2/K_3$	-0.36 °C
$T_c = T_a - \Delta T_1$	20.36 °C
$\Delta T_2 = T_{min} - T_a$	-60 °C
Preload Increase @ T_{min} (P_3)	$1.24\text{X}10^4$ N
Total Preload @ T_{min} ($P_1+P_2+P_3$)	$1.25\text{X}10^4$ N
Axial Contact Stress @ T_{min} (S_A')	$4.40\text{X}10^7$ N/m²
Safety Factor for Glass @ T_{min}	7.8
Lens Cross Section Area (A_G)	$3.26\text{X}10^3$ mm²
$\Delta T_3 = T_{minop} - T_a$	-50 °C
Preload Increase @ T_{minop}	$1.01\text{X}10^4$ N
Total Preload @ T_{minop}	$1.02\text{X}10^4$ N
Lens Axial Bulk Stress @ T_{minop}	$6.26\text{X}10^6$ N/m²
Birefringence @ T_{minop}	$1.16\text{X}10^2$ nm/cm
Cell Cross Section Area (A_M)	$2.67\text{X}10^3$ mm²
Cell Axial Bulk Stress @ T_{min}	$3.97\text{X}10^6$ N/m²
Cell Radial Stress @ T_{min}	none
Cell rss Stress @ T_{min}	$3.97\text{X}10^6$ N/m²
Safety Factor for Metal @ T_{min}	87

* Equivalent to 50 lb-in.

The rate of increase in preload with temperature (K_3) is now $2.05\text{X}10^2$ N/°C; almost 7 times less sensitive than with the aluminum cell. After allocating preload to withstand acceleration at T_a, we have practically none left to combat cell expansion under temperature rise so T_c is only 21.36°C. The contact stress at the minimum survival temperature is reduced to $4.40\text{X}10^7$ N/m² so the safety factor for the glass is 7.8. The birefringence at minimum operating temperature is reduced by a factor of about 7 from that for Case 1 to 116 nm/cm. No radial stress due to rim compression occurs at the low survival temperature so the safety factor for the cell is large (87). The thickness of the cell wall might well be reduced in this case.

Since the thermal expansion properties are fairly well matched in this design, the axial and radial gaps at maximum survival temperature (71°C) are very small. Gap_A is 0.7 µm while Gap_R is increased from 0.020 mm to 0.028 mm. The uncertainties in lens location and orientation due to these clearances are quite acceptable for the particular design investigated here. Injection of an elastomeric sealant around the rim of the lens would tend to hold the lens in place at elevated temperatures. One of several radially-directed holes provided for this purpose is shown in Fig. 11.

7. CONCLUSIONS

We have described here an approach to making engineering estimates of various mounting interface effects upon optical components such as lenses due to assembly preload and some environmental conditions. For simplicity, the equations given here do not include the effects of shear forces within materials or the inherent flexibility of mount components other than the mount (cell) wall. Omission of factors such as these cause the results to be conservative, i.e., to overestimate the magnitudes of the mounting effects. The degree of overestimation has not yet been determined. Further studies comparing results obtained by the methods outlined here to those obtained by more precise methods such as finite element analysis would help quantify this conservatism.

8. REFERENCES

1. Willey, R.R., "Maximizing both production yield and performance in optical instruments through effective design and tolerancing", Proc. SPIE CR43 (1992).

2. Shand, E.B., Glass Engineering Handbook, 3rd. Ed, McGraw-Hill, New York (1958).

3. Delgado R.B. and Hallinan, M., "Mounting of lens elements", Opt. Eng 14, S-11 (1975).

4. Hopkins, R.E., "Lens Mounting and Centering", Chapter 2 in Applied Optics and Optical Engineering, Vol VIII, R.R. Shannon and J.C. Wyant, eds., Academic Press, New York (1980).

5. Yoder, P.R. Jr., "Axial stresses with toroidal lens-to-mount interfaces", Proc. SPIE 1533, 2 (1991).

6. Rourk, R.J., Formulas for Stress and Strain, Table XIV, McGraw-Hill, New York (1954).

7. Yoder, P.R., Jr., Opto-Mechanical Systems Design, 2nd. Edition, Marcel Dekker, New York (in press, 1992).

8. Hatheway, A.E., private communication (1992).

9. Richard, R.M., private communication (1992).

10. Bayer, M., "Lens barrel optomechanical design principles", Opt. Eng. 20, 181 (1981).

11. Yoder, P.R. Jr., "Lens mounting techniques", Proc. SPIE 389, 2 (1983).

12. Kowalskie, B.J., "A Users Guide to Designing and Mounting Lenses", Digest of Papers, OSA Workshop on Optical Fabrication and Testing, North Falmouth (1980).

13. Valente, T.M. and Richard, R.M., "Analysis of elastomer lens mountings", Proc. SPIE 1533, 21 (1991).

14. Yoder, P.R., Jr., "Opto-mechanical designs for two special-purpose objective lens assemblies", Proc. SPIE 656, 225 (1986).

Optimization of the Optomechanical Interface Employing Diamond Machining in a Concurrent Engineering Environment

David J. Erickson, Roger A. Johnston and Anthony B. Hull
OCA Applied Optics, 7421 Orangewood Avenue
Garden Grove, CA 92641, USA
Telephone: 714/895-1667, FAX: 714/891-4356

ABSTRACT

Modern Single Point Diamond Turning (SPDT) technology offers new opportunities to address a large number of traditional problems in optical assembly and alignment. Traditional optical manufacturing rules-of-thumb have resulted from a lack of control over the Optomechanical Interface (OMI) . Because of the extreme accuracy associated with SPDT, conventional alignment methods can be replaced with deterministic and cost effective methods. These SPDT techniques have been demonstrated at OCA to be compatible with athermalization strategies, and with design-to-unit-production-cost (DTUPC) considerations.

1.0 INTRODUCTION

Over the past decade, the optical community has experienced an incredible growth in optical design capabilities. This has been driven in part by the increased pressure from more stringent system design requirements. It is also due in part to the ever-increasing power of computers and optical design codes. In an attempt to meet the often contradictory requirements of multi-spectral performance, severe light-weighting, athermalized operation, high-G environments and tight packaging constraints, the opto-mechanical designer has used every trick available. Armed with an optical design code and greatly enhanced CAD capabilities for mechanical layout, the designer has produced designs which employ multiple, general aspherics, often in non-axisymmetric configurations, with extremely tight fabrication and alignment tolerances.

This increase in both complexity and precision challenges the optical manufacturing community, even without considering the parallel pressures to reduce costs to minimum levels within very ambitious schedule constraints. Most of the challenges associated with converting these designs into functional hardware concern verification of the optical surface and control of the optomechanical interface.

2.0 THE OPTOMECHANICAL INTERFACE

The readers of this paper may have different perspectives on what are the crucial aspects of the optomechanical interface (OMI). Our perspective is that of an optical engineering/manufacturing facility engaged in the specification, opto-mechanical design, fabrication and test of limited quantity, high performance, optical subassemblies and systems. This environment provides the most thorough interaction between the design and fabrication disciplines.

An optical designer working outside the manufacturing environment views the OMI primarily as a constraint defined by a standard list[1,2,3,4] of achievable manufacturing tolerances within which he must optimize his design. The mechanical engineer who packages the optical design creates the OMI as a means of locating and constraining elements within a tolerance band defined by the optical designer, while coupling the mechanical and thermal properties of the chosen optical materials to those of the system platform.

The mechanical engineer's understanding of the optical design may be only sufficient enough to translate the optical designer's prescription into element drawings. More thorough, usually, is his understanding of requirements relating to manufacturability. However, subtleties of the OMI as relates to higher precision levels may be missed.

The manufacturing engineer has broad knowledge of "real world" issues that relate to the OMI and views control of the OMI as one key to the successful assembly and alignment of the optical system. However, distance from the design process limits the extent that manufacturing factors influence design tradeoffs.

For moderate to low tolerance optical designs, these design functions can be successfully performed serially with limited interaction between disciplines. However, as design complexity and precision levels increase, close interaction between the design functions becomes more critical. At the extreme of complexity and precision, a concurrent engineering approach becomes essential.

We are addressing systems with aperture diameters on the order of 1 to 40 cm , and our discussion may not apply equally to larger or smaller systems. Table 2.1 represents the various precision regimes encountered in optical fabrication.[1,2,3,4] First, observe the precision levels that are produced in conventional optical fabrication. It can be seen that in the context of conventional optical fabrication methods, the OMI problem can be characterized by the need to control positioning of optical elements possessing relatively high surface accuracy with mechanical housings machined to one or two orders-of-magnitude lower precision. The housing machining is less accurate than the optical surfaces both because lesser precision is needed, and because lesser precision is available by conventional methods. If the interface could be produced to optical surface accuracy, errors associated with the interface would be negligible in most cases. However, conventional methods result in interfaces that usually have an influence on optical performance. This influence is a result of the imprecise methods used to produce the interfaces, and the assembler's inability to deal with multiple degrees of freedom in alignment. Thus, implementation of the interface is as stressing to optical performance as is implementation of the optical surfaces. We will consider interface factors in both conventional and diamond turning applications.

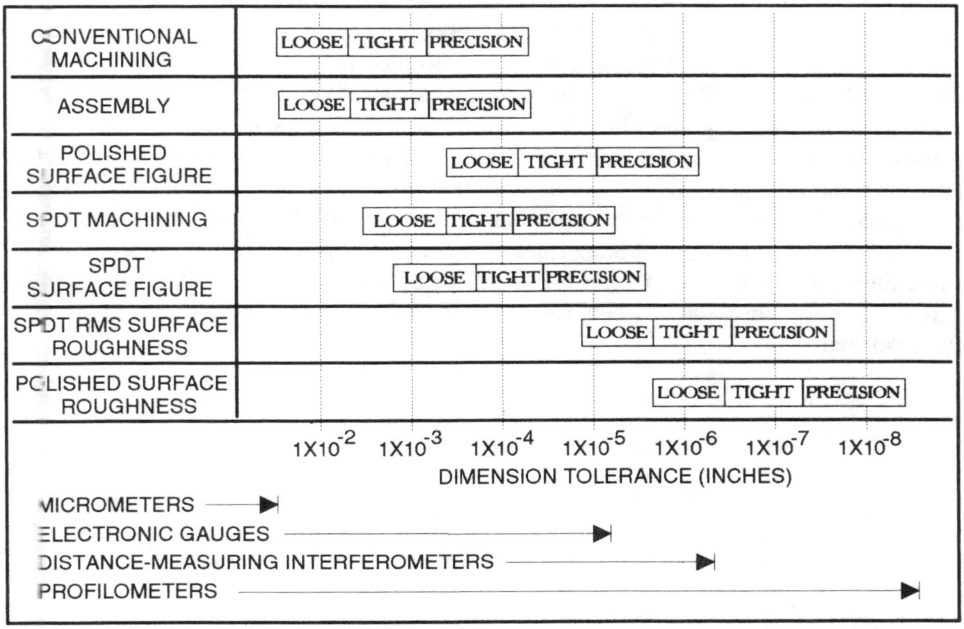

TABLE 2.1 : Precision regimes, comparing conventional machining and optical fabrication with Single Point Diamond Turning (SPDT).

3.0 CONVENTIONAL APPROACH TO INTERFACE CONTROL

Because of this disparity (See Table 2.1) between the high surface figure precision which can be achieved by conventional optical fabrication processes and the relatively poor dimensional control of interface surfaces, a number of techniques have been utilized to compensate for this deficiency. Among these are:

- mechanical indication of elements during assembly
- lapping and shimming for spacing control; "file-to-fit"
- monitoring of system aberrations or performance during element adjustment
- computer-aided alignment schemes
- lathe assembly
- matching of cells to glass after recomp for manufacturing actuals
- and mix-and-match techniques for large quantity applications.

As the precision and complexity levels increase, these approaches often become heroic in scope, usually becoming tooling and/or labor intensive.

3.1 Degrees of freedom

Even conventional approaches to simplify the OMI succeed because the design of mounting interfaces takes advantage of all available precision to minimize the number of adjustments that must be made and verified by the assembler. A non-axisymmetric optical element can be misaligned with respect to any or all of its 6 degrees of freedom: 3 axes of translation and 3 rotations. Rotationally symmetric elements have 5 degrees of freedom. Fig. 3.1-1 defines the cartesian coordinate system that we will refer to in establishing the OMI, and in defining the errors in the interface. This discussion is limited to optical surfaces that are either figures of revolution, or are off-axis segments of figures of revolution. Although the techniques we are discussing are applicable to the relatively infrequent inclusion of anamorphic elements, to dispersive elements such as diffraction gratings and prisms, and to optics including discrete binary sub-components, we have chosen not to generalize discussion to include such optical surfaces. In this discussion, when 5 degrees-of-freedom are controlled, the interface is fully defined.

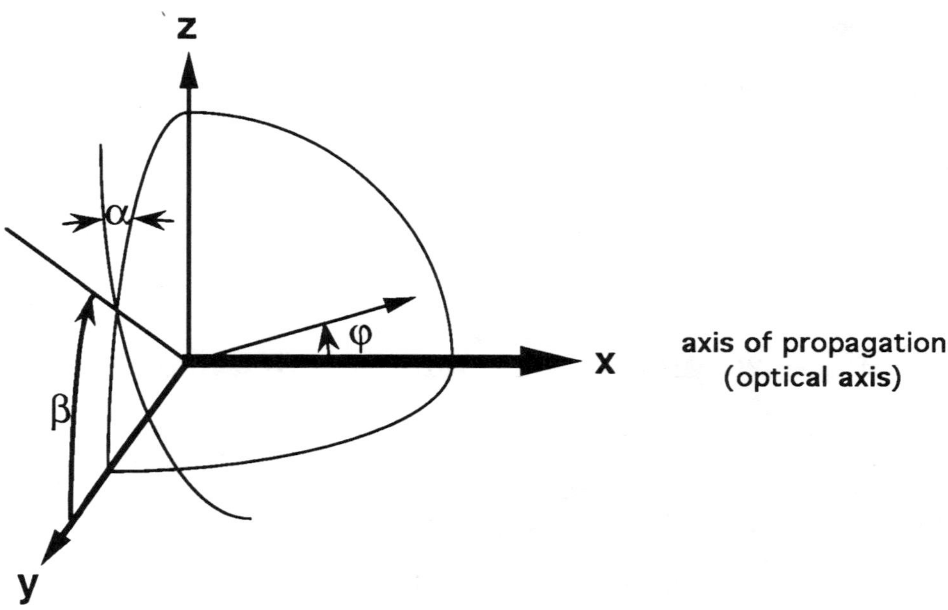

Figure 3.1-1 Cartesian Coordinate System. X defines the direction of propogation along the optical axis, and Y, Z define the perfect element rim plane. Any error in mounting an element into the perfect element rim plane will result in error in the surface normal ϕ as shown. The tilt will be angle α, and the axis of tilt is angle β . The six degrees of freedom that we discuss are the displacements x, y, and z, plus the angles α, β, and γ (not shown). γ is the degree of freedom corresponding to rotation about the x axis. We have limited this discussion to on-axis and off-axis figures of revolution.

Most conventional approaches of constraining the number of degrees-of-freedom are still left with 2 or 3, i.e. axial spacing and decentration. This is usually due to the shortcomings of conventional machining and/or glass shaping processes.

It would appear that without improving the precision of the OMI and optical element interface (C.T., O.D, Wedge, etc.), we are approaching the limits of system complexity and precision achievable through conventional manufacturing techniques.

4.0 THE SINGLE POINT DIAMOND TURNING MACHINE

4.1 Precision of SPDT Optical Surfaces and Interfaces

Table 2.1 demonstrates that with SPDT there is an immediate gain of at least an order of magnitude in the precision of mechanical features. Balancing this is some loss of optical surface figure and finish with respect to conventional polishing techniques. However, for infrared systems, this tradeoff can be quite acceptable. And by the appropriate choice of materials, figure and finish can be improved for visible applications by post-polishing. As we will see, the advantages gained by this improvement in mechanical feature precision are well worth these limitations.

Fig. 4.1 Moore M-18 AG Single Point Diamond Turning Machine

4.2 Properties of the Single Point Diamond Turning Machine

The SPDT machine is a CNC lathe which obtains its great improvement in precision over conventional lathes through the enhancement of some of its features.[5] The spindle which holds the element to be turned has been replaced by an air-bearing or hydrostatic bearing because of its low friction properties and microinch runouts and stability. The cutting tool is now precision-lapped and of diamond for its hardness and heat transfer properties. The diamond tool is translated in two axes by a hydrostatic or precision roller bearing stage which exhibits very high straightness of travel. Whereas the conventional CNC lathe has an optical or mechanical encoder to transduce position, the SPDT machine utilizes a Distance Measuring Interferometer (DMI) accurate to microinch levels. The CNC controller has changed the least, except perhaps in the number of bits of accuracy encoded. In order to cut aspheric curves on complex hardware, the development of customized software is usually required to augment the functions provided by the controller. The SPDT machine is usually built over a massive frame (the OCA Moore M18 weighs approximately 4200 lbs) to minimize loss of precision due to flexure and vibration and thermal inertia. The room which houses the SPDT machine is optimally controlled for temperature, humidity and vibration, the result of which is a very critical long-term stability.

Table 4.1 lists some of the operating parameters of a typical SPDT machine (Moore M-18 Aspheric Generator). The OCA M-18 AG also utilizes a third rotary axis, on which the toolpost is mounted, which rotates under program control to always position the diamond normal to the curve being cut. This feature also allows easy access to the sides and backs of parts, facilitating single-set-up turning.

Spindle speed	0-2000 RPM
Spindle Axial Error	2 μin.
Tool Feedrate	0 - 4 in/min
Minimum programmable step	1 μin.
Z-axis travel	9 in.
X-axis travel	16 in.
Max. part dia.(over rotary table)	13 in.
Max part dia. (no rotary table)	22 in.
Max part dia. (with risers)	33 in.

Table 4.1 Moore M-18 AG SPDT machine operating parameters

The precision levels we are seeking fall within the parts-per-million CTE regime of most turnable materials, as well as a regime coinciding with strain magnitudes expected from mounting and centrifugal forces. In a well designed element, these may be the limiting factors to achieving total control over the OMI.

4.3 OMI control of element features with SPDT technology

The accuracy built into the SPDT machine suggests some obvious uses, such as producing optical curves, precision diameters, etc. However, of more importance to this discussion is how this capability can be used to configure the optomechanical interface to make assembly efforts more efficient and accurate. It is this exciting area of SPDT technology that allows one to turn modern optical designs into realizable hardware.

As described in Section 4, control of the OMI condenses down to eliminating degrees of freedom that need to be controlled at the assembly stage. There are very basic concepts that can be applied to designing elements for diamond turning that in many cases reduce assembly degrees of freedom to zero. These principles, single-setup turning, align-and-turn, and stack-and-turn, will be briefly described here and clarified through examples in Section 6.

4.3.1 Single Set-up Turning

The most basic concept of OMI control with SPDT hinges on the principle that any feature cut on a lathe in one set-up shares the same rotational axis. Add to this the microinch control of axial and lateral positioning inherent in SPDT machines and this principle expands to insure comparable accuracies on diameters and axial heights of faces.

4.3.2 Align-and-Turn

Not all elements can be created entirely in a Single Set-up configuration. In fact, it is often advantageous to have some surfaces already prepared by diamond flycutting or conventional polishing to provide references. By allowing for alignment of the element to the SPDT spindle face during its mounting, one is provided with a much larger range of possibilites for OMI control.

4.3.3 Stack-and-Turn

By combining the two principles above, and sequentially adding elements to the spindle without removing previously turned hardware, one can extend the inherent accuracies of the SPDT machine to whole assemblies. This represents the pinnacle of OMI control. By utilization of pinning or kinematic mounting techniques, assemblies can be turned, disassembled, conventionally worked (plated, post-polished, coated, painted, etc.) and reassembled without losing these micro-inch accuracies.

4.4 OMI control during system alignment

Many optical systems employ a hybrid mix of elements, some of which are appropriate for SPDT manufacturing techniques and others (beamsplitters, prisms, gratings, fold mirrors, etc.) which still must be aligned to extremely tight tolerances. A class of optical elements for assembly which we call Metering Mirrors (see Section 5.7), are readily fabricated with SPDT techniques. These metering mirrors are designed to simulate the OMI of a previously diamond-turned element. The metering mirror may contain multiple reflective surfaces whose centers of curvature can represent various system datums (i.e., image plane location, element vertex or center of curvature location, or pupil locations). By locating these centers of curvature by interferometry or autocollimation, optical axes can be propagated through very complex folded paths. Section 6 includes more thorough descriptions of this powerful technique.

4.5 OMI and the optical test

By coupling the metering mirror concept with the well-known aspheric reflective null surface, the control of test conjugates for interferometric testing of aspheric elements or subassemblies is readily achieved.[6]

5.0 ILLUSTRATIVE EXAMPLES OF OMI CONTROL

5.1 Concurrent engineering: the cornerstone

So many of the examples to be given in the following section could not have been accomplished without the close interaction between the optical designer, the optomechanical designer, the manufacturing engineer, and the operators of the SPDT facility. The success of the approaches given hinged quite often on the understanding of the engineering team of specific SPDT machine capabilities and even idiosyncrasies. As system complexities increase and tolerances continue to tighten, the importance of more complete communication of machine capabilities will favor those endeavors which can control the whole design-through-fabrication cycle.

5.2 Designing for SPDT methods

The designer who can work with the technology of SPDT is far less constrained than the designer employing conventional technology. Sanger[5] has discussed many factors involved in producing SPDT optical surfaces. Clearly, unobscured mirrors and non-symmetric, irregularly-shaped and very fast aspheric optics can be efficiently produced with SPDT. For long-wave IR systems made from SPDT-compatible materials, there is no incentive to constrain the optical design to conics, nor is there a need to constrain the optical surface to be continuous. Similarly, unusual and efficient designs can be packaged by establishing the OMI with SPDT. Furthermore, the accuracy of this SPDT interface is so excellent that tolerances usually associated with alignment error may frequently be allocated to other system parameters. Although it is preferable that both the optic and the housing be produced by SPDT, we note that SPDT techniques used on a housing may interface efficiently with conventionally produced and lapped elements on optics, adjacent housings and optical benches.

SPDT methods are compatible with the most demanding spaceborne, military and scientific instruments. Examples below are representative of actual hardware produced at OCA incorporating SPDT to establish the OMI. Each example will show the application of a principle or principles discussed in Section 4.2. We start with a very simple (but significant) application of the principles, and in subsequent examples shall accumulate extensions to these principles, thus building to a description of implementation of a relatively complex system. The examples are also intended to be indicative of the range of application of the SPDT OMI methods, and suggestive of further extensions and developments.

5.3. Example 1: Fabrication and assembly of a 9-inch diameter germanium dome

Figs. 5.3-1, 5.3-2, and 5.3-3 show views of a germanium dome produced for a military application. The quarter-meter diameter, 6 mm thick germanium element extends beyond a hemisphere as a cylinder which serves as the mounting interface. The OMI is characterized by requirements for centrations of the germanium dome with respect to its titanium attachment ring reference surfaces to <0.001 in. both axially and laterally. These, as well as a <0.0003 in. germanium thickness uniformity requirement and wavefront requirement, are driven by optical performance considerations. The CTE values for the dome and ring materials were dissimilar enough to require 0.018 ±0.001 in. RTV thickness to control stresses over the operational thermal profiles. The titanium ring did not lend itself to being produced by SPDT methods; conventional lathe tolerances needed to be adequate. Developing production-level assembly techniques was a factor in the design effort.

It was realized that a variation of single-setup-turning could be developed to achieve all the dome requirements. We purchased the germanium generated to near-net-shape and went immediately to SPDT methods for curve generation. Fig. 5.3-1 shows the convex surface waxed to a turning fixture which is held on the SPDT faceplate by vacuum. This curve is cut from the cylindrical diameter to the vertex utilizing the 3-axis programming capabilities of the Moore M-18 AG. The convex surface serves as the reference surface for all future tooling. Optical performance is strongly influenced by the radius and figure control attained at this step.

Figure 5.3-1 Generation of Convex Surface of Germanium Dome

The concave surface of the dome is produced by a variant of stack-and-turn where a vacuum chuck (similar to the assembly chuck shown in Fig. 5.3-2) is mounted on the SPDT machine and a spherical surface is turned on this chuck which matches the radius of the convex dome surface. Machine coordinates at this point uniquely define the center of curvature of both surfaces of the dome as well as the vacuum chuck. The dome is then merely installed into the chuck and the program run to cut the concave interior, the cylindrical inside diameter, the end face and the protective bevels. In a 50-unit production phase, this sequence was automated to a turn-key operation with a superb yield rate.

The same principles are applied to the assembly phase tooling. Fig. 5.3-2 shows a schematic of the assembly tooling. Again, SPDT machining was used to insure precision. The Spacing Tool was diamond turned at the chuck interface end to produce a face and outside diameter. The chuck was then turned in one setup to produce a concave surface matching the dome convex radius as well as a diameter and face to match the Spacing Tool features. Without removing the chuck from the SPDT spindle, the Spacing Tool was installed on the chuck and the length (minus allowance for graduated spacers) and outside diameter of the Alignment Mandrel interface surfaces were finalized. The outside diameter was matched to the Alignment Mandrel by trial during final cuts.

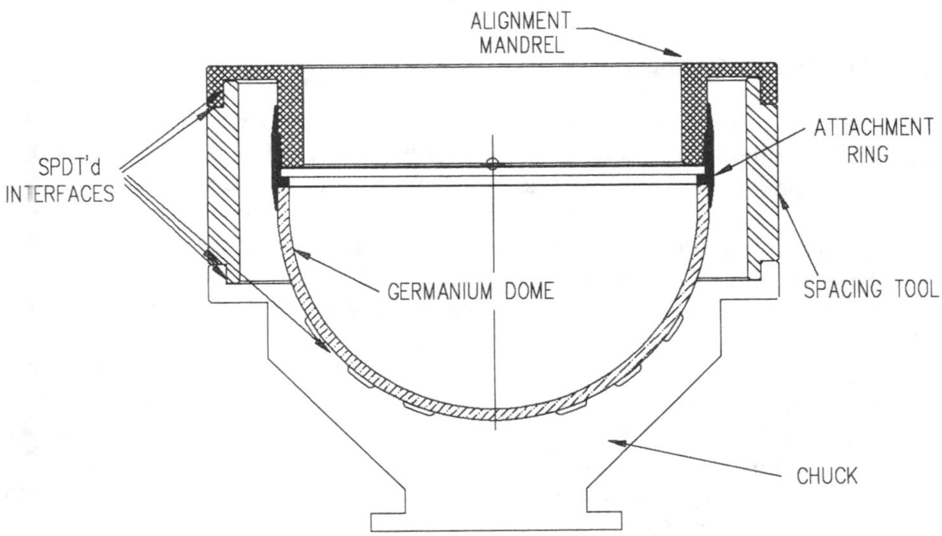

Figure 5.3-2 Germanium Dome Assembly Tooling

The final result was a dome assembly process which, after machinable spacer selection to compensate for Attachment Ring height variations, was essentially "drop-in" to <0.0007 in. TIR at a rate of one assembly per day. Fig. 5.3-3 shows a finished assembly mounted in test tooling.

Figure 5.3-3. Germanium dome after final assembly

5.4. Single Point Machining of Metal Mirrors

5.4.1 Introductory comments

Aluminum mirrors, copper mirrors, beryllium, titanium and metal matrix composite mirrors clad with electroless nickel, and other material combinations can use some or all of the techniques that we shall describe below. It matters little if the optical surfaces are flat, spherical, or general aspherics with great departure from the nearest sphere. It also is only a small complication to mount off-axis segments of full mirrors. The underlying techniques are the same. We have selected several examples to illustrate specific OMI control methods we have found effective in mounting reflective elements.

Precision machining to establish the alignment is best done if the constraints of alignment are considered at the design stage. Alignment should constrain mirror fabrication on the SPDT machine. Should the optical sensor operate at either visible or short wave infrared wavelengths, the design should also consider the constraints of post polishing (nickel plating, polishing surrounds, etc.) Integral to the design of the optical assembly is the design of fixtures required for fabrication and alignment.

The viability of precision machining to establish the alignment requires that the design consider the pressures, centripetal forces, and stresses that both the mirror and the interface will experience during fabrication, and have a first resonant frequency higher than potential excitation frequencies found in fabrication. The single set-up turning of the optical interface is desired, but not always possible. Special cases include cutting off-axis mirrors, supporting thin mirrors on plates, and post polishing considerations. In all cases, the mounting conditions selected for fabrication processes will affect the performance of the mirror. The following are selected examples of optimizing the OMI in the production of metal optics.

5.4.2 Example 2: A Two Mirror Off-Axis Beam Expander

We selected this example (see Fig. 5.4.2-1) because it illustrates the fundamental technique of accurate transfer from the mirror-manufacturing interface to an assembly interface. The beam expander of this example employs two confocal parabolas and is required to operate in the visible and far infrared simultaneously. Because of its use in the visible, post polishing is required to remove the periodic residuals of SPDT.

The OMI is characterized by requirements for visible operation, fixed package size, $\lambda/12$ wave RMS (HeNe) collimation, and entrance and exit beam geometries. In addition, this off-axis afocal beam expander was designed to be low cost and direct replacement for several thousand existing systems. Our manufacturing methods needed to be tailored to accommodate production rates of several hundred units a month, a rate we could not satisfy with conventional techniques.

In its finished state the OMI of the primary mirror is a plane surface normal to the axis of the parabola and 0.10 in. below its vertex. This surface serves to reduce the possible degrees of freedom of the primary to in-plane translations. This translation is further constrained by a pair of pins. The mirror is attached to the housing with six screws.

The OMI of the secondary mirror is a plane surface normal the the optical axis and 0.15 in. from its vertex. The motion of the secondary is constrained to translation along this surface. Oversized through-holes for the three screws that fasten the secondary, allow for small adjustments to system wavefront at final assembly.

The OMI of the housing consists of two SPDT parallel surfaces whose separation meters the assembly airspace. The OMI surface supporting the secondary serves the additional functions of boresight reference and next assembly interface.

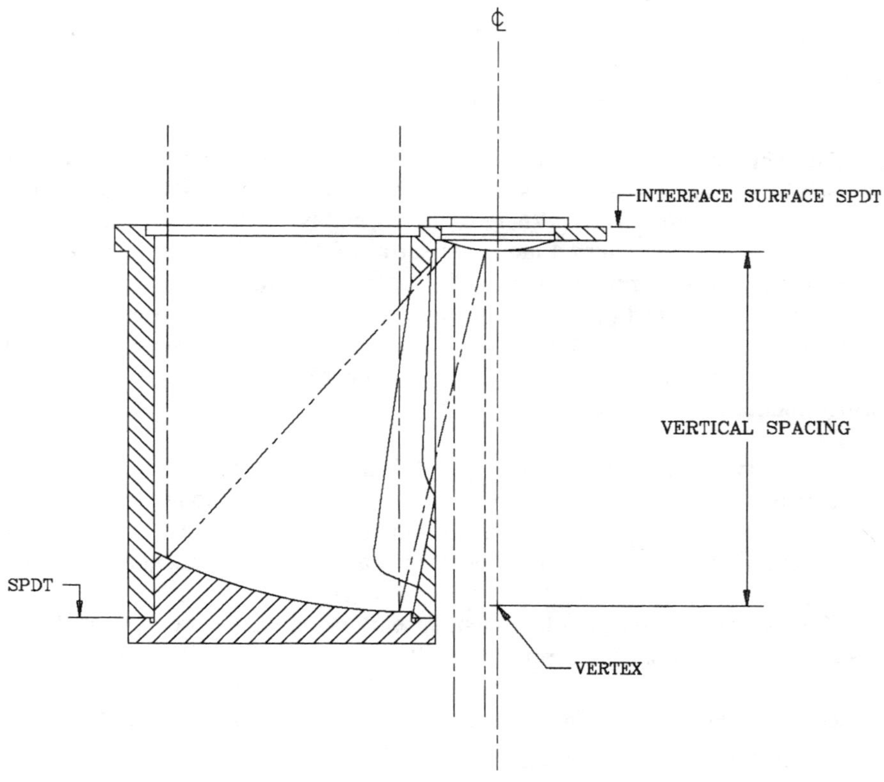

Figure 5.4.2-1 Two-mirror afocal telescope

5.4.2.1 Descripiton of SPDT process

The aluminium off-axis primary mirror was rough machined and stress relieved as were a surround and base plate. Although polishing single off-axis parabolic mirrors is within the capabilities of skilled opticians, it is more efficient to produce several in a multiple block and post-polish with a surround to fill in the spaces between mirrors. This efficiency is amplified when SPDT is used to generate the surface profile. The three elements were assembled and the parabolic surface was final machined on a conventional CNC lathe. The primary and surround were then plated with electroless nickel to a thickness of approximately 0.005 in. Fig. 5.4.2-1 shows the primary mirror installed in its surround.

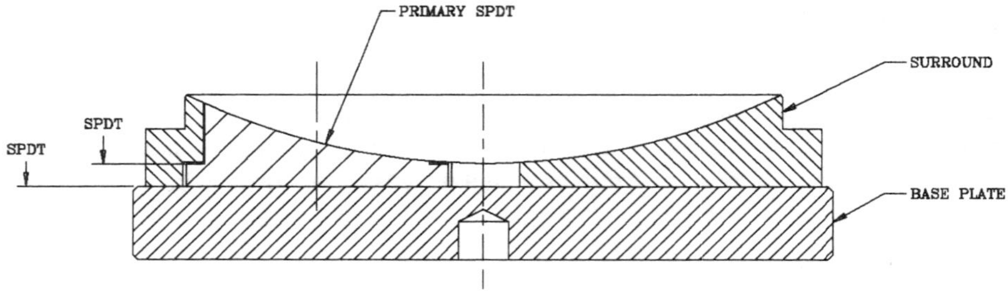

Figure 5.4.2-1 Primary mirror in polishing surround.

It is advisable to support the mirror during this manufacture with the same OMI used for assembly. Sometimes this desire cannot be realized and an additional OMI must be provided for manufacturing. Whereas the assembly OMI is the plane facing the secondary mirror, the preferred mounting surface for manufacture is the back plane of the primary. This requires a separate SPDT operation in which the back of the mirror is turned optically flat and parallel to the assembly OMI. It is appropriate to note here that these diamond turned reference surfaces can be used as alignment aids during in-process testing and are even more useful for interferometric monitoring of warpage during mounting operations.

The secondary mirror for this telescope, although used off-axis, was designed to be mechanically symmetric about its optical axis and oversized enough to serve as its own polishing surround. The OMI therefore could be turned in the same setup as the aspheric curve, thus allowing control of spacing and tilt of the optical axis.

After fine-tuning the process-dependent spacing parameters, assembly of these afocal telescopes consisted of monitoring of the flatness of exposed SPDT'd surfaces during torquing of the primary mirror screws and lateral adjustments of the secondary mirror to balance coma and boresight requirements. No shimming was required to achieve the $\lambda/12$ RMS (HeNe) collimation requirement.

5.5 Example 3: Fabrication and assembly of a monometallic cassegrain telescope

Figs. 5.5-1, 5.5-2, and 5.5-3 show a schematic and views of elements and the assembled cassegrain telescope for a spectrometer. The telescope has an 8-in. diameter primary mirror and is made entirely of 6061 aluminum to provide uniformity of thermal properties. The OMI is characterized primarily by a requirement that the alignment of the telescope remain stable throughout a temperature range from ambient to liquid nitrogen temperatures. It was decided that utilizing precision machined interfaces would provide the most stable assembled structure. Other constraints included low cost, ease of assembly and alignment, and a desire to quantify relative motions of the individual mirrors and telescope stucture with respect to the focal plane. Fig. 5.5-1. shows a schematic of the mono-metallic telescope. The principal elements are the primary mirror, secondary support, secondary mirror, focus spacer, and retaining hardware. Fig. 5.5-2 shows these elements in exploded configuration after SPDT operations.

This telescope was designed with the intent of utilizing as many of the advantages of SPDT as possible. The primary and secondary were designed with integral reference spheres whose nominal optical centers of curvature coincided with the system focal plane. Several concurrent engineering design iterations were required to effectively incorporate these metrology features into the optomechanical design. The specific order of turning operations also affected the design. Incorporation of diameters and faces which could be utilized when flipping the mirrors for back side preparation was necessary. Finally it was felt that to achieve cryogenic performance, all surfaces should be subjected to the same machining operations.

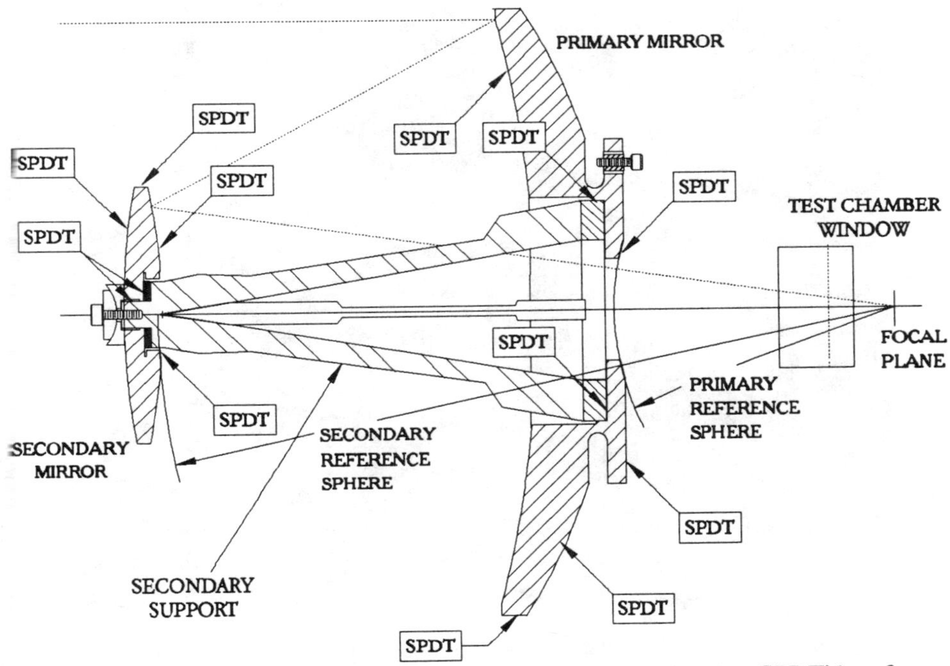

Figure 5.5-1. Schematic of SBIR mono-metallic telescope showing SPDT'd surfaces

After conventional machining operations to prepare the blank to a near net shape, the secondary mirror back surface was machined first. This established a precision mounting surface for the seocndary mirror aspheric turning operation. The secondary was then flipped and held to the SPDT spindle by vacuum. The outside diameter, inside diameters, general aspheric optical curve, and the interface surface for the focus spacer were then turned, taking advantage of the accuracies afforded by the single-step-turning principle. See Fig. 5.5-1.

The SPDT operations on the primary mirror also proceeded from its back. In one operation, the back mounting plane, the primary reference sphere, the spherical back of the primary mirror, and its outside diameter were turned. The part is then flipped and secured to the SPDT faceplate by its precision turned mounting plane. Axial locations of the front face with respect to back features are controlled by utilizing SPDT machine coordinates of the faceplate at the time of its last face-off. Lateral alignment of the primary mirror is assured by centering the primary to its outside diameter (align-and-turn) at which time the general aspheric primary mirror optical surfaces are cut. The interface plane for the secondary support is cut at the same time, at a precise distance from the prmary vertex. Finally the radially-defining diameter is cut to match-fit the conventionally-machined outside diameter of the secondary support base. (See Fig. 5.5-1)

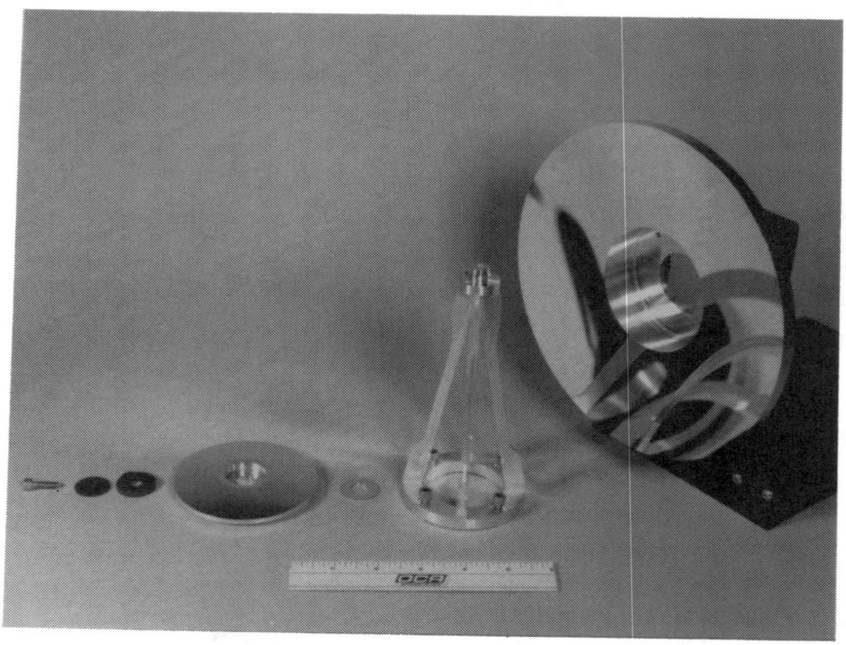

Figure 5.5-2. Exploded view of mono-metallic telescope elements

Without removing the primary mirror from the spindle, the secondary support was assembled to the primary with screws. The secondary mounting interface, a diameter and plane, are then turned on the end of the secondary support structure. Because of the accuracy of the stack and turn process, the optical axis of the primary is essentially transferred precisely to the mounting inside diameter of the secondary. After removal of the primary from the spindle, the focus spacer is machined to nominal thickness and parallelism and the secondary mirror installed.

Fig. 5.5-3 shows the final assembly photograph of the mono-metallic telescope. Without any need for spacer thickness or wedge modification, this telescope could be disassembled and reassembled, and without any adjustments, produce a wavefront error of < $\lambda/4$ P-V (HeNe). Also fringes from the two metrology reference spheres can be observed at the same time.

Figure 5.5-3. Assembled mono-metallic telescope

Thermal tests have not been run at this time, but assuming proper control of thermal gradients, this telescope should perform as designed or at least be easily monitored for element spacing and curvature changes.

5.6 Single point turning of refractive elements

5.6.1 Introductory comments

Although fabrication differences for refractive elements compared to reflective elements may be no more significant than modification of the cutting parameters to accommodate the ductility and other cutting properties of the material, mounting methods for refractive and reflective elements vary significantly. Schemes for mounting lens elements generally result in over-constraint and even the best schemes, employing slotted and compliant retainers, distort elements. Also, adequate mounting methods for lenses may not be appropriate for mirrors.

One of the tasks facing the element fabricator is providing an OMI for subsequent assembly. For the simple case of a bi-convex lens, the OMI is composed of the optical surfaces and the rim of the lens. This rim defines the mechanical axis of the lens and may coincide with the optical axis if the latter was properly aligned during the edging operation[7] (See Fig. 5.6.1). For the condition where the mechanical and optical axis coincide within the allowed tolerances, the lens is "centered".

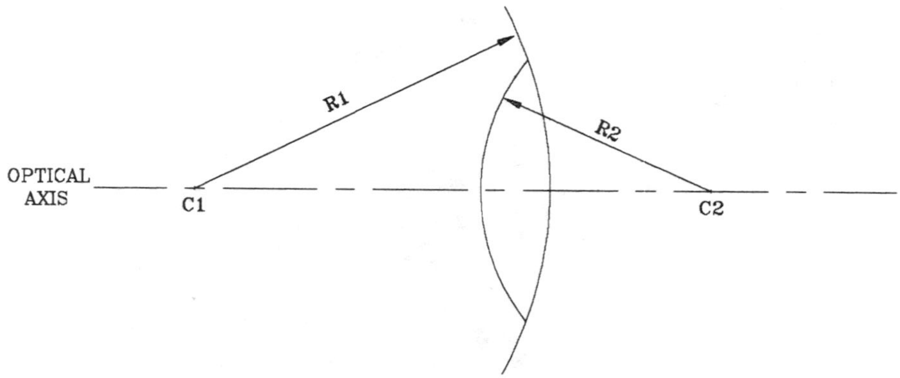

Figure 5.6.1. Condition for a perfectly centered element. The optical axis passes through both centers of curvature

Conventional lens manufacturing requires that polishing and edging operations be performed at separate times on different machines. The edging process requires the alignment of finished optical surfaces followed by grinding the rim. Today's optical designs demand precision, and are taxing the limits of conventional processes. Requirements for TIR <0.00005 in. are becoming common. Alignments to this precision level require great skill, are time consuming, and place delicate optical surfaces at risk of damage.

The separation of surface fabrication and edging is not necessary when SPDT methods can be employed. As illustrated in Table 2.1, SPDT enables direct machining to optical tolerances. The gap that existed between optical polishing and machining has been blurred, and the separate operations of surface finishing and OMI machining can now be accomplished on a single machine. The use of SPDT is limited somewhat by the selection of compatible materials. SPDT is ideally suited for infrared applications, where moderately loose figure requirements are complemented by moderately tight alignment requirements.

The infrared transmitting materials Ge, Si, ZnS, and ZnSe are readily turned using well established SPDT methods. Their turnability allows both the surface profile and OMI to be produced in a single setup. In addition, the programming, and positional accuracy of SPDT machines makes it no more difficult to produce a general aspheric profile than a spherical profile. This in itself may result in major simplifications of the optical subsystem.

5.6.2. Example 4. Refractive element production by SPDT

As an illustration of single-setup SPDT, consider the volume production of a ZnSe apsheric element. Our discussion will begin with the conditions and tooling required for the alignment of an optical surface to a chuck. Here the process is common with both conventional edging and the SPDT. From this common point, the SPDT process departs with a discussion of the single point generating followed by optical surface fabrication. We will demonstrate, how in a single-setup , the aspheric surface, center thickness (C.T.), and diameter are machined.

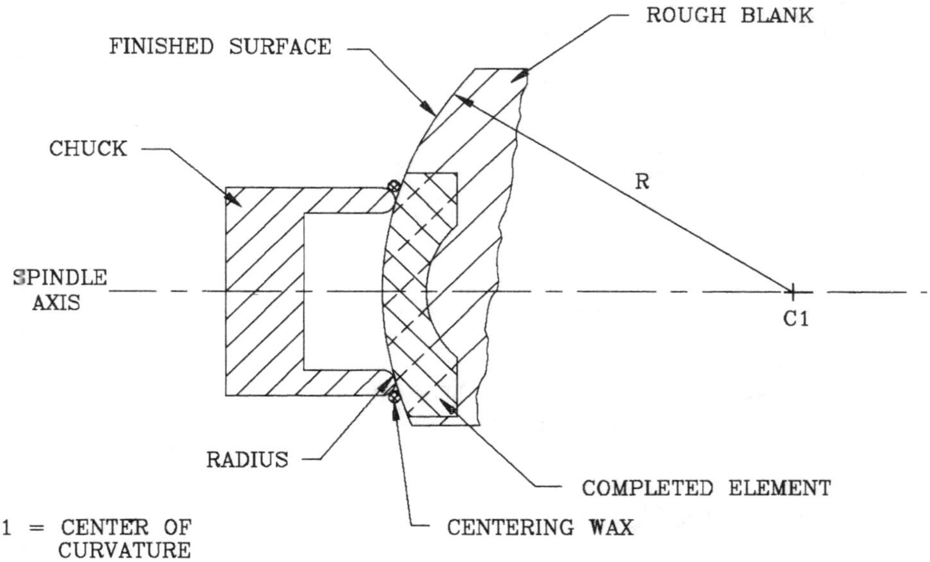

Figure 5.6.2-1. Element aligned for Single Point Generating

5.6.2.1 Element centering

Our process begins with a polished surface on an otherwise rough blank. As illustrated in Fig. 5.6.2-1, the polished surface of the element is placed firmly in contact with the radiused surface of the chuck and retained with centering wax. When this single spherical surface and chuck are in firm contact, motion is constrained to rotation about its center of curvature. Any decenter is accompanied by a compensating tilt, and centration is preserved. This greatly reduces the task of precision alignment. One need only assure the finished surface is in firm contact with the chuck and that sufficient material is available to yield a finished element. The TIR of this surface is monitored with an electronic gauge to accuracies of ±0.00005 in. while the element chuck ensemble is rotated on an air bearing. A process of heating, measurement, and manipulating the lens is repeated until the desired TIR is achieved. In the case of precision lenses, TIR's < 0.00005 in. are not uncommon. Once cooled, a probe is inserted in the axial hole of the chuck, and contacts the vertex of the optical surface for measuring its location. This data is essential in calculating the position of the cutting tool in relation to the desired center thickness. The element is now in alignment with the chuck references and the task of precise transfer of this alignment from centering to generating to SPDT finishing is made easy.

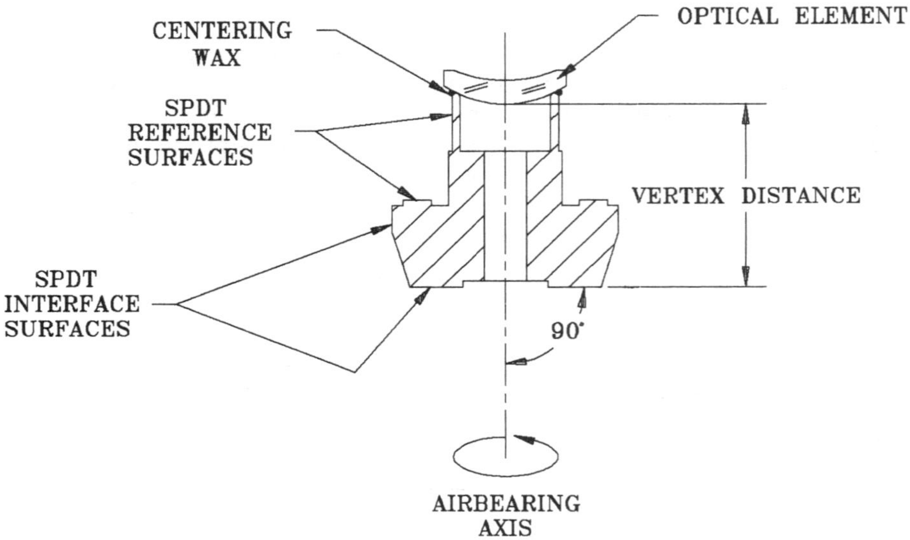

Figure 5.6.2-2 The transfer chuck provides single element alignment for many operations

5.6.2.2 SPDT chuck considerations

The quality of the element is dependent on the quality of the chuck, and therefore consideration must be given to the chuck's design and manufacture. Each chuck has built-in reference and alignment features to facilitate alignment prior to the centering and turning operations. Therefore, great care must be taken to assure these references are concentric to the chuck-radius supporting the element. One method of assuring this critical concentricity is to machine all surfaces in a single-setup.

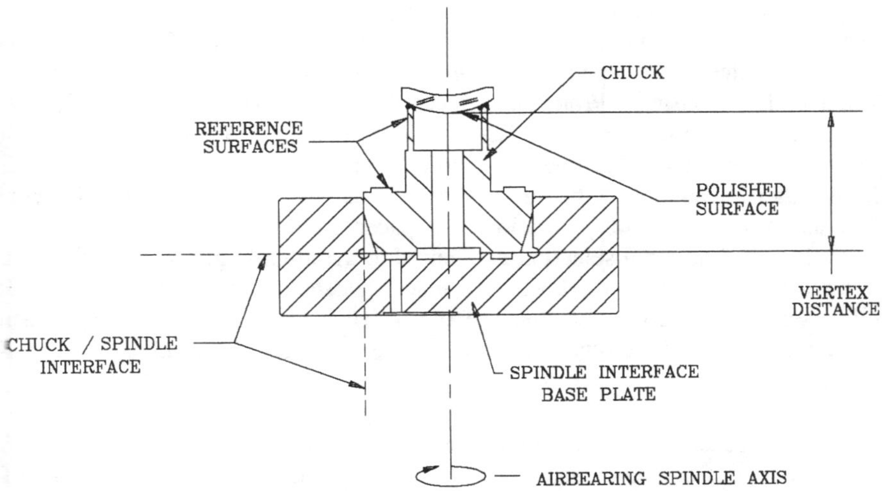

Figure 5.6.2-3. Chuck and airbearing spindle interface.

5.6.2.4 Single Point Generating (SPG) and SPDT finishing

With the chuck aligned to the spindle axis of a SPDT machine the process of generating can begin. Our element is an oversize rough blank. A single point tool consistent with high cutting rates is selected, and the programming of the SPDT machine is modified to allow for reciprocal operation. Modest cuts of 0.001 in. are advised while the surface is irregular to avoid knocking the element off of the chuck. As successive cuts remove these irregularities and the surface becomes more continuous, much larger cuts can be taken. We have removed as much as 0.010 in. in a single pass on ZnSe. With the completion of the single point generating (SPG) operation the optical surface and OMI have been established. Sufficient thickness, usually <0.003 in., remains on the element for the SPDT finishing operation that follows. Starting with a single optical surface, an oversized blank, and the concept of single-setup turning precise alignment of the optical axis to the OMI is assured.

5.7 Example 5: Single Point Turning Of Lens Sub-cells And Drop In Assembly

5.7.1 Discussion of sub-cell concept

A natural extension of our single setup turning discussion above is its application to the manufacture of assemblies employing sub-cells. Sub-cells have been used for years in the production of precision optical systems, with many variations represented in the literature.[7] The discussion of sub-cells that follows is intended to illustrate the application of single-setup SPDT.

Sub-cells are typically cylinders whose rim and faces serve as the OMI for an optical element. In applications where sub-cells are employed, one or more are assembled into a another cell or housing. From the perspective of parts count the sub-cell method appears to complicate the assembly. However if in a single setup all interface surfaces of the sub-cell are machined, one can readily achieve sub-micrometer control of the OMI. Another advantage of the sub-cell can be realized when an assembly requires multiple assembly-disassembly iterations for spacer trimming. If this disassembly is anticipated careful material selection is required to avoid galling, and within the limits of the environmental stress, the housing contact area should be minimized to facilitate smooth assembly. Additionally, the use of sub cells allows a modular approach to system manufacture, where sensitive elements are aligned and assembled as a module outside the system housing.

There are many schemes for retaining elements into these cells ranging from swaging to elastomeric suspension. Our discussion is limited to the latter. Here the skill of an assembler is applied to the alignment of the optical element onto a chuck rather than into a cell, and SPDT provides a precise OMI that facilitates drop in assembly. By mounting the element to a chuck as the first operation the assembler can verify the TIR of each optical surface and thus precisely control centering. The precision of this alignment is limited only by TIR gauging and operator patience.

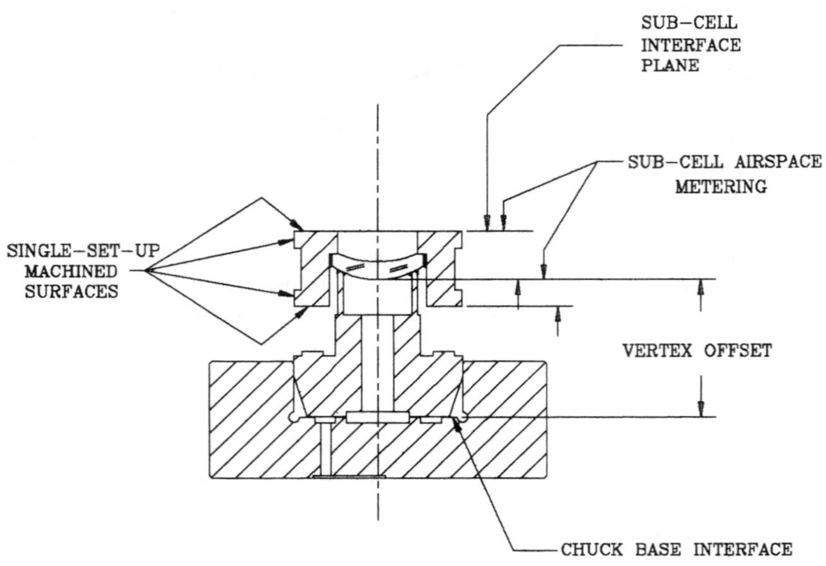

Figure 5.7.1-1 Oversize sub-cell ready for Single Point Machining

Once a lens is centered to the references of a chuck. An oversized sub-cell is bonded to the centered lens. (See Fig. 5.7.1-1) The lens, sub-cell, and chuck ensemble is set aside on a level surface while the adhesive cures. In most applications the sub-cell need only be aligned well enough to assure that OMI surfaces will clear during machining. In cases where the bond thickness uniformity is critical, and careful alignment is required, the sub-cell runout must be monitored during bonding.

With the sub-cell bonded to the element chuck ensemble, vertex offsets are measured. These offsets are used to establish the sub-cell airspace. (See Fig. 5.7.1-1) Machining of all interfaces is completed before removing the cell, lens, and chuck group from the SPDT spindle. Coolants used in SPDT will attack some adhesives and centering waxes and machining without coolant may be required.

In a single-setup, all interface surfaces of the sub-cell are SPDT'd and OMI accuracies of 1μm (4×10^{-5} in.) are readily achieved in production. Sub-cells of this precision allow drop in assembly. However, clearance between the sub-cell and housing must be provided, and this clearance is likely the single largest source of error. For assemblies under 3 in. in diameter, a clearance of 10μm (.0004 in.), is required for drop in assembly, worst case decenters resulting from this clearance are similar in magnitude.

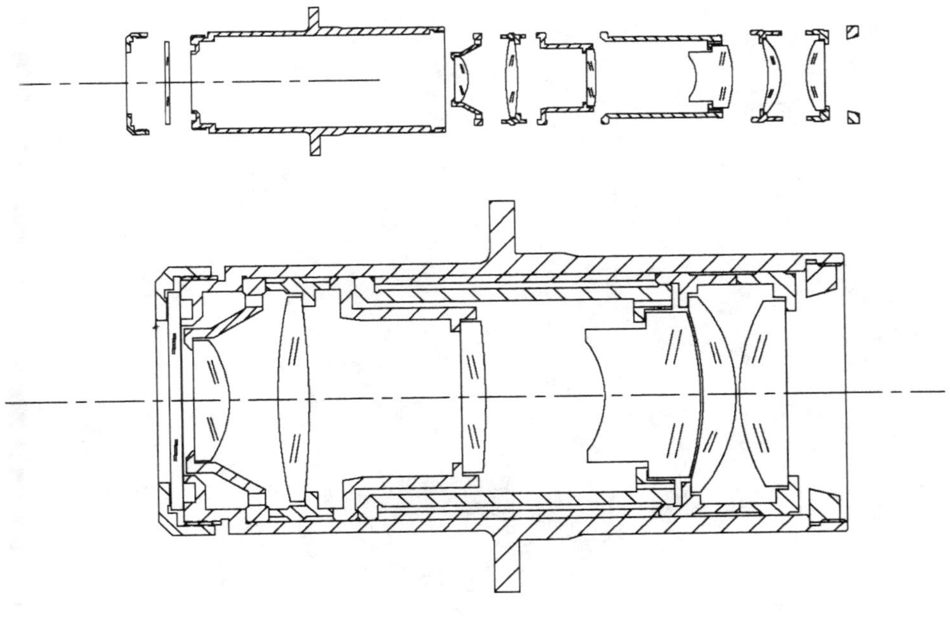

Figure 5.7.1-2 Schematic of a production drop-in assembly

Fig. 5.7.1-2 is a simplified sectional view of a production drop-in assembly. This low cost, controlled distortion, wide field lens was used for an infrared sensor. Its high performance required element wedge, tilt, and decenters of <10μm (<0.0004 in.). The image of several dozen completed systems were evaluated for rotational symmetry and none departed by more than 3μm (0.00012 in.). In conventional manufacture, this dispersion would typically be on the order of 30μm and would have been unfavorable.

5.8 Example 5: Metering mirrors

5.8.1 Metering Mirrors: optical elements to assist assembly

Folded optical systems with high density packaging present the greatest alignment challenges. Such systems are becoming increasingly common as minimal volume and mass are needed to satisfy the operation of many developmental military platforms. Such optical packages must accomodate gyros, encoders, dewars, cables, sensors and analog logic, all related in a robust manner. One parameter used to describe such systems is packaging density; the volume of all elements plus optical paths divided by the package volume. Packaging densities in excess of 90% are often sought, and this density limits the ability to either provide internal alignment features, or to have access to alignment points. Fortunately, the method of metering mirrors mitigates this circumstance. Fig. 5.8.1-1 represents such an optical system.

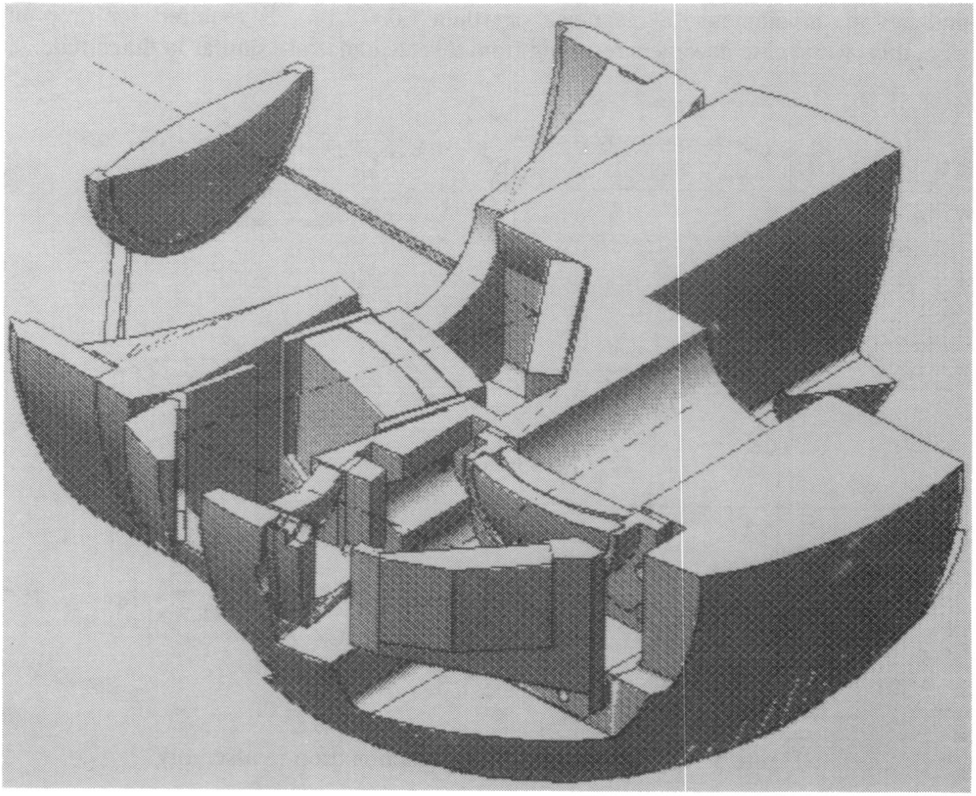

Figure 5.8.1-1 Representative Gimballed High Density Optical System

This system is packaged to a density of 93% with the aperture comprising 80% of the gimbal diameter, and the focal length comprising two times the gimbal diameter. Auxiliary methods are clearly necessary to establish the alignment of this system, and the metering mirror method described in this section has proved to be a practical auxiliary method.

5.8.2 Metering Mirror Principles

Fig. 5.8.2-1 shows the centered mirror concept, and Fig. 5.8.2-2 shows a metering mirror of the sort that is used to build up optical elements to establish an optical system in the high density gimballed system in 5.8.1. Mirror centration here is similar to that described in Section 5.6 where we discussed the mounting of refractive elements. It is apparent that the metering mirror is perfectly centered with respect to the optical axis when two conditions are satisfied.

1. By geometrical identity, the optical axis must be perpendicular to the intersecting plane of the mirror's rim.
2. The center of curvature of the mirror lies on the system optical axis.

Condition 1 establishes the OMI plane tilt, and thus constrains two degrees of freedom. Condition 2 establishes the OMI cartesian coordinates (position in-plane and along the optical axis), and thus constrains three additional degrees of freedom. Since the mirror has rotational symmetry, the sixth degree of freedom is irrelevant, and the OMI is completely constrained.

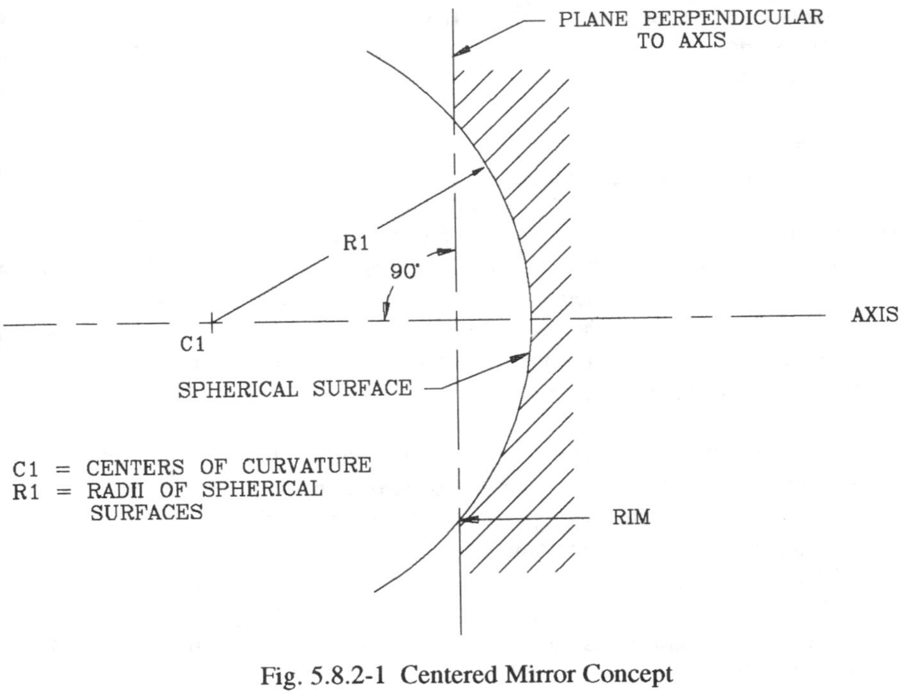

Fig. 5.8.2-1 Centered Mirror Concept

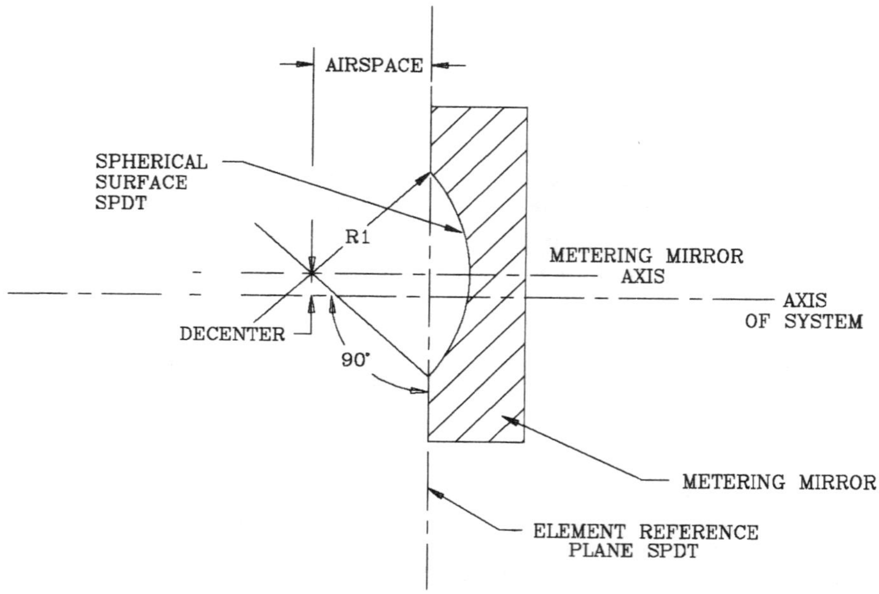

Fig. 5.8.2-2 metering mirror centration. The metering mirror is intentionally shown to be decentered from the system optical axis. Centration (translation) is sensed by interferometry, with central fringes resulting from decenter.

5.8.3 Use of metering mirrors to propagate the OMI

Metering mirrors are use to propagate OMI along the optical axis. To accomplish OMI propagation a fixed reference must be established against which all OMI are measured. A housing shoulder that precisely defines the optical axis serves this purpose and is used to support the reference interface plate. Refer to Fig. 5.8.3-1. Because it will not be removed until assembly is complete, attachment must provide rigid support without altering the reference surface figure.

The reference metering mirror is placed on the reference interface plate, which in turn is established at the primary datum of the optical housing. The reference metering mirror is not to be confused with the metering mirror. Note that the bore of the reference interface plate, and the diameter of the reference interface mirror are match machined by SPDT. Because the reference interface mirror will be taken in and out a number of times during alignment, the mirror surface S5 is relieved for minimal contact area to avoid binding during insertion. Binding could affect the registration or damage this critical interface.

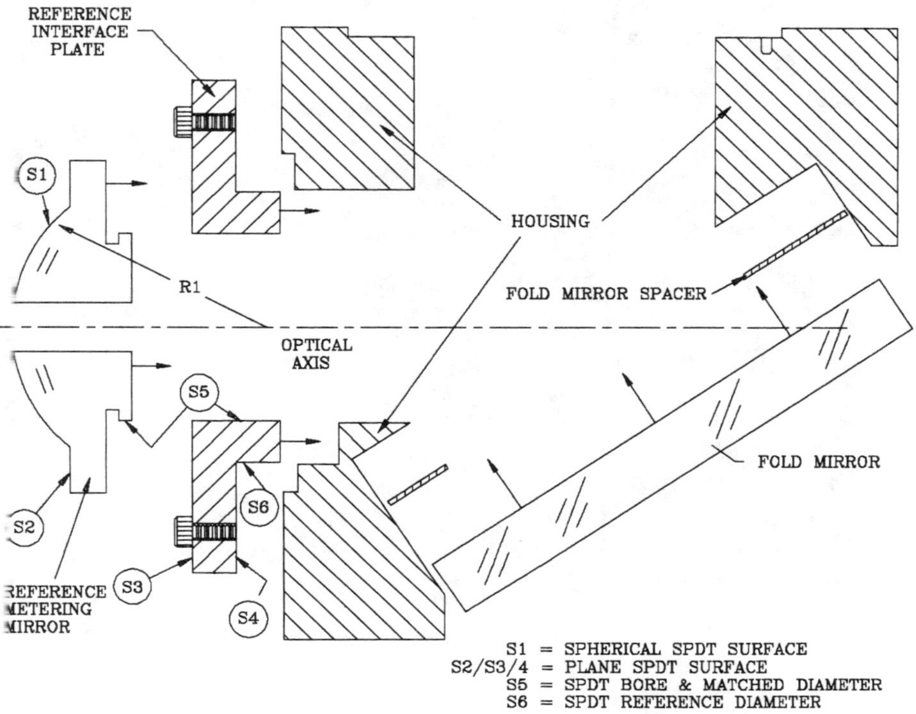

Fig. 5.8.3-1 Reference Metering Mirror and Reference Interface Plate. The optical axis is transferred precisely from the reference interface mirror at entrance of the housing to the reference metering mirror optical surface.

The metering mirror is quite straight forward to use. The assembler will need to have an interferometer of sufficient aperture to fill the full optical surface of the reference interface plate. Fig. 5.8.3-2 shows the set-up we use at OCA.

Fig. 5.8.3-2 Interferometric sensing of the metering mirror OMI transfer

It is of note that the accuracy of the method does not depend on the positioning of the interferometer in the set-up. The housing can be removed and replaced with no loss of accuracy. Two steps will establish the tilt, despace and decenter of the propogated OMI with respect to the datum. These two steps are as follows:

1. The reference metering mirror is removed as seen in Fig. 5.8.3-3. The tilt of the fold mirror is adjusted with respect to the metering mirror such that the plane wave front form the metering mirror is parallel to the plane wave front of the reference interface plate. This condition produces null fringes for each of these wavefronts. An example of the interferogram resulting form alignment is provided in Fig. 5.8.3-4.

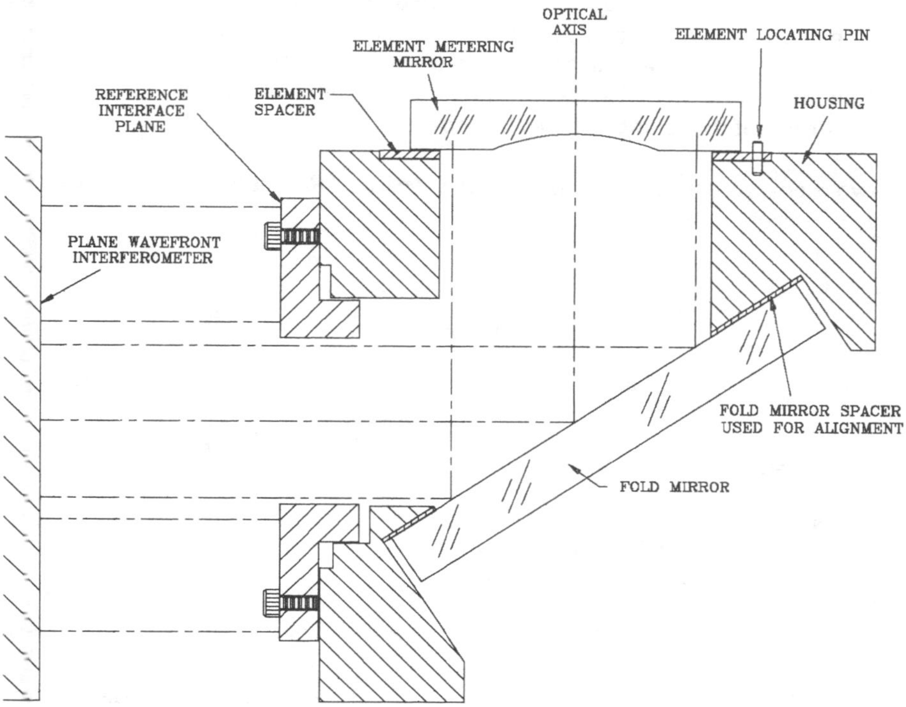

Fig. 5.8.3-3 Configuration for tilt adjustment. The reference metering mirror is removed.

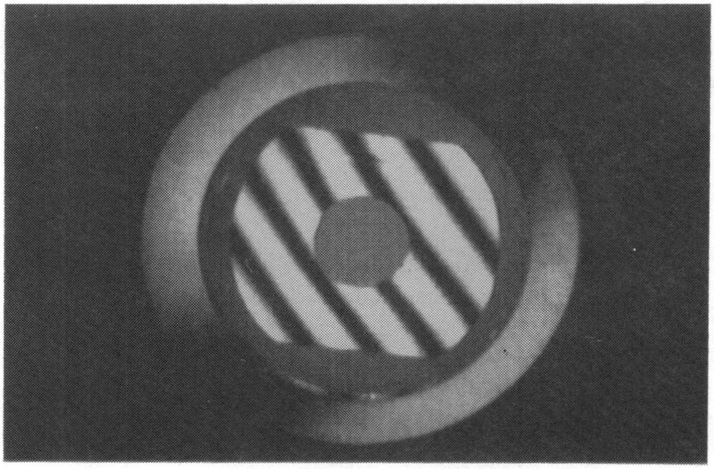

Fig. 5.8.3-4 Interferogram of adjustment for tilt. The difference between the two fringe patterns is 5 fringes (50 μin.) over a 3 in. diameter, which corresponds to a tilt of 21 μradians.

2. The reference metering mirror is inserted as seen in Fig. 5.8.3-5. The element metering mirror is adjusted until its center of curvature matches that of the reference metering mirror. This is termed the confocal position. This again will produce null fringes, however the null is with respect to the spherical surface of the reference metering mirror.

This two step process is repeated until the desired OMI alignment and despace is acheived for each element.

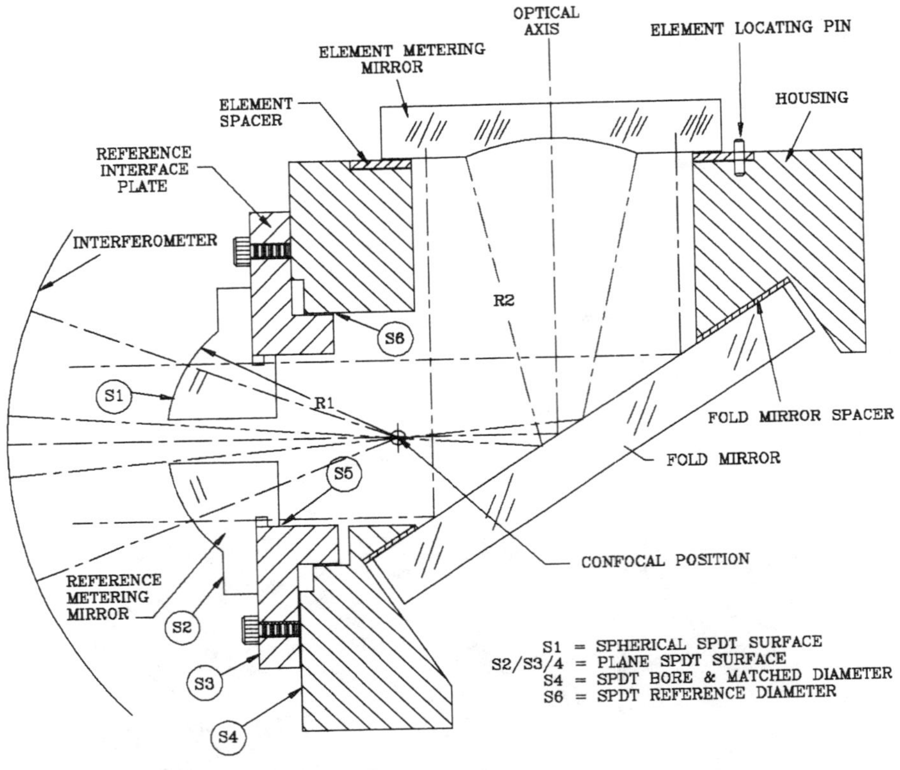

Figure 5.8.3-5 Configuration for despace and decenter adjustment. The reference metering mirror is installed.

5.8.4 Manufacture and validation of the metering mirror, and accuracy of the resulting OMI

The metering mirror, reference metering mirror, and reference interface plate are all produced as though they are optical elements. Carefully conditioned aluminum and SPDT single set-up techniques are used. Validation of the surface figure is easy to verify by interferometric testing.[6] A conventional interferometer set-up is used with a radius slide that includes a distance measuring interferometer.

Errors in using the metering mirror are dominated by interferometric determination of the optical axis and center of focus, and errors associated with the fabrication of the metering mirror are negligable with respect to this (typically 5μin. and 10 μradians). Fig. 5.8.4-1 expresses the relationship of errors in both despace and decenter to the F-number of the test. Here, λ is the wavelength in inches, the number of fringes is the difference between the fringe count for the reference metering mirror to that for the metering mirror. F/# is the F-number or radius of curvature of the metering mirror divided by the clear aperture of the metering mirror. Note that the radius of curvature is established by the distance of the OMI from the reference metering mirror. Note that, as the distance becomes long, the errors in spacing rapidly increase. The tilt is simply determined by the difference in number of fringes between the metering mirror and the reference interface plate, divided by the clear aperture.

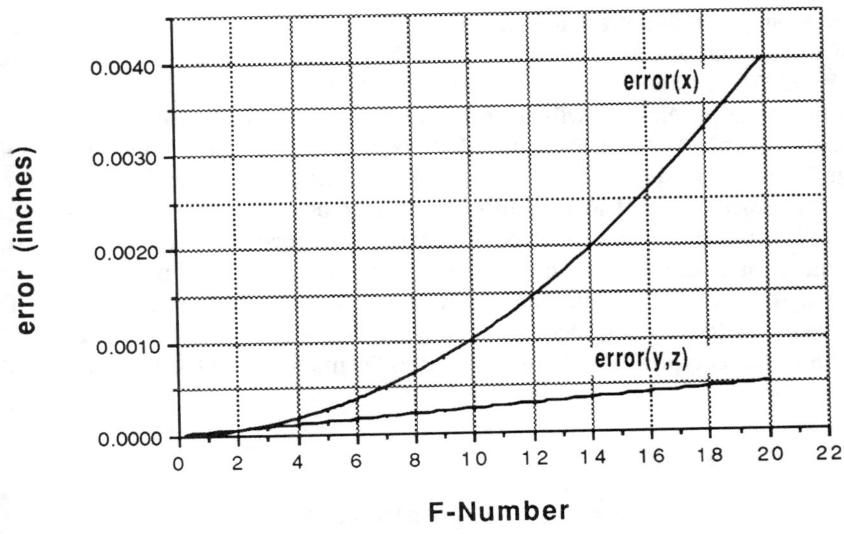

Figure 5.8.4-1 Errors associated with metering mirror alignments.

The despace error $\varepsilon_x = 4\lambda *$(estimated accuracy of number of fringes)$*F/\#^2$ following Smith.[8] The decenter error $\varepsilon_{x,y} = \lambda *$(estimated accuracy of number of fringes)$*F/\#$. The despace number of fringes is the difference in power derived from the interferogram of the reference metering mirror and the element metering mirror, each of which may be provided directly by the interferometric analysis software. The difference typically is accurate to better than ±0.1 fringe. The decenter error is set by the ability of the assembler to simultaneously null the fringes of both the reference metering mirror and the metering mirror, and we estimate this to be better than ±1 fringe.

5.9 Assembly of the optical element in identity to the metering mirror position.

Once the assembler is satisfied that the metering mirror is located to the requisite accuracy, this positioning must be accurately transferred to the optical element. There are several ways of achieving this. For example, the metering mirror may simulate each degree-of-freedom of the OMI. The offsets that are used to establish (shim) the position of the metering mirror are used to establish the position of the component, and the component is bonded in place. The accuracy is mitigated by several factors:

- ability to maintain clean surfaces
- minimization of degradation of the shims
- ability to duplicate the loading forces experienced by the metering mirror
- different flexure properties between the element and the metering mirror
- nonuniformity and stresses caused by the curing of adhesives

Degradation in a carefully controlled environment typically is <0.0005 in.

If an interferometer with a precision slide is available, and the lens to be mounted has a spherical optical surface, greater precision can be acheived. In this case, the element cell will be SPDT finished to act as an auxillary metering mirror reference interface, and the reflection from the element will act as an auxilliary metering mirror sphere. The two step process described above (Section 5.8.3) is repeated but now with the auxillary metering. This usually does not obviate using a metering mirror at this location because the metering mirror is used to establish fold mirror positions prior to mounting any elements. The limiting factors will be only those described in Section 5.8.4, and the nonuniformity and stresses caused by the curing of adhesives mentioned above. Degradation in a carefully controlled environment typically is indistinguishable from that provided for despace and decenter in Figure 5.8.4-1, and for tilt as discussed in Section 5.8.4 (tilt is typically controlled to 10 to 30 μradians, which to date is seldom required).

6.0 FUTURE DIRECTIONS

We have considered the implementation of current SPDT technology to efficiently establish the OMI for complex systems. For many cases, the OMI can be established to between one and two orders-of-magnitude better than with non-SPDT techniques. In others, the interface can be so simplified that producibility is greatly enhanced. We expect that future designs will increasingly exploit this improvement jump in the following manners:

- Design of more complex and densely packaged systems
- Reallocation of the tolerances to acknowledge the OMI accuracy available
- Design for simplification of the interface, and production drop-in assembly

In order to affect this exploitation, there will be a closer fusion between the optical and mechanical designers and the manufacturing operation. Fortunately, this is increasingly being done in the current industry drive for concurrent engineering. Designers will want to educate themselves on what can be accomplished, and not rely on old rules-of-thumb for assembly.

Simultaneously, capability to manufacture the OMI will be increasingly stressed. We are already seeing instances where designers are defining requirements for which the one-to-two order of magnitude improvement cited above is insufficient. Areas that may be productive in ameliorating these limitations over the next decades are apparent. As defined for this paper, we are primarily concerned with assemblies of characteristic dimensions of the order of 1 to 40 cm. Different factors would apply to either large assemblies, or to micro-miniature assemblies, and a list corresponding to divergent size regimes may be significantly different. Many of the potential developments we list in Table 6.0-1 will affect both the OMI and the optical surface.[9]

Table 6.0-1 Future directions to further improve manufacturing the omi

Better Position Controls:

- Improved position sensors, defining the relative location of the tool edge and work piece possibly including interferometric sensing of the diamond edge
- In-situ optical metrology (independent of the machine's metrology)
- Improved work piece mounting methods
- Stiffer machines
- Micro-damping through positional feedback
- Improved servo controls
- More adaptable control software and AI corrections from in-situ metrology
- More agile tool carriages
- Better environmental controls

Better understanding of materials and tools:

- Tribology of the tool, work piece interface, and elastic/plastic properties of micro-cuts
- Active temperature control (cooling or heating) of the tool
- Alloys with improved homogeneity, isotropy, hardness, grain structure and micromechanical properties (microyield, microcreep), and with minimum defects
- Improved tools
- Tools and methods to work strategic materials like beryllium, SiC and glass ceramics which are largely inaccessible for these techniques

Fusion with other techniques

- Laser milling
- Autonomous alignment verification methods using embedded binary optic and holographic elements

- Extension of these methods for establishing precision surfaces into ductile grinding techniques for materials not accessible to diamond tools.

7.0 CONCLUSION

The power of SPDT techniques to establish the OMI to an accuracy to one or two orders-of-magnitude better than that achieved by conventional alignment methods has been explored, and the practical manufacturing approach that OCA has been using has been described through a series of examples. The underlying principles of single set-up turning, align-and-turn, and stack-and-turn have been defined, and starting with the mounting of single refractive or reflective elements, we have systematically introduced concepts through examples. This has led to description of the more advanced metering mirror methods that are effective for a wide range of alignment problems including those associated with high density, folded systems including decentered off-axis aspheric elements. Such SPDT methods are available to new systems, and are being practiced in the optics manufacturing community with encouraging results. The authors have stressed the importance of concurrent engineering to achieve the potential precisions and efficiencies of these SPDT techniques to establish the OMI. Consideration is given to potential advances in techniques, materials and machines that may yield further precision and simplicity to establishing the OMI.

8.0 ACKNOWLEDGEMENTS

The authors wish to acknowledge our colleagues who have participated in establishing this technology at OCA, and have participated in discussions or provided comments relevant to this paper. Bert Fischer has helped develop these ideas, and has implemented them on the our SPDT machines. Ralph Scicchitano has assisted in preparing figures for this paper. Bill DeBoynton has challenged these OMI concepts with both increasingly stringent design demands, and with insights into what can be machined. Dan Barber has always been willing to woodshed new problems.

9. REFERENCES

1. R. H. Ginsberg, "Outline of Tolerancing (from Performance Specification to Toleranced Drawings)", Optical Engineering, Vol. 20(2), pp. 175-180, 1981.

2. P. Kuttner, "Trends and Limits in the Production of Optical Elements and Optical Systems", Advances in Optical Production Technology II, SPIE Procedings Vol. 163, pp. 24-30, 1979.

3. R. E. Parks, "Optical Component Specification", 1980 International Lens Design Conference, SPIE Procedings Vol. 237, pp. 455-463, 1980.

4. J. L. Plummer, "Tolerancing for Economies in Mass Production of Optics", Optical Systems and Components Specifications, SPIE Procedings Vol. 181, pp. 90-111, 1979.

5. Gregory M. Sanger, "The Precision Machining of Optics", Applied Optics and Optical Engineering, R. R. Shannon and J. C. Wyant, Volume X, pp. 252-390, Academic Press, Inc, 1987.

6. D. Malacara, Optical Shop Testing,John Wiley and Sons, New York, 1978.

7. P. R. Yoder, Jr., Opto-Mechanical Systems Design, Chapter 4, Marcel Dekker, Inc, New York, 1986.

8. W. J. Smith, Modern Optical Engineering, The Design of Optical Systems, Second Edition, Chapter 11.2, McGraw Hill, Inc., New York, 1990.

9. D. L. Decker, "The Future in Diamond-Machined Optics", Ultraprecision machining and Automated Fabrication of Optics, SPIE Proceedings Vol. 676, pp. 2-7, 1986.

Analytical Design Methods

Chairs
R. Barry Johnson
Center for Applied Optics,
University of Alabama in Huntsville

A Review of Finite Element Analysis Techniques: Capabilities and Limitations

Alson E. Hatheway

Alson E. Hatheway Inc.
Engineered Products and Systems

INTRODUCTION

In its most general form the finite element method is a computational procedure used for resolving tensor states in three dimensional continua based upon the conditions at the boundary. The procedure provides approximate solutions for continua of any shape or configuration, some of which may be very difficult or impossible to solve in closed form. This facility is provided by the fact that the procedure divides a continuum into smaller units, often many units, and the shape of the units are selected to be shapes for which solutions are either known or can be solved by approximate numerical methods. The elastic equations for each of the small units are assembled for simultaneous solution by matrix methods, thereby solving for the boundary conditions (displacements) of each small unit that are consistent with the conditions at the boundary of the continuum. The final step of the analysis is to apply a modified Ritz analysis method to determine the tensor state inside each small unit based upon the boundary conditions for the unit.

The finite element method is a very powerful tool for the analysis of extended field problems. It has been most widely used in structural mechanics, for which use it was originally developed. Portions have also been applied to the mechanics of heat, the mechanics of electro-magnetics and the mechanics of fluids (potential flows). Because of the computational demands of the procedure it has found the digital computer to be a natural companion but most of the capabilities and limitations of the method are legacies from an earlier age when the computation was labor intensive and assisted only by mechanical calculators and electrical direct analog computers.

The structural analysis codes are all safety related in that they are intended to calculate stresses in structural members so they can be designed to stay within the safe operating stress ranges. The methods used in structural mechanics derive just enough information to calculate the stresses; no more, since that would be wasteful of resources and no less, since the problem would then be indeterminate. This has left curious "holes" in the procedure and its results that may frustrate and bewilder other users of the data. One well known example is the method's failure to calculate rotational displacements as well as translational displacements. Another may be the computational singularity in which the method is unable to calculate stresses at a free-edge interface between two different materials.

Nonetheless, the finite element method has shown that it is possible to make very accurate calculations of very complicated structural systems involving many thousands, or hundreds of thousands, of equations. The correlation between a well formulated finite element model and a prototype in the real world is generally excellent. The finite element method has pressed analysts to improve their understanding of the nature of structures in order to make more faithful models for analysis. In the writer's experiences the difference between the results of a finite element analysis and the results of a physical test have been the faithfulness with which the analyst has been able to render the real world into the available mathematical models for analysis.

The method has proved to be very useful for disciplines such as optics which must rely on structures to support the instruments. By the method of analogies it has proven possible to incorporate into the structural analysis codes other disciplines such as heat transfer, control systems, fluid mechanics and optics. This allows assembling the equations into one matrix for simultaneous solution and greatly extends the utility of the method.

This paper addresses the capabilities and limitations of the finite element method based upon its historic development and its contemporary use. Emphasis is placed on structural analysis codes, the theories behind them and their utility to the optics community. There is no discussion of problems or limitations associated with computational engines (truncation, round-off, matrix conditioning, etc.).

The writer also outlines some difficulties he sees in the next twenty to fifty years and suggest possible solutions which would benefit the optics community.

WHAT IS THE FINITE ELEMENT METHOD?

The finite element method is a collection of mathematical techniques developed over the last three hundred and fifty years to allow the calculation of stresses in a solid body under known loading conditions. The driving force for the development has been public safety; to avoid injury and loss of life from the failure of large structures such as bridges, buildings, aircraft and sports arenas. It has also enabled designers to avoid economic loss from the failure of structures such as children's' toys, wrenches and tools, sports equipment, medical instruments, and scientific apparatuses of all kinds. Its success in these applications has fueled interest in applying it to other disciplines as well.

It has become the indispensable tool for design of large optical instruments ranging from relay lenses to astronomical telescopes. The demands on the instruments create structural support forms that often are not easily analyzed by the available closed form solutions for beams and shells and the required accuracy is very demanding. However, the analyst of optical systems requires deflections, not stresses (only rarely is strength an optical issue), and this requirement may be in conflict with the (strength based) majority of the finite element analysis community.

The finite element method is only the latest form (and the form currently being used) to analyze complex structures. The development has been going on for a long time, and continues today (Love, 1927; Bathe, 1976 and Boyer, 1989):

1638	Galileo	asked, "if one pushes on the end of a beam, how much will it move and when will it break?" (Galileo's Prob lem.)
1744	Euler	derived the equation for the elastic curve of a beam

		from the condition of minimum strain energy.
1811	Gauss	taught us to solve simultaneous linear equations.
1822	Cauchy	proposed most of the elements of a general elastic theory for solid materials.
1870	Maxwell	related the stress and strain components to a system of linear equations.
1909	Ritz	showed that the boundary conditions of a solid body may be calculated from the internal states of the body.
1943	Courant	inverted Ritz' method to calculate the internal state from the boundary conditions.
(1960	Clough	coined the phrase "Finite Element Method.")

The above abridged chronology does not credit the many physicists and mathematicians who developed the theories of beams, plates and shells and vibration and shock dynamics nor many others who contributed to the general theory. As you can see from the list, however, the major advancements have occurred at rather widely spaced intervals. It has been thirty-two years since Clough's seminal paper "The Finite Element Method in Plain Stress Analysis" and forty-nine years since Richard Courant made it all possible. It may be about time for another milestone step.

The elastic theory that Cauchy proposed was intended to determine the state of stress at any point in an elastic body. First he defined the deformation response at points in the elastic body using nine partial differential equations:

$$\epsilon_x = \frac{\partial u}{\partial x}$$

$$\epsilon_y = \frac{\partial v}{\partial y}$$

$$\epsilon_z = \frac{\partial w}{\partial z}$$

$$\gamma_{xy} = \frac{\partial u}{\partial y} + \frac{\partial v}{\partial x}$$

Strain Components

$$\gamma_{yz} = \frac{\partial v}{\partial z} + \frac{\partial w}{\partial y}$$

$$\gamma_{zx} = \frac{\partial w}{\partial x} + \frac{\partial u}{\partial z}$$

$$2\omega_x = \frac{\partial w}{\partial y} - \frac{\partial v}{\partial z}$$

$$2\omega_y = \frac{\partial u}{\partial z} - \frac{\partial w}{\partial x}$$

$$2\omega_z = \frac{\partial v}{\partial x} - \frac{\partial u}{\partial y}$$

Non-strain Components (Rotation Components)

He then stipulated that the stresses at points were related to the strains at the same points by the elastic properties of the solid material (Young's modulus, modulus of rigidity, Poisson's ratio, bulk modulus, etc.), only two of the properties being required to determine all the rest.

Unfortunately the direct solution of the Cauchy's equations proved exceedingly difficult except for simple geometric cases or cases where symmetry could be used to reduce the general three dimensional problem to one or two dimensions. Therefore the dream of being able to directly calculate the stresses at points inside elastic bodies has not been fulfilled and various approximate methods have had to be developed. This development work resulted in the finite element method as we know it today.

The convergence criterion.

The approximate methods include Ritz' analysis and numerical integration techniques that circumvent the need for closed form solutions. However, since the use of approximate methods implies the presence of errors and human safety may depend upon the accuracy of the analysis, the method must offer some criterion for analysts to apply so they may become confident in the accuracy of the resulting calculations.

In finite element analysis "convergence" has become the universal criterion for acceptability of an analytical procedure. The convergence criterion asserts that the computational result becomes more accurate as the discretization intervals (time and space) defining the finite size of the steps in the analysis become smaller. The process of reducing the discretization interval has become know as "refining the mesh."

1) Refining the mesh improves the accuracy and

2) If refining the mesh produces negligible changes (improvements) in the answers then the answers are acceptably accurate.

Note that all the computational procedures in the finite element method converge as the mesh is refined. This allows the analyst to test the user dependent variables (element selection, material properties, load application, etc.). Analysts who regularly use the finite element method often run test problems to verify the accuracy of their modeling techniques. These test runs are usually made on small problems with known results and allow the analysts to validate their methods in the specific medium (code) to be used on a larger problem. These test runs provide confidence to an analyst that he is achieving sufficient accuracy in large problems without having to "re-mesh" a large model. The increased expense of halving the mesh and running a sample problem may be an order of magnitude or more.

Typical convergence behavior is shown for a number of sample problems in Figure 1 (from MacNeal and Harder, 1984). The problems were proposed to the American Institute of Aeronautics and Astronautics as techniques to evaluate the quality of finite

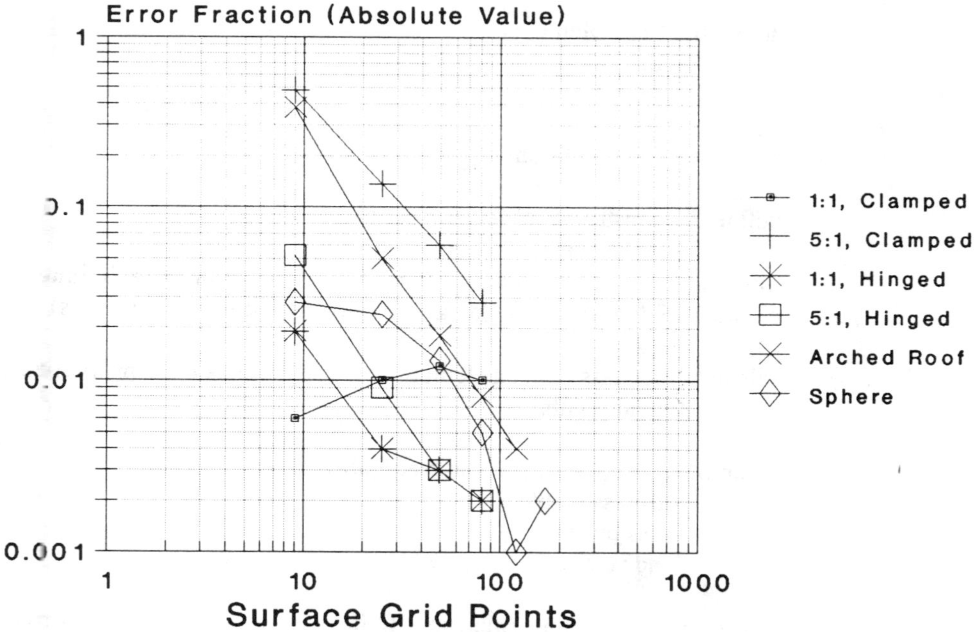

Figure 1. Displacement errors reduce without limit as the mesh is refined.

element formulations that may be proposed by various developers of such routines. The six configurations selected for presentation here represent shapes that may be of some interest to workers in optics; spherical shells, cylindrical arches and flat plates with various boundary constraint conditions. The models use similar quadrilateral shell elements. All the problems have known closed form solutions to permit assessment of the errors in the finite element formulations. The reader may see that as the number of grid points defining the surface is increased the residual error fraction tends to decrease. The square (1:1) plate with clamped edges exhibits anomalous behavior that would tend to discredit its use, except that its error tends to be one percent or less across the whole range and one percent error is considered acceptable in most structural analysis situations. The real puzzle is "Why does it perform so well in coarse models (six- tenths percent error in a model of nine grid points)?"

A similar convergence occurs in the time domain for transient analysis, as the time interval for integration is reduced the results converge to a more accurate answer.

The direct static response analysis sequence.

In structural mechanics the finite element method is used to calculate the "typical" state of stress in a defined body of finite size (an approximation of Cauchy's stress at a point). Let us recapitulate the steps required in such finite element analyses under the influence of statically applied loads:

<u>In static response analysis:</u>

define the finite elements
- geometry
- material properties

define loading conditions

define the constraints of the structure

numerically integrate the equations of elasticity through each finite element to determine its elastic influence coefficients (stiffnesses)

calculate the mass of each element and distribute its mass properties to each of its grid points

assemble the equations of static equilibrium
- stiffness matrix
- mass matrix
- load vector

partition the equations (matrixes) to determine the elastic solution set

solve equations for all the grid point displacements using Gauss elimination (or some permutation that provides a direct solution of the elastic equations):

$$[k][u] = [p] + [m][d^2u/dt^2]$$

use element boundary displacements to calculate the desired elements' states of strain and stress using Ritz' analysis techniques

Some of these steps involve approximation and others are exact so it is important for us to review them to understand where the approximations lie and why the method exhibits convergence. The quality of the finite element model depends to some extent on the experience of the analyst who defines it and his perceived need for accuracy in the results. But the wide spread use of the method testifies to the fact that method is intrinsically capable of producing valuable analyses.

Nearly all finite element formulations are based upon integrating through a volume bounded by straight sides or flat surfaces. The only exception to this is a curved beam element for which closed form solutions are available (obviating the numerical integration). The accuracy of a finite element model will depend upon how well the elements are distributed to simulate the intricate and often curved surfaces and edges in a structure. The process of defining the elements may therefore involve some approximations of the geometry of the structure but all such approximations converge with mesh refinement.

There is some uncertainty in the properties of materials used in structures but this is an uncertainty of material science, not the finite element method. The computational accuracy of the method is independent of the material properties used.

Loads are all applied at the grid points of a model, even those that are distributed such as pressure and acceleration. Codes that support distributed loads convert them to equivalent grid point loads for use in formulating the equations. The accuracy of distribution of loads improves with mesh refinement.

A single point load will not give accurate results in the vicinity of the grid point at which the load is applied. If results are required in the vicinity of a grid point load, elements must be added so that the element that will be used for stress recovery does not include the point at which the load is applied and grid points used for displacement recovery are not the same as the points where the loads are applied. The greater the distance (in elements) between the applied point load and the recovered data, the better the accuracy.

Constraints and supports for the model follow the same guidance as for loads; for accurate results in the vicinity of a point constraint, increase the mesh density.

The stiffness coefficients for the elements are determined by a numerical integration of the equations of elasticity through the volume of each element. The routines used in most codes are most accurate for element geometries that are equilateral, square or cubic (aspect ratios of 1:1). For skewed geometries (2:1 to 5:1) acceptable accuracies are usually achievable. If the aspect ratios exceed about 10:1 the integration accuracy may be unacceptable and the accuracy may be falling off rapidly. The accuracy of the stiffness coefficient is determined by the aspect ratio of the elements and the integration algorithm used. Badly skewed elements may converge with mesh refinement but to an erroneous answer.

The mass of the model is summed from the mass of each element in it. The mass of each element is determined from the volume of the element and it material properties (density). The mass of each element is then distributed among its grid points. There is no physical basis for this distribution and the quality of the mass distribution depends partly on the type of solution routine to be used. All methods of mass distribution converge with mesh refinement.

Partition of the equations and solution for the displacements involve a direct solution of the simultaneous equations by substitution. As such it is an exact solution of the equations incorporating the stiffness and mass coefficients. The organization of the equations into matrixes is shown in Figure 2. The terms of the stiffness matrix are the influence coefficients determined by numerical integration of each element. The displacement

$$\begin{bmatrix} k_{11} & k_{12} & k_{13} & k_{14} & \cdot & \cdot & \cdot \\ k_{21} & k_{22} & k_{23} & k_{24} & \cdot & \cdot & \cdot \\ k_{31} & k_{32} & k_{33} & k_{34} & \cdot & \cdot & \cdot \\ k_{41} & k_{42} & k_{43} & k_{44} & \cdot & \cdot & \cdot \\ \cdot & \cdot & \cdot & \cdot & \cdot & & \\ \cdot & \cdot & \cdot & \cdot & & \cdot & \\ \cdot & \cdot & \cdot & \cdot & & & \cdot \end{bmatrix} \times \begin{bmatrix} u_1 \\ u_2 \\ u_3 \\ u_4 \\ \cdot \\ \cdot \\ \cdot \end{bmatrix} = \begin{bmatrix} p_1 \\ p_2 \\ p_3 \\ p_4 \\ \cdot \\ \cdot \\ \cdot \end{bmatrix}$$

[Stiffness Matrix] [Displace-ment Vector] [Load Vector]

Figure 2. Matrix techniques provide convenient notation for solving the equations.

vector represents the displacements to be determined. The load vector is the sum of all loads applied to the grid points of the model and may include D'Alembert's inertia forces. The quality of the solution calculations is independent of mesh refinement.

Calculation of the states of strain and stress in selected elements depends upon a numerical integration through the volume of each of the elements. The accuracy of this integration depends upon the aspect ratio of the element as discussed above for determining the influence coefficients, as well as the accuracy of the displacements at each of the boundaries determined in the previous step. This calculation determines one value of strain and stress associated with each element and it should be considered as a "typical" value for the whole element. There is no fixed geometric point in the element to which this value applies in every case (except, possibly, in an abstract mathematical sense). If one wishes to determine the stress at a particular point he must determine the stress in the elemental volume that includes that point and satisfy himself that the elemental volume is small enough to represent the point with adequate accuracy. The accuracy of the stress determination converges as the mesh in the vicinity of the point of interest is refined (but more slowly than the displacements).

The only step in the entire process that is always approximate is the Ritz solution for the states of strain and stress.

The only step in the entire process that is always exact is the simultaneous solution of the elastic equations.

The accuracy of all the other steps depends upon the problem being solved. It is possible to define problems in which they are all virtually exact. In most cases however (complex structures with curved surfaces) they will not be exact but they will all converge to acceptably accurate answers if the element geometries are not too badly skewed.

> **The static response analysis converges to accurate answers (with mesh refinement) because the accuracy of the artificial allocation of distributed quantities among the grid points more nearly simulates a continuum as there are more grid points to be used.**

The direct dynamic response analysis sequences.

Now let us look at dynamic response analyses. There are two basic sequences, transient response analysis and frequency response analysis. Many of the steps in the procedures are similar to the static response analysis steps. The equation solution methods are similar also although the equations are formulated slightly differently. The solution steps are repeated, usually many times, in order to cover the entire time duration or frequency band.

<u>In direct transient response analysis:</u>

define the elements
  geometry
  material properties

define damping

define the constraints of the structure

define time dependent loading

define time intervals for integration

assemble equations of dynamic equilibrium
- stiffness matrix
- mass matrix
- damping matrix
- load vector

partition the equations (matrixes) to determine the elastic solution set

organize the elastic equations (matrixes) to set them up for iterated solutions

look-up the applied load vector for the current time step

solve elastic equations for acceleration at all the grid points using Gauss elimination (or some permutation that provides a direct solution of the elastic equations):

$$[k][u] + [c][du/dt] + [m][d^2u/dt^2] = [p]$$

integrate acceleration once to derive velocity and once more to derive displacements for current time step.

sum the displacement vectors from this and all previous time steps for use as displacement loads in next time step.

iterate for all time steps.

use element boundary displacements to calculate the desired elements' states of strain and stress using Ritz' analysis techniques

The reader may observe that the solution to transient response problems is similar to the solution to static response problem except that the direct solution is repeated a number of times using the integrated results from the previous solutions as part of the instantaneous applied loading. There are several methods for performing this integration and they are all approximate, like the Ritz' method. Most of the integration methods converge to the correct answers as the time intervals are reduced, a convergence in the time domain similar to the other convergences in the dimensional domain (mesh size).

The other difference from static response solution is the specification of damping. Damping is the removal of energy from the kinetic response of a dynamic system. The observed effect usually takes the form of converting kinetic energy to heat through some physical mechanism such as relative motion between magnetic domains in the material, migration of lattice defects in crystalline materials, viscosity effects, etc. Physicists have observed and measured very many damping effects in many materials and been able to relate them to material properties. For the most part these material property effects are small (equivalent to a viscous critical damping ratio, c/c_c, $< .0001$), very much smaller than the values frequently observed in real structures ($c/c_c = .01$ to $.1$). Some of these phenomena are physically simple but computationally awkward in large scale finite element models (viscosity and coulomb friction) while some computationally convenient methods are difficult to rationalize in physics (modal damping and mass damping).

It is fortunate for transient response analysis that the damping specification has only a very small influence on the response unless the damping is extremely high or the transient excitation contains a long duration (several cycles) harmonic motion component. In either of these cases the analyst must be very careful in specifying damping in order to get accurate results. He may need to rely on his experience and intuition (and perhaps do some research) since there are few physical models to guide him.

The amplitude and timing of the transient response depend somewhat upon the specification of damping. There is no convergence test for it. It is a property of the model's materials like density and Young's modulus and the analyst is expected to supply the correct specification. Damping in transient response calculations is often assumed to be zero; an assumption that may provide a worst case response and may be very close to the actual response if the structure is lightly damped.

> **Transient response analysis converges to correct answers as the time step is reduced as well as with mesh refinement as discussed under static response analysis.** The analyst should be sensitive to the uncertainties relating to the specification of damping in direct transient response analysis.

The cost of this analysis may be estimated based upon the cost for the analysis of the static response of the same model. The solution of the equations at each time step is essentially equivalent to a static solution so that if one hundred time steps are being analyzed it will be equivalent to one hundred static analysis solutions (one hundred of the equation solution steps, but none of the preceding steps necessary to formulate the equations which occur only once and may be half to two thirds the cost of performing a single static load case analysis).

In direct frequency response analysis:

- define the finite elements
  - geometry
  - material properties
- define frequency dependent loading conditions
- define the constraints of the structure
- define damping
- define frequencies for solution
- assemble equations for harmonic motion
  - stiffness matrix
  - mass matrix

damping matrix
load vector

partition the equations (matrixes) to determine the elastic solution set

organize the elastic equations (matrixes) to set them up for iterated solutions

solve the elastic equations for acceleration and phase at all the grid points using Gauss elimination (or some permutation that provides a direct solution of the harmonic motion equations):

$$\{[k] + 2i(pi)f[c] - 4(pi)^2f^2[m]\}[u] = [p]$$

derive velocities and displacements from the equations for harmonic motion and the acceleration amplitudes.

iterate for all frequencies.

use element boundary displacements at frequencies of interest to calculate the desired elements' states of strain and stress using Ritz' analysis techniques

The solution of the harmonic motion equations and the derivation of velocities and displacements from them is an exact procedure based upon the sequential elimination of the variables after the desired frequency is substituted into the equations. As a result, all the comments associated with static response analysis apply here as well. The single exception is the specification of damping.

Frequency response analysis is even more sensitive to the appropriate specification of damping than is transient response analysis, above. Without damping all resonant responses will be infinite and it is the resonant responses that in reality limit the utility of structural systems in a vibration environment. The analyst should apply damping to achieve reasonable results and agree as closely as possible to experimental data for similar structures.

Direct frequency response analysis converges to correct solutions with mesh refinement as discussed under static response analysis. The analyst must be sensitive to the potential uncertainties associated with the specification of damping in direct frequency response analysis.

The cost of direct frequency response analysis may be estimate from the cost of a static response analysis on the same model in a manner similar to direct transient response analysis. Note that the solution at each frequency interval is roughly equivalent to the solution at one time step in the direct transient response case.

The above three procedures are the basic constituents of the finite element method as practiced today. Each of the sequences is complete, meets the convergence criterion and has proven to give accurate results. All three converge toward exact solutions with refinement of the position mesh of the grid points in the model. The transient analysis also converges toward exact solutions with refinement of the time steps used for integration.

Reducing the problem size.

Analyses in the time and frequency domain as outlined above may be relatively expensive (by factors of 10 to 1000 or more depending upon the number of frequencies or time steps involved) compared to a static analysis. Much effort has gone into developing dynamic analysis techniques that are competitive with static loads analysis. Two methods have resulted from this effort:

- mode extraction and modal superposition
- matrix reduction
  - Guyan reduction
  - generalized reduction

Both of these methods of reducing the size of dynamics problems have proved themselves valuable in the calculation of stresses in structural components of systems. But, they do not fully satisfy the convergence criterion for accuracy.

Mode Extraction and Modal Superposition:

Modal superposition techniques in either the time or frequency domain require the initial extraction of the eigenvalue/eigenvector sets (modes or roots) of the characteristic equation of the system. If the system has a thousand equations of motion (displacement components or degrees of freedom) the characteristic equation will be of degree one thousand and will therefore have one thousand modes (solutions) each of which will be associated with a vibrational resonance mode of the model.

Not all of the modes are equally significant to the critical structural response quantities of the system. In dynamic stress analysis it has been demonstrated that the highest mounting base stresses are usually associated with the motion of the largest proportion of the mass of the system; and in turn, these motions are usually associated with the lowest resonant frequencies. For this reason the eigenvalue extraction techniques used in finite element analysis work emphasize the accurate extraction of the lowest "n" eigenvalues, where "n" is specified by the analyst. Or, if the analyst specifies a frequency band of interest, the routine determines the number of eigenvalues in the band of interest (using a Sturm sequence check) and sets "n" equal to this number.

Modal superposition techniques assume that the structural response can be completely characterized by the eigenvalue/eigenvector sets retained from the extraction process.

<u>In modal transient response analysis:</u>

define the elements
- geometry
- material properties

define damping

define time dependent loading conditions

define time steps for solution

assemble characteristic equation
- mass matrix
- stiffness matrix
- dmaping matrix

extract desired roots of characteristic equation (modes or eigenvalue/eigenvector sets)

apply load and damping to each of the modes

calculate the response of each mode as a single degree of freedom system under the influence of the loads for the current time step

integrate the response of each mode during the time step to determine the velocities and displacements

sum the contributions of each mode at each time step to calculate displacements at each time step

use these displacements as part of the loading for the next time step

repeat the calculation through all time steps

use element boundary displacements at times of interest to calculate the desired elements' states of strain and stress using Ritz' analysis techniques

<u>In modal frequency response analysis:</u>

define the finite elements
- geometry
- material properties

define damping

define frequency dependent loading conditions
define frequencies for solution

assemble characteristic equation
- mass matrix
- stiffness matrix
- damping matrix

extract desired roots of characteristic equation (modes or eigenvalue/eigenvector sets)

apply load and damping to each of the modes

calculate the response of each of the modes as a single degree of freedom system under the influence of the loads at each frequency

Sum the contributions of each mode at each frequency

repeat for each frequency of interest

use element boundary displacements at times of interest to calculate the desired elements' states of strain and stress using Ritz' analysis techniques

Eigenvalue extraction is of necessity an approximate process since it involves solving an algebraic equation of degree equal to the number of elastic equations. There is no way to solve algebraically for the roots of an equation of degree five or higher (the fourth degree equation solution is the highest available) and most characteristic equations are of very high degree.

There are a number of different methods available for the analyst to solve for eigenvalues and eigenvectors, each with its own benefits and limitations. All are approximate and most converge with mesh refinement. Some allow the analyst to specify or adjust a convergence parameter value to improve the accuracy of the solutions (at some additional cost). Extraction methods are evolving continuously, mostly to find economic methods for extracting a few eigenvalues from a characteristic equation of very high degree and to resolve closely spaced or even congruent modes.

There are two principle reasons for extracting solutions to the characteristic equation,

1) to determine the resonant frequencies of a structural model in anticipation of setting-up the direct solution of its dynamic equations and

2) to reduce the number of equations carried through to subsequent dynamic analyses using modal superposition techniques.

The adequacy of the eigenvalue extraction technique used depends upon the anticipated application. If the eigenvalues are to be used to determine the solution frequencies (for frequency response) or the integration time step (for transient response) in a direct solution of the full set of equations, then the analyst only needs to verify the number of resonances (eigenvalues) in the frequency band of interest (by a Sturm sequence check) and obtain eigenvalues that are accurate with respect to the frequency intervals or time steps to be used in the subsequent analysis. These comments apply to reason 1) above.

If modal superposition techniques are to be used to calculate the response then the analyst must be much more concerned with the quality of the eigenvalue/eigenvector sets (orthogonality, etc.) and the accuracy of the eigenvalues in the frequency domain. He must also assure himself that the responses of the "features of interest" are included in the eigenvalue/eigenvector sets that he has extracted. In the stress analysis of mounting bases only the lowest frequencies may be required based upon sound physical reasoning. However, in general dynamic analysis there may be no physical reasoning that can lead the analyst to a proper selection of eigenvalue/eigenvector sets. These comments apply to reason 2) above.

Both modal solution sequences will converge to a fixed value as the position mesh is refined and as the size of the time step is reduced (modal transient response analysis), but the fixed value is not the exact answer:

> **both sequences result in a fixed residual error associated with information in the eigenvalue/eigenvector sets not used in the solution.** The analyst is left to determine whether the residual error is acceptable.

Matrix Reduction:

Matrix reduction allows the analyst to reduce the size of his original matrixes (the number of equations to be solved) by a dramatic figure. In Guyan reduction the analyst specifies the components of motion that are to be retained in the final matrix and the Guyan reduction routine reduces the stiffness and mass matrixes to connect only these components of motion (degrees of freedom). In general the analyst attempts to capture the components of motion that comprise the principal masses in the model since these are usually associated with the lowest resonant frequencies, and the lowest resonant frequencies ordinarily produce the highest stresses in the mounting base of the instrument. Guyan reduction has been shown to be equivalent to Gauss' elimination (the conventional method for solving statics problems) of the undesired degrees of freedom and if used in the static solution sequence will give answers equal to a solution of the full (unreduced) set of equations (at essentially the same cost as the unreduced solution). It is therefor rarely used in static solution sequences.

It has been shown that Guyan reduction does not perform quite so well in dynamic analysis. This is because although there is a physical basis for the algorithm for reducing the stiffnesses of the structure to the retained degrees of freedom (grid points) there is no physical basis for an algorithm that assigns the mass of the model to the retained degrees of freedom. For this (and several other) reason(s) an analyst may choose to retain three or four times as many degrees of freedom in his model as would be strictly required to extract the desired number of resonant frequencies; the additional degrees of freedom help to better distribute the mass in the reduced model, improve the chances of capturing the lowest resonant frequencies and thereby improve the quality of the response calculations.

There are generalized reduction techniques that are designed to attempt to guarantee the analyst that he captures the lowest resonant frequencies in his model as he defined it and to assure the accuracy of the resulting eigenvalues and eigenvectors. Typically the routines may achieve an accuracy of ninety-nine percent (one percent error) in the resonant frequency with only 1.5 to 2.0 times the desired number of modes (considerably fewer than required by Guyan reduction). The cost of generalized reduction and eigenvalue extraction techniques is much greater than that of Guyan reduction but this cost is often recovered in the subsequent modal superposition analyses since fewer modes are required for accuracy. This is because the generalized routines use additional "generalized" degrees of freedom to which they reduce the mass and stiffness matrixes and this is thought to give much better distribution of the mass and stiffness among the retained "generalized" degrees of freedom.

In generalized reduction the analyst specifies either the number of resonances to be used or the frequency range of interest. The reduction algorithms reduce the mass and stiffness matrixes to generalized coordinates of motion that are sufficient to cover the frequency range of interest or the number of resonances. Note that the generalized coordinates are not physical points or degrees of freedom that the analyst defined in the model but rather computational degrees of freedom required for the reduction process and defining the characteristic equation. Once the extraction process is complete the generalized coordinates and related matrixes are discarded. Only the extracted roots of the characteristic equation are retained for use in modal superposition solution sequences. The reduction algorithm normally uses a Sturm sequence check to verify the number of resonances in the frequency band.

Guyan reduction techniques reduce the matrixes to real components of motion defined in the model (grid points) and such a reduced model may be used in either direct or modal analysis solution routines.

On the other hand generalized reduction techniques do not produce reduced matrixes but rather eigenvalues and eigenvectors (and transformation matrixes) which characterize the structure. This form of reduction can only be used in subsequent modal analysis solution routines.

At this point it should be clear to the reader that matrix reduction techniques and modal analysis techniques are nearly equivalent methods for reducing the size of some problems by eliminating what is thought to be unnecessary detail. The distinction between the two is that Guyan reduction techniques may subsequently be used in both direct solution sequences and modal superposition solution sequences whereas automatic reduction and mode extraction may be used only in modal superposition techniques.

If one uses Guyan reduction on a problem prior to extracting the roots of the characteristic equation the eigenvectors and eigenvalues will contain the uncertainties from both the reduction process and the root extraction process. All analyses using reduced models will converge to a value which includes a fixed residual error depending upon the modes retained and mass distribution algorithm (for modal superposition solutions) or the errors in mass distribution in the reduced mass matrix (for Guyan reduction) or both if Guyan reduction and modal superposition are used in the same solution.

Both matrix reduction solutions will converge to a fixed value as the position mesh is refined and as the time step is reduced, but the fixed value is not the exact answer:

> **matrix reduction methods result in fixed residual errors associated with information in the roots not used in the solutions (generalized reduction), mass distribution errors in the mass matrixes (Guyan reduction) or both (if Guyan reduction is performed on the equations before the mode extraction is performed).** The analyst is left to determine whether the residual error is acceptable.

One way analysts may check the accuracy of a reduced matrix dynamic analysis is to run occasional cross-checks with the full unreduced model in a direct solution sequence. These checks may be at selected frequencies or time steps (with known initial conditions) and will indicate the error to be expected in the reduced model results. The reader should note that the finite element method is capable of computing results for any of these models in direct solution sequences without mode extraction and superposition or Guyan reduction. Such solutions will satisfy the convergence criteria in both the dimensional mesh and time domains, i. e., **there is no more accurate calculation possible for the constitutive equations of the problem than that assembled by the analyst in his finite element model and solved in unreduced form in a direct solution sequence.**

One reason for running reduced models is "conservation of computational resources" (direct solutions may be expensive). If the analyst cannot find sound physical reasoning to verify the accuracy of his results (as has been done for mounting base strengths in stress analysis) he always has recourse to the full equation set in a direct solution sequence as a computational baseline.

Compensating for the residual error.

Stress analysts often compensate for the residual error in reduced models by increasing the mass (and stiffness) of the model so the zero frequency peak response is equal to the equivalent steady acceleration response from a static solution. This is known as the "residual stiffness" or "residual mass" technique for stress analysis and is occasionally used in safety related analyses. Note that if the mass and stiffness are changed uniformly in the same proportions the eigenvalues and eigenvectors will be unchanged.

The residual stiffness technique can only be applied rigorously to a single response quantity in the model, say one of the stresses in a particular finite element or one of the displacement components at one grid point. It is not possible with this technique to correct the responses of several quantities, let alone a complete model, so a residual error of unknown magnitude will persist in the remainder of the model. Also, it is theoretically valid only at zero frequency (or the frequency at which the correction is made).

The magnitude of the fixed errors cannot be predicted. Residual stiffness corrections assume the error is negative (calculated responses are smaller than actual responses). This assumption is valid only for the stresses near the fixed mounting base and at zero frequency. Other responses may be greater or smaller. The writer knows of no way to bound the error (set an upper and lower limit on its magnitude).

From the above discussion it should be clear that the finite element method is a very highly developed tool for performing stress analysis to assure the safety of structures. The codes may be expensive to run, especially for large problems and the codes have been optimized to produce the most accurate stress results for the least cost (computer run time).

Resolving stresses.

The finite element method is not an approximate method in the sense that the true solution is approached iteratively and the analyst may specify a greater number of iterations to improve the accuracy. The finite element method calculates the true solution with the most direct computational techniques available. Some of these techniques are necessarily approximate since exact closed form solutions are absent. Only in eigenvalue/eigenvector extraction is a convergence parameter sometimes available.

The finite element method does not calculate stresses at points, it calculates the stresses typical of a finite sized region of the continuum. If one wishes to determine the stress at a point he must be sure that the finite shape that surrounds the point is representative of the point, i.e., that the stress gradients across the element are small with respect to the accuracy with which the stress must be determined. The finite element method is not capable of calculating the stresses at the points that bound the finite shape (grid points), i. e., at edges or on surfaces.

Many software codes will print out estimates of the stresses at the points that bound the finite elements. These estimates can be no better than the element values since the algorithms that are used to generate boundary point stresses are principally extrapolation functions and they are at best only partly derived from the equations of elasticity.

One way to assess the quality of these algorithms is to observe the values of the grid point's stresses calculated by each of the elements that share the grid point. Differences in the grid point's stresses may be due to errors in the element stress calculation or errors in the extrapolation function or both. It is also possible to have off-setting errors in the two calculations so that one observes apparent good agreement among the grid point stresses when in fact the two errors may be significant, of the same magnitude but opposite sense, so the effects of the errors nearly cancel each other.

To further aggravate the problem many codes will generate contour plots of the grid point stress data. In these "post processors" the stress at a point is assumed to be the average of the grid point stresses in each of the elements that is connected to the point in question. If the point in question has four stresses associated with the grid point stresses of each of four shell elements, the grid point stress will be the numerical average of the four stresses, assuming that each of the elements is attempting to approximate a unique value of stress at the point.

But consider that a unique value of stress may not exist at the point in question if the shell elements are not all the same thickness; a dual value of stress would be appropriate at each such point in the case where the thickness is discontinuous and two thickness are involved. In fact there should be as many stress values at a point as there are thickness for the shell elements connected to the point. The stress contours should be discontinuous across such a structural discontinuity. Most software post-processors that perform contour plotting of stresses ignore structural discontinuities and simply plot the average stresses at the points as described above.

The finite element method as currently used cannot calculate stresses at the grid points of a model. Rather, the finite element method has been optimized to accurately calculate the stresses in a finite body. These may be uniquely determined by using only six equations (direct strain, three equations and shear strain, three equations). Knowing the x, y and z displacements of each corner of a rectangular parallelepiped it is believed that one may calculate the internal state of strain (and stress) in the element. Considering the cost of running any of the larger finite element codes and the competitiveness in the market for them it is reasonable that the proprietors of the codes would attempt to minimize the number of computations necessary to achieve their goal: determine the stresses in an element.

However, if the full set of nine equations (including the three rotational components) is used then the three rotations at each point would also have to be used in the solution, thereby doubling the size of every problem run in the code.

A Computational Singularity.

There is at least one other difficulty in the finite element method's calculation of stresses. If one attempts to calculate the stresses in the vicinity of a free edge between two different materials the results may be very wrong.

Figure 3 shows a finite element model of a plane section through an adhesive bond line between a rigid base and an aluminum plate (Hatheway, 1989). The writer was interested in the strength of the adhesive bond under the influence of a bending moment. The moment caused tensile (peeling) stresses in the adhesive, this stress being highest at the end nearest the applied bending moment. Attempts to determine the stress at the end of the bond line and next to the aluminum (in the interface stress plane) were disappointing, the results being unreasonable. Furthermore, as the mesh of the model was refined the unreasonableness was aggravated. The convergence criterion was violated.

A plot of the stresses in the bond line next to the aluminum, Figure 4, shows reasonably good behavior some distance from the end of interest but at locations approaching the free edge of the adhesive (the end where the moment is applied to the aluminum plate), the stress values begin to oscillate (with a cyclic period equal to two elements). Furthermore, as the mesh is refined in the model the cyclic oscillations in the bond line stress become more pronounced (diverge), with the cyclic period shortening to agree with the new element dimensions. Figure 4 is based upon grid point stresses but the element stresses showed the same behavior.

A regression analysis of the stress of interest shows (Figure 5) that as the mesh dimension was halved the stress value nearly doubled, indicating that as the mesh is refined the calculated stress increases without limit, an unrealistic conclusion.

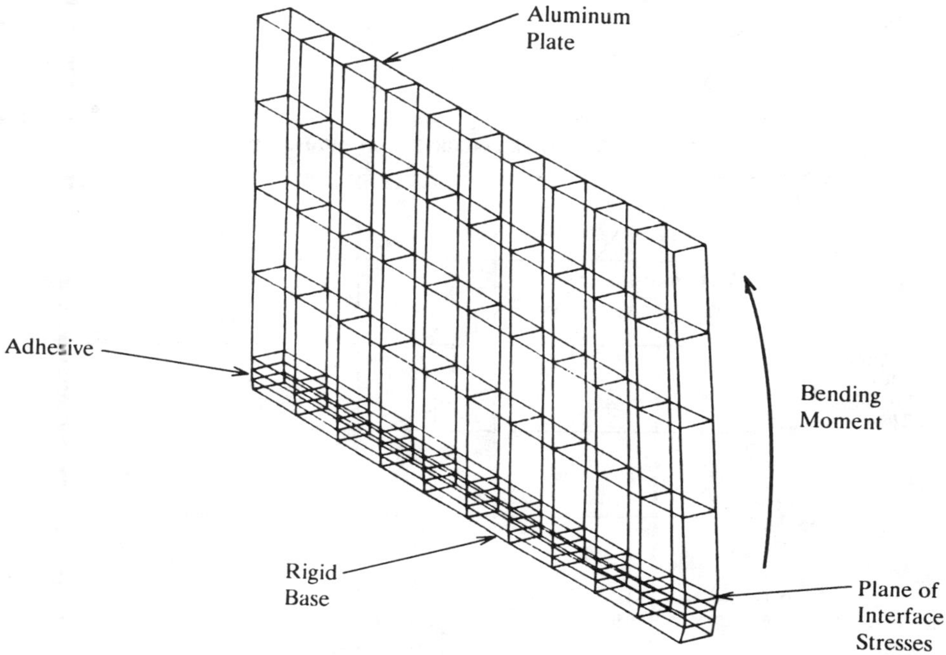

Figure 3. A finite element model of a plane section through a portion of an adhesive bond joint.

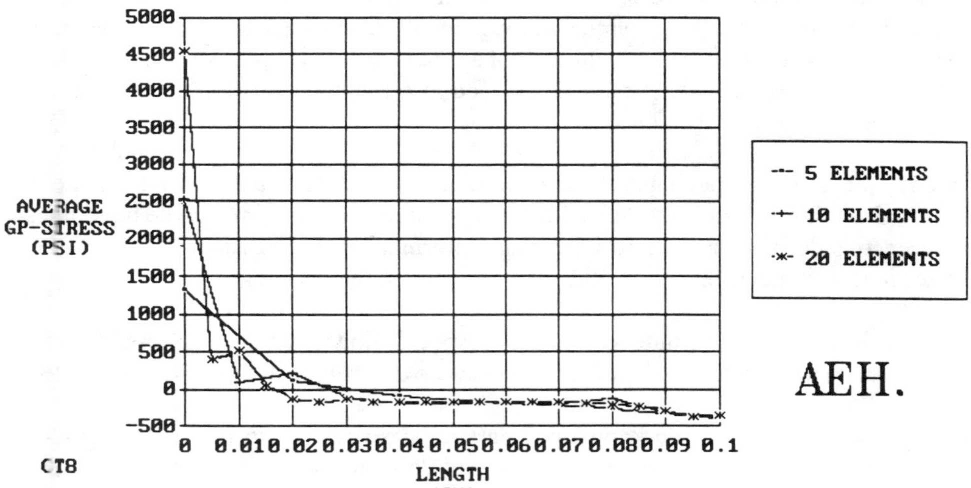

Figure 4. Stresses at the loaded end of the model tend to <u>diverge</u> with mesh refinement..

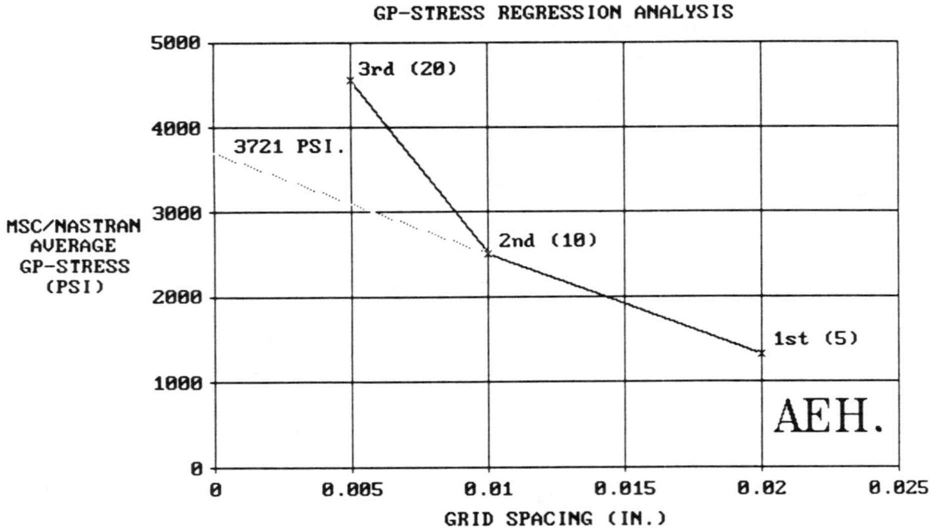

Figure 5. The stress increases without limit (<u>diverges</u>) as the mesh is refined.

On the other hand however, if one plots the displacements of the points on the bond line, a smooth monotonic curve results, as one might expect. Further if one performs a regression analysis on the displacement of the point at which the stress is desired he finds that it converges and approaches a finite limit as the mesh approaches very small values (Figure 6). The displacements can also be used to estimate the stress in the element, which appears to be a reasonable level.

What appears to be happening in the computational sequence is that the finite element code is calculating the translational deflections properly for the points on the bond line. However, this does not appear to be enough information for the Ritz solution of the stresses in the elements, especially the element nearest the free edge.

The difficulty appears to be independent of the specific finite element code used for the solution; the writer has run the identical model in two different codes with similar results.

This is an example of a little-known computational singularity in the finite element method, and appears to be sensitive to the value of Poisson's Ratio; if two materials have different Poisson's Ratios at a free edge, the method cannot calculate the stresses (apparently).

This computational problem may be related to the missing rotational degrees of freedom.

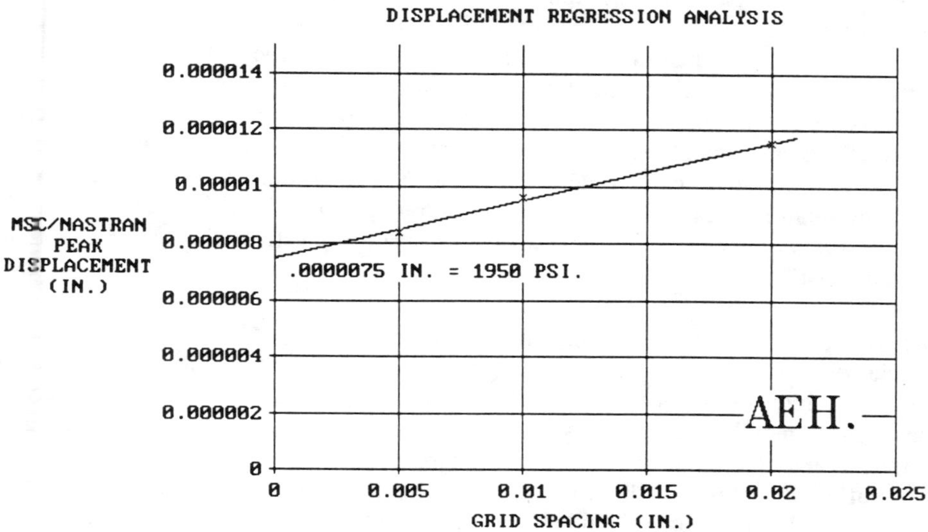

Figure 6. The displacement approaches a finite limit (converges) as the mesh is refined.

If the finite element solution uses only the first six equations of elasticity (relating to direct and shear strains) and ignores the three "rotation components" (since they are not ordinarily needed to determine the internal state of stress or strain) it may have insufficient information on the boundary of the element in the problem described above to perform a non-singular Ritz solution for the interior states of the element. On the other hand if the "rotation components" are included in the formulation then one would have additional equations in each element to reduce the singularity. Moreover, he could almost certainly calculate not only the state of strain for the element but also the state of the three strain gradients as well. (Note that this would also provide an algorithm, based upon elasticity, for calculating states of stress and strain at the boundary points of an element, i. e., grid point stresses.)

It is true that this would increase the size of a problem, but the increased accuracy in calculating the grid point stresses would allow some reduction in the mesh density, partly compensating.

Summary of the finite element method's capabilities and limitations.

The finite element method is an economical and reliable tool for the analysis of structures, especially tailored for strength and safety analysis of civil structures, i. e., bridges, dams, buildings, aircraft, etc. It has been shown to converge to the exact solution of the equations as the dimensional mesh is refined and as the time steps (in transient analysis) are reduced. It offers reduced cost dynamic analysis through various matrix size reduction procedures. The residual error in reduced matrix solutions can always be assessed, if not by physical reasoning then by spot checks against the full equation set. On the other hand, the method appears to be incomplete as presently implemented in commercial codes. This incompleteness is manifested in the omission of rotational degrees of freedom from the solution for displacements in two and three dimensional elements where the historical development of the closed form solutions became exhausted. The omission of these rotational degrees of freedom is tantamount to asserting that the rotations do not exist, a claim that is controverted by the evidence readily at hand. The rotational degrees of freedom exist in the physical world, they are included in the equations that define elasticity and they in fact occur implicitly in the behavior of finite element elastic models.

The sense of incompleteness is augmented by the existence of at least one computational singularity; a situation in which the codes have insufficient equations to solve for meaningful interior states of strain and stress in the elements. One such is described above, there may be others. It is possible that the computational singularities are a symptom related to the missing rotational degrees of freedom. Whatever the source or the solution, it seems that there is still something amiss in the finite element solution method and that it is not yet a fully kinetic solution to the equations of elasticity.

The ambiguity and uncertainty surrounding damping specifications cannot really be called a shortcoming of the finite element method. But the absence of a physically sound theory for the dissipation of the kinetic energy in structures is a very real and serious limit on the finite element method's ability to analyze dynamic systems. In safety related strength analysis this shortcoming has been overcome by agreeing to "conservative" values of the dissipation to expect in specific applications and then demonstrating that a design has "safe margin" under the "conservative" assumptions.

In other disciplines however, such as optics, where the instantaneous magnitude of the response quantities is important for predicting the performance of an instrument, the uncertainty surrounding the specification of damping may be overwhelming, especially in vibration related analyses. It is not uncommon for the "conservative" specifications of damping to be a factor of two or three greater than that justified by subsequent test data. Such a large uncertainty is a very large challenge to the designers of instruments relying on structural analysis to define their performance capabilities and limits.

So, although damping is not really a part of the theory of elasticity, it has a major influence on the utility of elastic theory in dynamic analyses. It is probable that the uncertainty concerning damping specifications will persist for some time. Even if some physical basis for damping can be agreed to, it is not clear that it would be computationally compatible with the finite element method as presently formulated; two examples, 1) viscous flow damping might be based upon the Navier-Stokes equation which is non-linear and therefor insoluble using conventional methods, or 2) simpler viscous damping terms might be used but they often generate unsymmetrical matrixes which multiplies the cost of solution by a sizeable integer.

APPLYING THE CAPABILITIES OF THE FINITE ELEMENT METHOD.

The reliability and economy of the finite element method has led to applying its techniques to many other disciplines; heat transfer, control systems, fluid mechanics, electro-magnetics and others. In fact many existing structural analysis codes, such as NASTRAN, already include some explicit capabilities (by analogy to structural mechanics) in heat transfer, control systems and fluid mechanics in order to investigate the interactions between these disciplines and structural mechanics (thermo-elastic analysis, servo-elastic analysis and aero-elastic/flutter, analyses).

There has been an attempt to couple the finite element codes to other codes used in other disciplines in order to perform sequential analyses in multi-disciplinary systems. A typical analysis in such a system might perform a thermal analysis in SINDA and then send the temperatures to NASTRAN to be use as a load in a thermo-elastic analysis; or NASTRAN's eigenvectors may be sent to $MATRIX_x$ for the design and analysis of a structural servo control system; or surface displacements from NASTRAN might be sent to CODE V for optical evaluation of the surface distortions.

This is a system of analysis called "Integrated Analysis" (Figure 7). Integrated Analysis has not performed well, largely because it has not developed an adequate physical basis

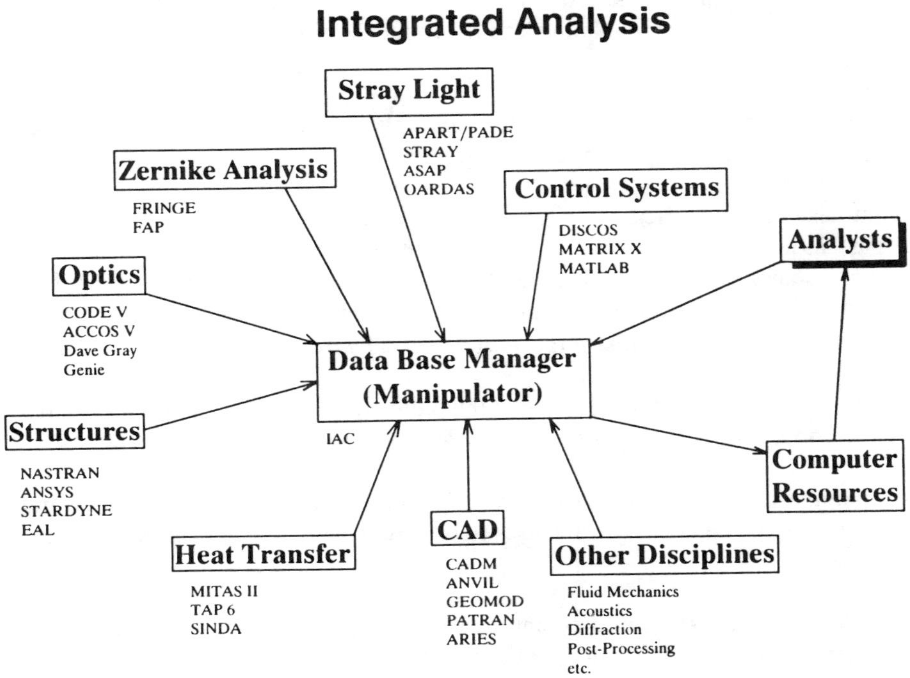

Figure 7. Integrated Analysis exchanges data (input and output) among the various disciplines' analysis codes.

for the interpolations and transformations necessary to interface the codes and disciplines:

1) if one takes the results of a two hundred node thermal analysis in SINDA (a large thermal model) and applies it to a two thousand grid point structural model (a moderate size structural model), **there is no physical basis or rule for the expansion of the two hundred temperatures to two thousand temperatures** and one may expect sizeable uncertainties in the thermo-elastic analysis;

2) if one takes a two thousand grid point structural model and reduces it to one hundred eigenvectors for use in design of a control system for an optical structure **there is no way to be sure that they include sufficient high frequency response in the vicinity of the actuators or optical element support structures;**

3) if one takes the distortion of ten optical surfaces in a finite element code and reads them into an optical design code for evaluation **he may loose accuracy of between seven and twenty-five percent per surface, due to interpolation,** a root-mean-square loss in accuracy of between twenty-two percent and eighty percent for a ten surface optical system.

Without a strong foundation in physics it should not be surprising that the many translations, interpolations and transformations involved in Integrated Analysis create cumulative errors of overwhelming size in all but the simplest, most mundane sample problems.

But the question remains, is there a way to apply the principles of the finite element method to broader multi-disciplinary analyses?

The success of the explicit analogies in the commercial codes has spurred a modest "cottage industry" among analysts that use the more capable codes to define new "implicit" analogies in order to increase the number of disciplines that may be submitted to the rigor of the finite element method. These implicit analogies are occasionally reported at the colloquia and symposia sponsored by the software code suppliers.

Paxson has investigated the use of the finite element method to design controllers for both thermal control systems (1988) and laser beam pointing systems (1989);

Genberg has analyzed the illumination in a copier's light box (1983) and the sensitivity of plastic optical lenses to humidity (1987), both using a finite element code;

Cifuentes (1989) created a combined elasticity/viscosity shell element for use in a finite element code to analyze the elasto-hydrodynamics of high speed magnetic tape gliding across a static tape head;

Mera (1986) successfully coupled hexahedral acoustic elements to elastic shell elements to predict the cabin noise generated by the jet engines on a new transport aircraft;

the writer (Hatheway, 1984) has developed analogies that facilitate the element by element ray tracing of optical systems (including evaluation of image motion and quality) in the finite element method.

Figure 8 is a schematic representation of the equations of each discipline set up in matrix form for solution in a single finite element code, an approach called "Unified Analysis."

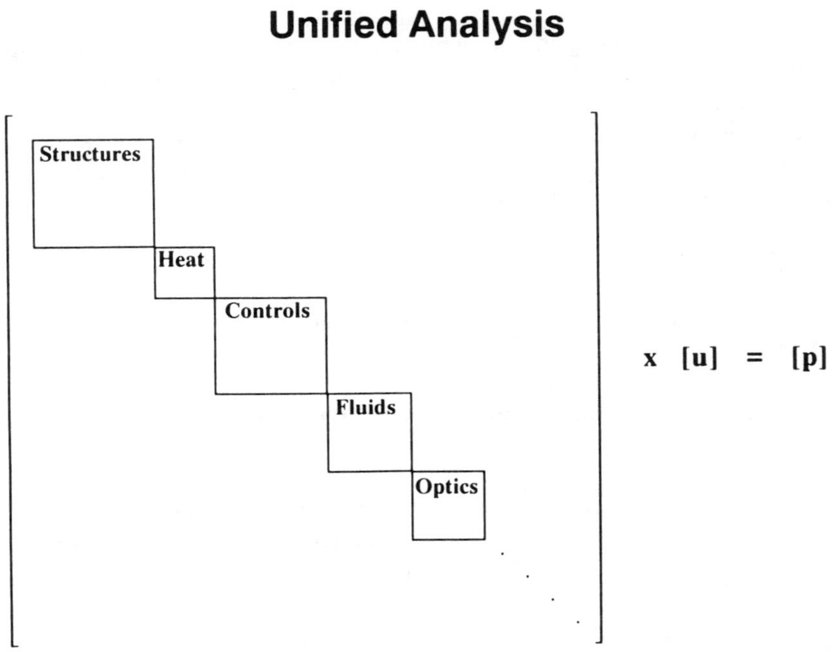

Figure 8. Unified Analysis organizes the equations of all disciplines for simultaneous solution in one software code.

After a decade of disappointment using Integrated Analysis techniques, the use of explicit and implicit analogies has set the analytical arts on the road to affordable and accurate computational results.

This past century saw fragmentation of the field of mechanics into dozens of disciplines. It now appears that the finite element method may be a unifying force, bringing them back together. By organizing the equations of each discipline for simultaneous solution in the same matrixes, Unified Analysis offers the fastest, most economical and the most accurate computation system available.

The challenge of applying Unified Analysis techniques is in organizing the equations of each discipline for solution. To this end it would be very helpful, especially in the analysis of optical systems, if elastic finite elements could be developed that included the rotational degrees of freedom in the solution. Slope changes on the surfaces of optical systems are the biggest contributors to loss of performance in a system. Presently the analysts need to "fool" the code into providing this information by adding structural elements to the model. The results are reasonably good (much better than interpolation) but the technique is only semi-physical and does not produce the exactly "right" values for the slope changes.

WHAT ARE THE PROSPECTS FOR THE FUTURE?

One way to gather a feeling for the demands that are going to be placed upon the engineering arts is to attend a gathering of astronomers and astrophysicists who are discussing future plans. The writer was fortunate to be able to attend the ASTROTECH 21 Optics Technology Workshop in March of 1991. The meeting was a gathering of NASA and industry scientists, hosted by Cal. Tech.'s Jet Propulsion Laboratory, to plan the next twenty to fifty years of projects in space optical physics.

All of the projects discussed at the gathering shared three aspects in common;

> the instruments are going to get bigger, typified by twenty to fifty meter apertures in imaging systems;
>
> the structures will be unconventional, such as a filled Fizeau telescope with a thirty meter "cross" dilute aperture; and
>
> the experiments will demand extraordinary precision; diffraction limited performance in the visible and near infrared and visible light interferometers with reference baseline lengths of 10^1 to 10^7 kilometers.

The writer is certain that each of these three aspects would be echoed by any group of scientists gathering their plans together for the next couple of generations. Terrestrial astronomy, particle physics, gravitational research, fusion power; all their experimental apparatuses will be bigger, more innovative and more precise than anything we've seen before

Just how will the designers of such instruments and systems satisfy themselves that they can be made to work; that the complicated interactions among all the portions of the apparatus support the experiment rather than interfere with it?

Size, innovativeness and precision do not come together for free. We need to start planning, and investing in, the tools necessary to design and construct these truly amazing instruments. We need to view each of these projects as if each of us be builders and constructors with a vested interest in making them work, otherwise we will not bring to the planning activity sufficient emotional commitment to successfully press for the funding and other resources necessary to have the tools in place when they will be needed.

1) The finite element method must become fully kinetic, offering accurate solutions for all the response quantities needed by the various disciplines that rely on it for more than safety related stress analysis.

We cannot count on the structural mechanics community to undertake this effort or guide it along all by themselves. This community is happy with the method as it is today and sees no need to extend it. Similarly, the community of software writers and proprietors who supply the finite element analysis codes sees no commercial advantage in these proposed enhancements since they will be useful to only a small community of analysts and the market for analysis codes is very competitive.

2) Damping effects must be integrated into the computational procedure with a rigor and accuracy comparable to elasticity.

The quality of the results from the new optical instruments will be driven by dynamic effects, not static effects; witness the growing number of instruments that rely on servo systems to control and stabilize their behavior. If computationally compatible damping theories are not possible, the true physical phenomena must be related to the available computational procedures in such a way as to guide analysts in the appropriate application of damping principles to dynamic systems analysis. Only by tying computational methods to physical principles can reliable dynamic analyses be made.

In this area the structural mechanics community may be of some support. They are aware of the limitations placed on their analyses by the poor theoretical basis of damping. They have some procedures that assist them in the stress analysis of structural bases but they are also aware that as their problems become more complex, their dynamic analyses rest on much less rational assumptions about the dynamic behavior of structures.

3) A new computational (finite element) vehicle should be developed which fully supports all the disciplines of interest and in which all disciplines will feel comfortable.

It is clear that Unified Analytical techniques can support the interdisciplinary efforts required by the planned projects but the, seemingly arbitrary, limitations of the present codes (the combined result of lack of foresight, commercial considerations and budget constraints) need to be lifted and adequate lucid documentation must speak easily to the users. There are some codes that currently exist, one of which might provide an excellent starting point for this effort; there is no reason to rewrite solution routines that all ready exist, and most of them do.

Again, however, the existing communities of users and providers of finite element codes will probably not be interested in making this happen. The instrument designers, will have to take the initiative.

4) A new computational engine may be needed by the end of the decade, even if the software developments described above materialize.

The large space instruments will need very large finite element models to capture their behavior because their sparse structures will require fine mesh models to capture the full band width of their dynamics. Solution of a few hundred thousand to a million simultaneous equations may be necessary, assuming an evolutionary extension of the current Unified Analysis methods to models of the advanced systems discussed above.

A revolutionary approach should also be considered; extending the finite element method to the solution of Maxwell's equations for electromagnetic radiation. Such an extension would eliminate the (current) need for analogies for including the optical phenomena in the structural analysis codes, and open the field to rigorous and complete simultaneous computation of diffraction effects, scattering effects, stray light suppression and other phenomena which are today the province of separate highly specialized codes. Modeling a one meter class instrument operating in the visible wave length band in such a code would generate matrixes of the order of 10^{21}, a monumentally large number by today's practice (10^4 to 10^5). Numerical methods and software may be able to ease the computational load some but the number of equations to be solved will still be larger than anything currently contemplated.

REFERENCES

ASTROTECH 21, Optics Technology Workshop, Pasadena, CA, March 1991.

Bathe, Klaus-Jürgen and Wilson, Edward L., Numerical Methods in Finite Element Analysis (Englewood Cliffs, New Jersey: Prentice- Hall, 1976) pp. xiii-xv.

Boyer, Carl B.; Merzbach, Uta C., A History of Mathematics, 2nd. ed. (New York: John Wiley & Sons, 1989).

Cifuentes, Arturo O.; Bock, Timothy L. and Coppolino, Robert N., "A Finite Element Technique for Tape-Head Interaction Problems in High Speed Recording: the Steady-State Clause," The MSC 1989 Word Users Conference Proceedings (Los Angeles: the MacNeal-Schwendler Corporation, 1989).

Genberg, Victor; Oinen, Don and Fronheiser, Sharon, "Diffuse Illumination with MSC/NASTRAN," MSC/NASTRAN Users' Conference Proceedings (Los Angeles: The MacNeal-Schwendler Corporation, March, 1983).

Genberg, Victor, "Optical pathlength calculations via finite elements," Structural Mechanics of Optical Systems II, Volume 748 (Bellingham: SPIE, January, 1987).

Hatheway, Alson E., "Finite element methods for evaluating optical system performance," Optical System Engineering IV, Volume 518 (Bellingham: SPIE, October, 1984).

Hatheway, Alson E., "Evaluating Stresses in Adhesive Bond Lines," The MSC 1989 World Users Conference Proceedings (Los Angeles: The MacNeal-Schwendler Corporation, March, 1989).

Love, A. E. H., A Treatise on the Mathematical Theory of Elasticity, 4th. ed. (New York: Dover Books , 1944 (1927)), pp. 1-31.

MacNeal, Richard H. and Harder, Robert L., "A Proposed Set of Problems to Test Finite Element Accuracy," Finite Element Validation Forum, (Washington, D. C.: American Institute of Aeronautics and Astronautics, 1984).

Mera, A.; Yantis, T. F.; SenGupta, G. and Landman, A. "MSC/NASTRAN Implementation of Coupled Structural-Acoustic Response for Aircraft Cabin Noise Prediction," MSC/NASTRAN Users' Conference Proceedings (Los Angeles: MacNeal-Schwendler Corporation, March, 1986).

Paxson, Ernest B., Jr. "Simulation of Room Temperature Control System (RTCS) with MSC/NASTRAN," The 1988 MSC World Users' Conference Proceedings (Los Angeles: The MacNeal-Schwendler Corporation, March, 1988).

Paxson, Ernest B., Jr. "Simulations of Small Structures-Optics- Controls Systems with MSC/NASTRAN," The 1989 MSC World Users' Conference Proceedings (Los Angeles: The MacNeal-Schwendler Corporation, March, 1989).

CREDITS

MSC/NASTRAN is a registered trade mark of the MacNeal-Schwendler Corporation, Los Angeles, CA.

NASTRAN is a registered trade mark of the National Aeronautics and Space Administration, Washington, D. C.

Non-FEM Analysis and Modeling in Opto-Structural Mechanics

Pravin K. Mehta
Hughes Danbury Optical Systems, Inc.
Wooster Heights Road, Danbury, Connecticut 06810-7589

Abstract

In spite of the widespread and mushrooming use of FEM (Finite Element Method) based modeling and analysis by the optical community in recent years, other methods of analysis and modeling are still viable, useful, cost-effective and sometimes necessary in some areas of opto-structural mechanics. This paper provides a review of some of this writer's applications using closed-form solutions and mathematical modeling based on the equations of elasto-mechanics to three topics in opto-structural mechanics; namely, (1) the distortion of circular mirrors under operational and environmental loads, (2) design/analysis of deformable mirrors and related performance prediction, and (3) polishing of optical surfaces (analysis/design of flexible polishing laps, material wear rate and smoothing). Solutions of Kirchoff's flat plate equation and E.Reissner's spherical shell equation were derived and used for the first two topics of application. A modified flat plate equation, which included, besides bending, the transverse shear and lap compressibility, was solved for the flexible lap pressure distribution. Using this, Preston's wear rate equation was solved for material wear and smoothing during optical polishing.

Where available, comparative data are presented in this paper for finite element analyses and test measurements corresponding to the closed-form solutions. Excellent correlation is observed among the three. The paper also includes reviews of (1) the flexural rigidity characteristics of open and closed back light weighted mirrors, (2) the meshing characteristics of curved deformable mirrors, and (3) related publications of other writers in the field of opto-structural mechanics.

1.0 Deformation of Circular Mirrors

A design and analysis process can be established for the development and engineering of large optical mirrors for use in systems that typically have very diverse applications, environments, and optical performance criteria. The process, however, remains analytically more or less common for such diverse applications, since it addresses static, dynamic, thermal, fabrication, and in-process metrology requirements, with direct relation to meaningful optical effects. Also, the design of the mirror substrate and mounting system is normally driven by similar goals of structural efficiency (light weighting), minimization of mount complexity, determinacy of support influence, reasonable cost and schedule parameters, and a list of candidates to be drawn from a rather limited number of optical quality materials.

In pursuing a design, preliminary calculations, trade-offs and sizing analysis are conducted by the use of design curves and scaling laws for structural deformation under various loading and boundary conditions, basic thermal laws, experimental factors including information from prior experience, and closed-form solutions. Once the basic design parameters have been established, further detailed analysis may be conducted by detailed finite element and/or boundary element models. It is evident that the closed form analysis has a substantial contribution to make to the overall design/analysis process.

1.1 Applied Loads

Before discussing some closed form solutions for the distortion of circular mirrors, it is useful to review typical applied loads, which may be divided into two groups on the basis of their significance to mirror surface distortion and strength, The distortion group is comprised of loads based on operational, metrology, ground testing and figure correctibility (for active mirrors) considerations. Those for the strength group are based mainly on launch survivability factors. The loads for each group are identified in Table 1.

Table 1. Applied Loads for Large Mirrors

<u>**Surface Distortion**</u>

- Operational
  - Thermal
    - Orbital
    - High Energy
  - Tracking Dynamics
  - High Energy Device Random Vibration
  - Mount Constraints
  - Operating Scenarios
- Metrology
  - 1-g release
- Ground Testing
  - Self-weight
- Correctibility (Active Mirrors)

<u>**Strength**</u>

- Launch Load Factors
- Launch Vibration Loads
- Handling and Transportation

The differential equations for the bending of flat and curved plates are analytically tractable with most of the above type of surface distortion loading functions. Of the three major types of loads, the self-weight/1-g release load is self-explanatory. We will describe the other two, namely, the thermal and orbital tracking loads, in some more details.

1.1.1 *Thermal Loads*

Thermal loads on a mirror can be derived from the temperature distribution in the mirror by using the following formulas[1] for thermal moment and thermal force.

$$M_T = \int_z E \left(\int_{T_o}^{T_s} \alpha_T \, dT \right) z \, dz \qquad (1)$$

$$N_T = \int_z E\left(\int_{T_o}^{T_z} \alpha_T dT\right) dz \tag{2}$$

where

M_T = thermal moment at a given point (x,y) or (ρ,θ),
N_T = thermal force at the same point,
x,y = cartesian coordinates of above the point,
r,θ = polar coordinates of the same point,
ρ = normalized radius = r / a,
a = radius of the mirror,
E = Young's modulus,
α_T = temperature dependent coefficient of thermal expansion,
z = distance of a point on mirror cross section from the neutral surface,
T_o = reference temperature at which the mirror is free from thermal stresses and deformation,
T_z = temperature at a distance z from the neutral surface of the mirror.

Unlike E, the coefficient of thermal expansion α_T may be temperature dependent as in the case of Corning's ULE glass. Assuming that α_T = f(T) is some function of temperature and is integrable in the temperature range of interest, we have

$$\int \alpha_T dT = \int f(T)\, dT = \varepsilon(T), \tag{3}$$

where ε = thermal strain. Then

$$\int_{T_o}^{T_z} \alpha_T\, dT = \varepsilon(T_z) - \varepsilon(T_o) = \varepsilon[T_z(\rho,\theta)] - \varepsilon(T_o) = \frac{\Delta L}{L} \tag{4}$$

The integral in eq.(4) represents free thermal strain at any point (ρ,θ,z), or ΔL/L as commonly known. Substitution for the thermal strain integral from eq.(4) into (1) and (2) yields the following.

$$M_T = M_T(\rho,\theta) = \int_z E\varepsilon(T_z)\, z dz - E\varepsilon(T_o)\int_z z dz \tag{5}$$

$$N_T = N_T(\rho,\theta) = \int_z E\varepsilon(T_z)\, z dz - E\varepsilon(T_o)\int_z dz \tag{6}$$

The second integral in (6) vanishes for cross sections that are symmetric with respect to the neutral surface. If $T_z = T_z(\rho,\theta)$ is known, say, from a heat transfer analysis, then thermal moment and force distributions can be calculated from formulas (5) and (6).

1.1.2 *Tracking Loads*

Loads caused by the tracking operation are inertial and generally nonuniform. The calculation of tracking loads is based on the geometry in Figure 1. The intensities of centrifugal and tangential

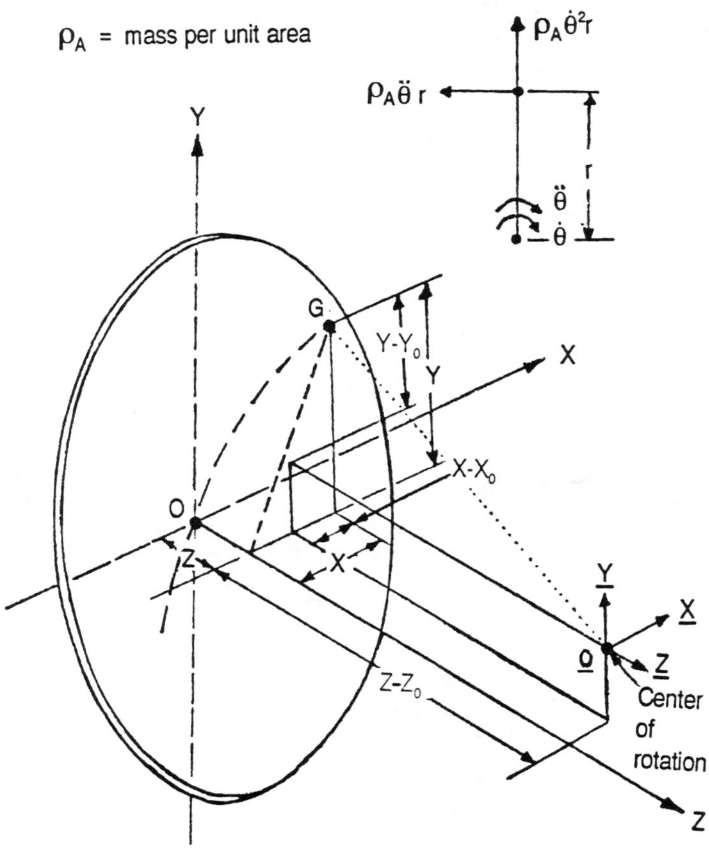

Figure 1. Configuration for computing loads due to tracking rate and acceleration

inertial loads experienced by a point G on a body which is rotating about an axis through a point $\underline{O}$ are $\rho_a\dot{\theta}^2 r$ and $\rho_a\ddot{\theta}r$, where $\dot{\theta}$ is the angular velocity and $\ddot{\theta}$ is the angular acceleration. Applying this to the mirror, the radius vector from $\underline{O}$ to G has three components : $x - x_o$, $y - y_0$ and $z - z_o$, where (x_o, y_o, z_o) are the coordinates of the center of rotation. This results in two components of linearly varying pressure loads due to velocity and two more due to acceleration about each of the two axes of rotation - in all eight components shown in eqs. (7) and (8) with the following notation.

$\dot{\theta}$ = angular velocity,
$\ddot{\theta}$ = angular acceleration,
p = intensity of induced pressure,
ρ_A = mass per unit area of mirror.

The subscripts for θ and θ in (7) and (8) as well as the second subscript for the pressure p indicate the direction of rotation of the mirror. The first subscript for p indicates the direction of pressure.

Rotation about the $\underline{x}$ axis :

$$\left.\begin{array}{llll} \ddot{\theta}_{\underline{x}} : & p_{zx} = -\rho_A \ddot{\theta}_{\underline{x}} (y - y_o) & \dot{\theta}_{\underline{x}} : & p_{zx} = -\rho_A \dot{\theta}_{\underline{x}}^2 (y - y_o) \\ & p_{yx} = -\rho_A \ddot{\theta}_{\underline{x}} (z - z_o) & & p_{yx} = +\rho_A \dot{\theta}_{\underline{x}}^2 (z - z_o) \end{array}\right\} \quad (7)$$

Rotation about the $\underline{y}$ axis :

$$\left.\begin{array}{llll} \ddot{\theta}_{\underline{y}} : & p_{zy} = +\rho_A \ddot{\theta}_{\underline{y}} (x - x_o) & \dot{\theta}_{\underline{y}} : & p_{zy} = -\rho_A \dot{\theta}_{\underline{y}}^2 (x - x_o) \\ & p_{xy} = +\rho_A \ddot{\theta}_{\underline{y}} (z - z_o) & & p_{xy} = +\rho_A \dot{\theta}_{\underline{y}}^2 (z - z_o) \end{array}\right\} \quad (8)$$

1.1.3 *Concentrated Loads*

This includes support reaction as well as actuation loads for point force actuators and moment actuators. Introducing the following notation,

P = a concentrated force,
M = a concentrated moment,
a = mirror radius,
c = radial coordinate } for P or M
φ = angular coordinate } for P or M

these loads can be represented by a Fourier series as follows[2].

$$P = \frac{P_c}{\rho_c \pi a}\left(\frac{1}{2} + \sum_{n=1}^{\infty} \cos n\phi\right) \quad (9)$$

$$M = \frac{M_c}{\rho_c \pi a}\left(\frac{1}{2} + \sum_{n=1}^{\infty} \cos n\phi\right) \quad (10)$$

1.2 Mirror Bending Equations

The closed-formed deformation problem of circular mirrors is formulated by using appropriate load functions in an appropriate differential equation for plate and shell bending. By considering only the plate and shell bending equations here, it is not implied by any means that other equations of elasto-mechanics are or have not been used in mirror analysis. A good many examples of the later type are available in literature. However, the bending of mirrors under the type of loads described in section 1.1 is relatively common, and more literature is available in that field than what can be accommodated in this review article. Accordingly, our discussion will be confined to bending. Further, the equations are applicable to isotropic materials.

Kirchoff's flat plate equation :

This equation[3] governs the bending of thin flat mirrors without including the transverse shear deflection. It may be approximately applied to mirrors of moderate curvature (high f-numbers). For light-weighted mirrors, the transverse shear deflection may be estimated separately and added on.

$$\nabla^2\nabla^2 w = \frac{p a^4}{D} + \frac{a^2}{D(1-\nu)}\nabla^2 M_T \tag{11}$$

where

$$\nabla^2 = \frac{\partial}{\partial\rho^2} + \frac{1}{\rho}\frac{\partial}{\partial\rho} + \frac{1}{\rho^2}\frac{\partial}{\partial\theta^2} \tag{12}$$

and

r	=	radial coordinate,	D	=	flexural rigidity of the mirror,
θ	=	angular coordinate,	p	=	applied pressure load function,
ρ	=	normalized radial coordinate,			
	=	r / a,	M_T	=	applied thermal moment function,
a	=	mirror radius,			
w	=	mirror deflection function,	ν	=	Poisson's ratio.

Modified flat plate equations :

A system of three simultaneous differential equations govern the bending of a thin flat plate with transverse shear deformation[4, 5, 6]. In these equations, w, Φ and Ψ are generalized displacements, related to the deflection and rotations of plate cross sections.

$$\left.\begin{array}{l} \left.\begin{array}{l} \nabla^2\Phi = -\dfrac{p}{D} \\[2ex] \nabla^2 w = -\Phi - \dfrac{p}{D_s} \end{array}\right\} \rightarrow \nabla^2\nabla^2 w = \dfrac{p}{D} - \dfrac{\nabla^2 p}{D_s} \\[6ex] \nabla^2\Psi - \dfrac{2D_s}{(1-\nu)D}\Psi = 0 \end{array}\right\} \tag{13}$$

where

w = combined deflection due to bending and transverse shear
Φ = div ψ, } , where ψ = vector of plate rotations,
Ψ = curl ψ, }
D_s = Shear rigidity.

All other quantities have been defined above in connection with Kirchoff's flat plate equation. It may be noticed that in eqs.(13), the first two second order equations are combined to form a fourth order equation, which together with the third second order equation forms the governing system of differential equations for this problem. This system of equation requires satisfaction of three boundary conditions at each boundary in contrast to Kirchoff's equation, for which only two conditions have to be satisfied correspondingly. This is accomplished in Kirchoff's theory by approximately combining two of the three conditions into one condition.

Eric Reissner has derived similar equations for anisotropic flat plates[7].

Reissner's shallow spherical shell equations :

These two fourth order differential equations[8, 9, 10, 11] govern the combined bending and inplane stretching of shallow spherical shells and may be applied to curved circular mirrors which fall within the mathematical limit of spherical shallowness as defined by E. Reissner. This limit is given later.

$$\left.\begin{aligned} D\nabla^4 w + \frac{1}{R}\nabla^2 F &= -p - \frac{2\psi}{R} - \frac{\nabla^2 M_T}{1-\nu} \\ \nabla^4 F - \frac{[(1-\nu^2)K]}{R}\nabla^2 w &= -(1-\nu)\nabla^2\psi - \nabla^2 N_T \end{aligned}\right\} \qquad (14)$$

where

$$\nabla^2 = \frac{\partial}{\partial r^2} + \frac{1}{r}\frac{\partial}{\partial r} + \frac{1}{r^2}\frac{\partial}{\partial \theta^2} \qquad (15)$$

and

r = radial coordinate,
θ = angular coordinate,
w = mirror deflection function,
F = membrane stress function,
R = radius of curvature,
D = flexural rigidity ,
K = extensional or membrane rigidity,
ν = Poisson's ratio,
p = applied transverse load function,
Ψ = applied inplane load potential,
M_T = applied thermal moment function,
N_T = applied (inplane) thermal force function.

The mathematical definition of shallowness, within which eqs. (14) are applicable, is given by

$$\frac{H}{2a} = \frac{a}{4R} = \frac{1}{16f_n} \leq \frac{1}{8}, \qquad or \qquad f_n \geq \frac{1}{2} \tag{16}$$

where

a = outer radius of mirror,

f_n = f-number of the mirror,

H = geometric sag of the mirror.

This range of shallow shell theory includes perhaps the optically fastest conceivable mirrors of the present state of the art and the near future. Further, the application of eqs. (14) is not limited to only spherical mirrors. The geometric difference between aspheric mirrors and the best fit spherical mirrors is very small form the view point of elasto-structural modeling and analysis in contrast to that relative to flat mirrors. So the shallow spherical shell theory can be well applied to aspheric mirrors. This theory, like Kirchoff's thin flat plate theory does not account for the transverse shear deflection.

1.3 Applications

Kirchoff's flat plate equation is the simplest of the three mirror bending equations described in the preceding section. Consequently, solutions of Kirchoff's thin flat plate equation (11) for problems that have applications to circular mirrors are more abundant in literature than other equations.

1.3.1 *Mirrors on multi-point supports :*

Of reoccurring interest is the problem of plate (mirror) deformation with a multi-point support configuration. A number of papers on this subject are available in literature, including two[12, 13] by Bassali prior to known interest and application in the optical industry. Bassali's paper also serves as a useful source of reference on related prior publications.

Among the earliest work in the optics community on the subject of multi-point supported mirrors is that by Williams[14, 15]. Any number of support points are permitted on a single support circle concentric with the circular plate and located anywhere on the plate. The applied load is axisymmetric. The method and results are presented in terms of real variables, and the solution to the problem is obtained by a superposition of three applicable known solutions. Extensive tabulated numerical data for deflection coefficients are provided for a variety of support numbers, normalized support radius locations and normalized grid point locations on the mirror. Further, the experimental data provided shows good correlation between the analytical data and test measurements.

Jerry Nelson and his associates have described[16] a method of calculating (a) optimum support point location locations for a variety of support point topologies based on solutions of Kirchoff's flat plate equation for circular and infinite plates, and (b) the maximum diameter of ensuing optical image blur for optimized support systems from the maximum slope, which is related to the rms deflection and to an "effective" length between support points.

A problem of interest in supporting optical mirrors is that of multi-level kinematic mount, which is required when a single level 3-point kinematic mount is not adequate for limiting the stress and/or deflection level to design specifications. It may then be necessary to employ a multi-level kinematic mount, also known as a Hindle (or Wiffle tree) mount, with some integer multiple of 9 points, such as 9, 18, 27, 36, 81, etc. The number of levels may be two or more. A closed form analytical approach to this problem has been demonstrated by the present author in a recent paper[17]. In this paper, a **direct** closed-form solution of Kirchoff's equation is first derived for m-point ($m \geq 2$) supports, equi-spaced on a concentric circle, and then applied to a 9-point hindle mount, illustrated in Figure 2. There are six support points on the outer support circle and three support points on the inner support circle. Nominal design curves are developed for determining nominal locations of the nine support locations, associated RMS deflections and support location sensitivity. Also several cases of the 9-point Hindle mount solution were compared with independent Nastran based finite element solutions. Excellent correlation between the two was obtained in all the cases. An illustrative correlation plot is presented in Figure 3.

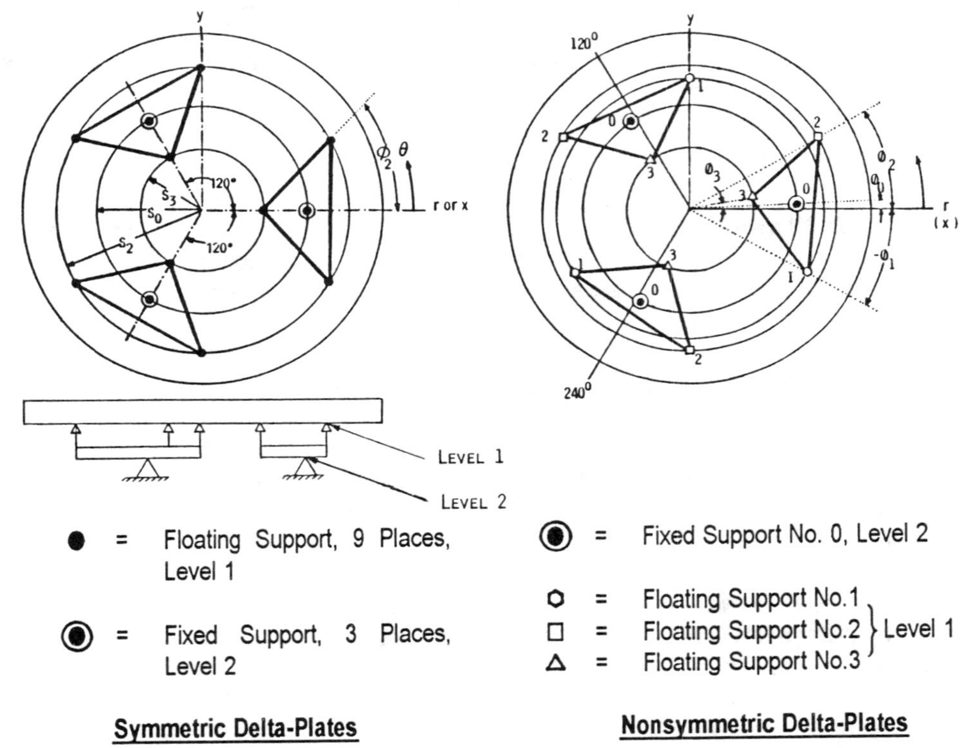

Figure 2. A schematic representation of a 9-point Hindle mount

Given required resources, the same approach for a multi-level kinematic mount (Hindle mount) can be applied to curved mirrors via Reissner's shallow spherical shell equations, as well as to the modified flat plate equations with shear deformation.

One of the first applications (in the optics community) of the modified flat plate equations to circular mirror analysis is by R. A. Ross and R. G. Mady[18]. They used the point matching technique for solving the multi-point support problem. In this method, the governing differential equations are satisfied exactly, while the boundary conditions are satisfied at discrete points on the boundary. The applied load was assumed to be represented by an infinite trigonometric power series.

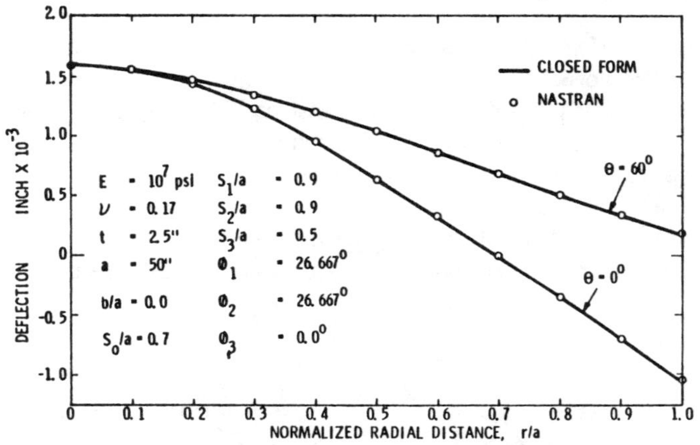

Figure 3. Comparison of closed form and Nastran solutions for a 9-point Hindle mount application

The present author has applied the modified flat plate equations to derive the pressure variation under flexible polishing laps used to smooth high frequency surface errors during optical polishing with a full size lap. This process is known as "Bridging". More details on this work will be given later in section 3.2 .

1.3.2 *Thermal deformation of circular mirrors*

A topic of continued interest in mirror design/analysis is that of thermal distortion. Expanding the thermal moment and thermal force functions into trigonometric power series of the following form

$$M_T = \sum_{m=o}^{M_m} \sum_{n=o}^{\infty} (\mu_{mn} \cos n\theta + \xi_{mn} \sin n\theta)\, \rho^m \tag{17}$$

$$N_T = \sum_{m=o}^{N_m} \sum_{n=o}^{\infty} (\upsilon_{mn} \cos n\theta + \zeta_{mn} \sin n\theta)\, \rho^m \tag{18}$$

where

μ = thermal moment coefficient, cosine series

ξ = thermal moment coefficient, sine series

υ = thermal force coefficient, cosine series

ζ = thermal force coefficient, sine series,

m = radial index of thermal load distributions,

n = azimuthal index of thermal load distributions,

closed form solutions of both Kirchoff's equation and Reissner's equations have been obtained for nonuniform thermal loading of flat and curved circular mirrors by the present author[19, 20]. A variety of nonsymmetric functional forms of thermal moment and thermal force distributions can be constructed from the assumed forms of eqs. (17) and (18). From these solutions, normalized performance curves are developed for nonuniform thermal bowing and ensuing RMS figure error. Several test cases of the closed form solutions were compared with Nastran based finite solutions. Excellent correlation was obtained in all cases.

In one series of test runs, the 50-inch diameter mirror was assumed to be made of six sectorial panels fused together along their straight edges as shown in Figure 4. The sectors were assumed

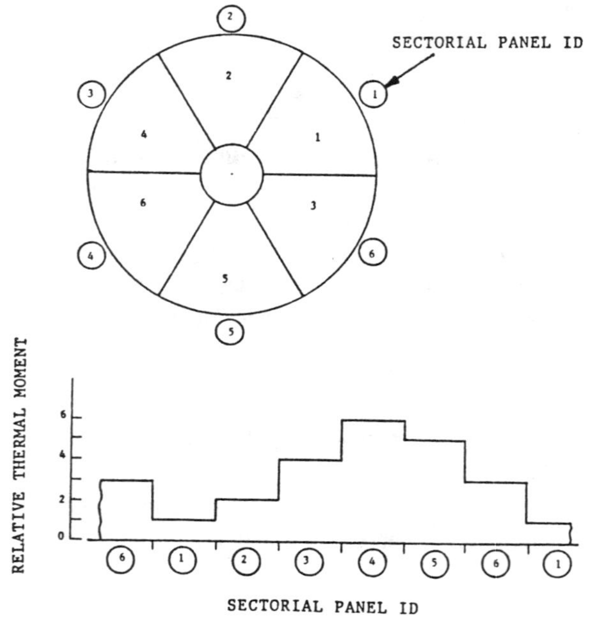

Figure 4. Relative azimuthal distribution of thermal loads for a test run

to be made of the same material but with different spatially averaged coefficients of expansion. The resulting thermal moments and forces in the panels were, therefore, different, although they were all exposed to identical temperature distributions. Figure 4 depicts the resulting relative variation of thermal loads along the azimuthal direction. For radial variation of thermal loads, it was assumed that the integrated value of the load over a panel area remained invariant regardless of the radial variation. Considering the thermal moment, for example, its integrated value over a panel is given by

$$\frac{2\pi}{6}\int_{\rho_b}^{1}(\mu_{mn}\rho^m)\,\rho\,d\rho = \frac{\pi}{3}\mu_{mn}\frac{1-\rho_b^{m+2}}{m+2} \tag{19}$$

where ρ_b = normalized inner radius of the mirror = b/a, b = inner radius of the mirror, and a = outer radius of the mirror. Similarly, for m = 0, the integrated value of thermal moment over a panel is given by

$$\pi \mu_{0n} \frac{1 - \rho_b^2}{6} \tag{20}$$

By the above invariance assumption, the integrated thermal moment values given by (19) and (20) are equal. Equating the two and solving the resulting equation for μ_{mn}, we have

$$\mu_{mn} = \mu_{0n} \frac{(m + 2)\ (1 - \rho_b^2)}{2\ (1 - \rho_b^{m+2})} \tag{21a}$$

Similarly,

$$\upsilon_{mn} = \upsilon_{0n} \frac{(m + 2)\ (1 - \rho_b^2)}{2\ (1 - \rho_b^{m+2})} \tag{21b}$$

Corresponding to any assumed values of μ_{0n} and υ_{0n}, the thermal moment and force for any value of m > 0 can be calculated from formulas (21a) and (21b) for the test runs. Two illustrative correlation plots of thermal distortion are shown in Figures 5 and 6. Many more plots for this and other test runs are given in the two papers on thermal distortion[19, 20].

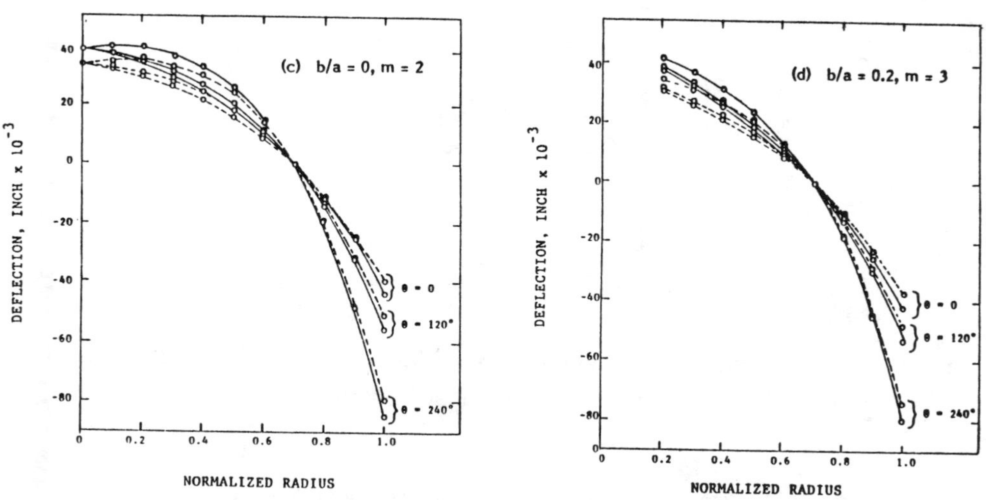

——— Closed form, fully heated aperture, $0 \leq \rho \leq 1$

- - - - - - Closed form, partially heated aperture, $0.2 \leq \rho \leq 0.9$

O Nastran

a = 50", b = 0 or 10"

E = 10^7 psi, ν = 0.17

D = 13.4 x 10^6 lb. in.

Figure 5. Test run correlation between closed form and Nastran solutions for a flat circular mirror subjected to the thermal moment distribution of Figure 4.

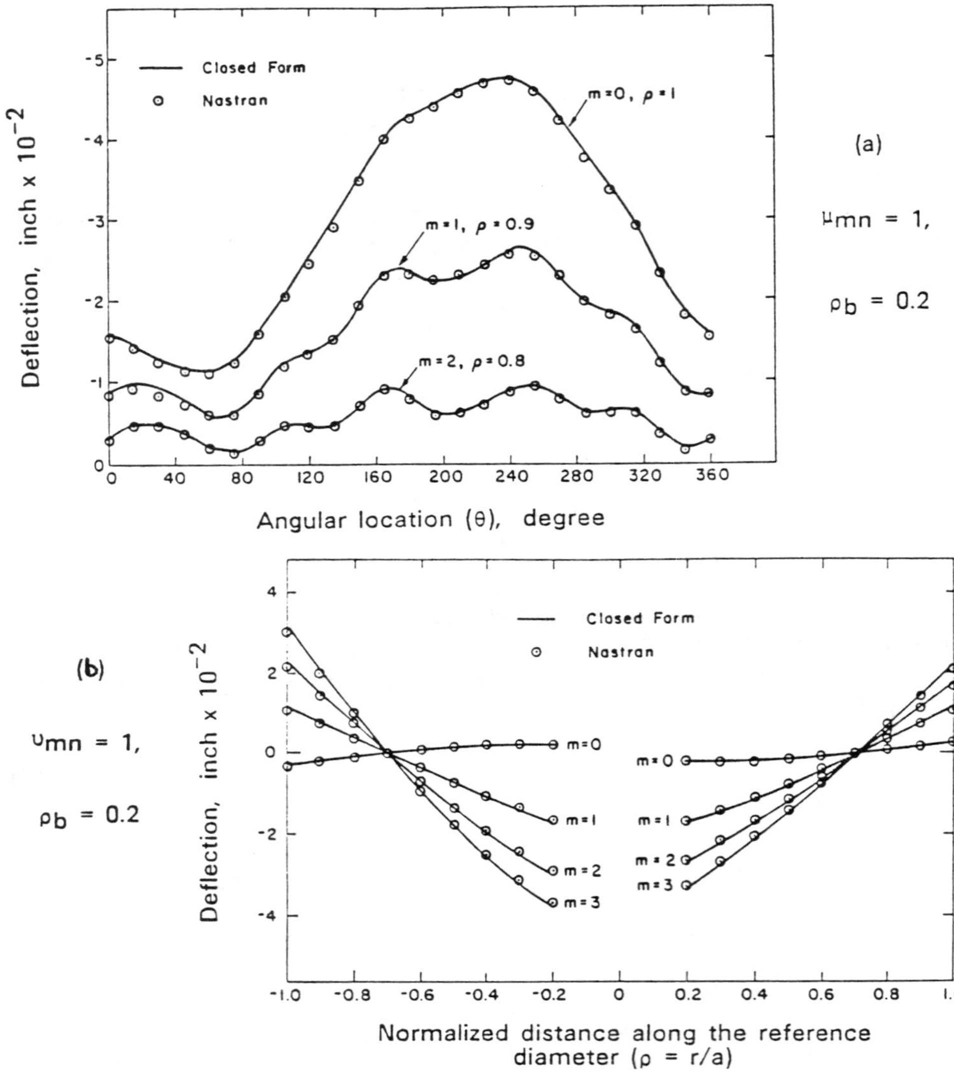

Figure 6. Test run correlation between closed form and Nastran solutions for a curved circular mirror subjected to the thermal moment distribution of Figure 4.

It is a common practice in the optics community to assume, at least at the feasibility study phase, a uniform thermal moment distribution obtained from a known or assumed temperature gradient across the cross section of a mirror, and then compute thermal bowing. This gives only the power term or defocus and is not of much use in many systems, which are provided with a focus correcting capability. It is more useful to know about astigmatism, coma, and higher order aberrations. The closed form solutions of these two papers provide such a capability without resorting to finite element models. The equations can be easily programmed for any small desk top computer for the convenience of the analyst.

Both the flat and curved mirrors in these solutions are assumed to be kinematically mounted, so that the support reactions under thermal loading are negligible or absent. Under these circumstances, the transverse shear deflection is relatively unimportant, and the solutions may be reasonably well applied to constant thickness light weighted mirrors with ribbed structures (sandwich or open back).

2.0 Deformable Mirrors

2.1 Various Actuation Approaches

Actuators for deformable mirrors can be broadly classified into two types : (a) force/displacement or piston actuators with the line of actuator either parallel to the optical axis, or normal to the optical surface, and (b) moment actuators with the line of action either perpendicular to the optical axis or parallel to the optical surface, as schematically shown in Figure 7.

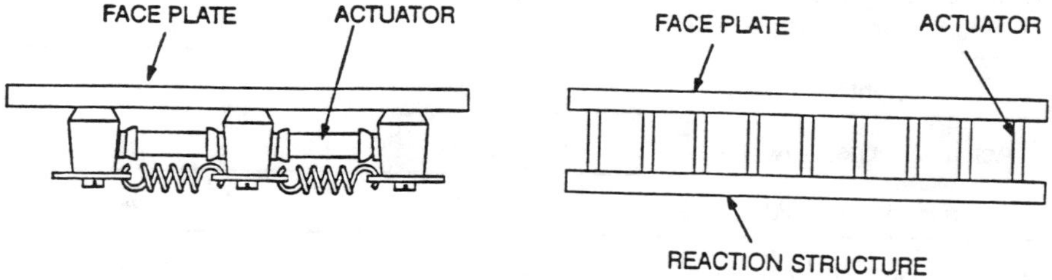

(a) Deformable mirror with moment actuators (schematic) The actuators are parallel to the mirror and/or perpendicular to the optical axis

(b) Deformable mirror with piston actuators (schematic) The actuators are parallel to the optical axis and/or normal to the mirror

Figure 7. Two different actuation approaches for deformable mirrors

There are several other important differences between these two types. Table 2 gives a summary of the generic differences between them. For a moment actuator, a single device (such as a piezo-electric stack) can be made to produce two equal and opposite forces at its ends parallel to the optical surface, thereby creating two self-balanced moments on the substrate. Therefore, there is no need for a reaction structure unlike in the case of piston actuators. This results in several important advantages for moment actuation approach : (a) The reaction structure induced distortion of the mirror (in response to environmental changes, for example) is absent; (b) a kinematic mount of the mirror is feasible; (c) the coarse and fine controls for the aberrations can be decoupled with separate actuators for each type; and (d) the overall weight of the system is reduced,which is usually a critical consideration in outer space applications.

A very important difference is the actuator connectivity in the two approaches - a series connectivity for moment actuators in contrast to a parallel connectivity for piston actuators. Since the series connectivity results in a cumulative action, the dynamic range of the mirror with moment actuators can be several times that of a single actuator. This can be of a significant advantage in some applications over the piston approach, for which the dynamic range of the mirror is limited to

Table 2. A generic Comparison between Moment and Piston Actuation Approaches

Characteristic	Actuation Approaches	
	Moment	**Piston**
Line of action	Perpendicular to optical axis or parallel to surface	Parallel to optical axis, or normal to surface
Reaction structure	Not required	Required
Kinematic Mount	Yes	No
Coarse/Fine control	Uncoupled (using separate actuators)	Coupled (using duel-mode actuators)
Coupling of reaction/support distortion with mirror	No	Yes
Weight	Less	More
Actuator connectivity	Series	Parallel
Dynamic range of DM	Variable	Fixed
DM resonant frequency	Less	More
Actuator density	Less	More
Influence function	Global	Global for force actuators Pseudo-zonal for displacement actuators
Control system matrix	Coupled	Coupled for force actuators Pseudo-diagonal for displacement actuators

that of a single actuator.

Among the disadvantages for a moment actuation approach relative to the piston actuation approach are : (a) reduced resonant frequency caused by actuator mass attachments to the back of the substrate and loss of stiffness augmentation of the substrate, which exists with displacement actuators; (b) reduced actuator density due to relatively more packaging space required per actuator. These considerations are not necessarily always critical and should be evaluated for a given application in the light of spatial and temporal frequency requirements of the control system.

2.2 Moment Actuation Approaches

The most common moment actuation approach centers on applying variable bending moments around the outer boundary of the mirror. This is known as the "Edge moment control". The generated shape depends on the location and orientation of the actuators and the thickness variations of the substrate. Only quadratic surfaces (defocus, astigmatism, etc.) can be obtained with constant thickness mirrors. For high order surfaces, nonuniform thickness variations are required. In principle, any desirable shape can be generated by using an appropriate thickness variation. The edge moment control has two disadvantages : (a) Only a fixed radial shape can be produced for a given thickness function, and (b) for curved mirrors, the deflection due to an edge moment attenuates towards the interior of the mirror, causing a deterioration in correctibility.

The constant thickness modal mirror[21] with three pairs of edge moment actuators, developed by Perkin-Elmer's Electro-Optics Division (now Hughes Danbury Optical Systems), was one of the earliest known applications of this approach. This 6-inch diameter circular mirror was 5/16" thick, had the first measured resonance at 1,950 Hertz, was made out of molybdenum, and had an elliptic clear aperture. The residual rms error for defocus (Zernike polynomial P_4) correction was 0.9 %, while that for astigmatism (P_5 and P_6) ranged from 1.1 % to 1.4 %. In a recent application of this approach, Lawrence Livermore National Laboratory[22] designed and built a 6-inch diameter mirror with nonuniform thickness to generate a Gaussian shape.

In an approach, which is more versatile than the edge moment control, the moment actuators are placed in a suitable two-dimensional array all over the back of a constant thickness mirror and not just at the edge. Many a variety of shapes can be generated with this approach. Consequently, the mirror is tunable in operation in contrast to the two edge moment control approaches discussed above. The first known study of this approach was published by R. M. Scott[23] (using influence functions generated by the present author by means of a Nastran FEM model). His study gives results for generating shapes up to Zernike P_{10} along with a speculation for shapes up to P_{15}. However, all actuator placement concepts reported in his paper are applicable to shapes up to P_6 only. Using our versatile approach, a deformable mirror has been developed[24, 25] with 30 moment actuators arranged in twelve radial (or six diametral) arrays, as schematically shown in Figure 8. This

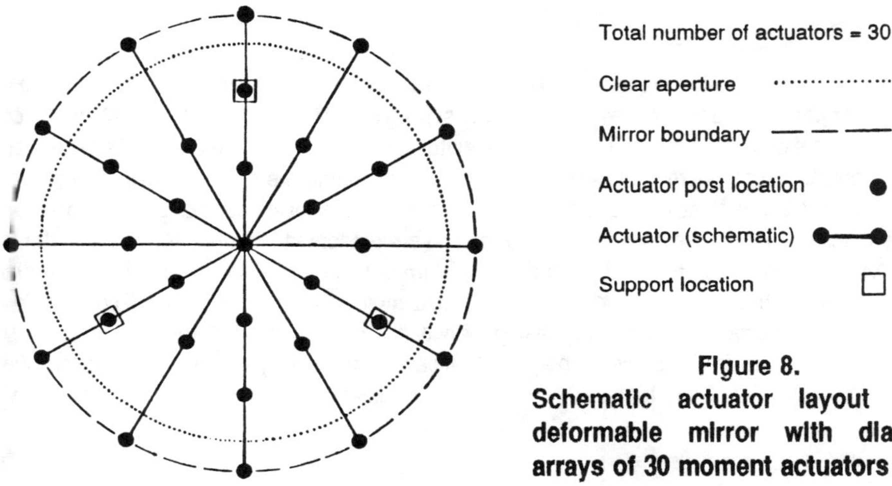

Figure 8.
Schematic actuator layout for a deformable mirror with diametral arrays of 30 moment actuators

mirror has a diametral cross section similar to that shown in part (a) of Figure 7, but with six moment actuators rather than two. It is capable of generating shapes up to Zemike P_{11} with an average rms residual error of about 2 %. This mirror has substantial capability beyond $P_{11.}$

2.3 Moment Actuator Influence Function

The influence function of an actuator in a deformable mirror is required to determine its capability of wave front or aberration correction, which is usually accomplished by determining the optimum set of actuator actions (force, displacement, moment, voltage, etc.) that yields a minimum residual wave front error based on a "least square fit" algorithm, given by

$$V = [A^T A]^{-1} A^T w \tag{22}$$

where V = optimum actuator action, A = actuator influence matrix and w = the desired shape vector.

For a moment actuator placed parallel to the mirror between two end posts at arbitrary locations A and B, as shown in Figure 9, the actuator generates two equal and opposite forces at A and B, and these forces create two equal and opposite bending moments on the mirror at the post locations.

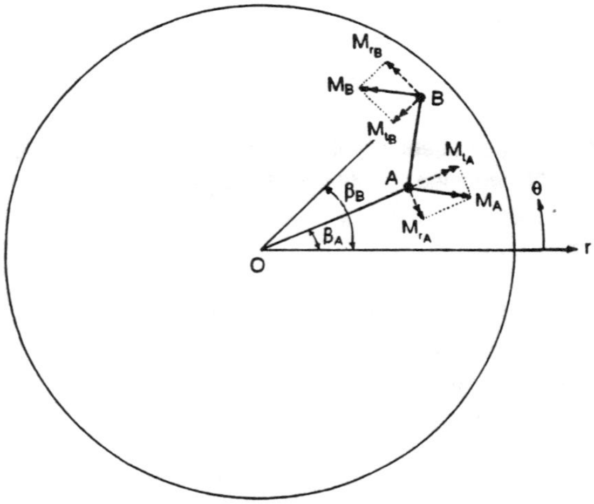

M = moment
M_r = radial component of moment
M_t = tangential component of moment
A = subscript for end A of actuator
B = subscript for end B of actuator
For the influence function, $M_A = M_B = 1$

Figure 9
Arbitrary placement of a moment actuator and related notation

A closed-form influence function corresponding to this arbitrary placement configuration of a moment actuator has been derived by the present author[26] by solving Kirchoff's thin flat plate equation. Since the end moments generated by such a moment actuator are self balanced, no reaction is generated at the support locations. Many test runs were made for comparing the results of this closed form solution with those of a Nastran based finite element analysis. Excellent correlation was observed in all the runs. Two illustrative cases for this correlation are presented in Figure 10. The actuator is located for each case between points A and B as shown in Figure 10. The first plot is for a radial moment actuator, while the second one is for a tangential moment actuator. A vectorial combination of the two in desired proportions gives us the capability of arbitrary orientation, provided that the post locations A and B are in the same respective locations for the two types. It is evident that the correlation between the closed form and Nastran solutions is excellent.

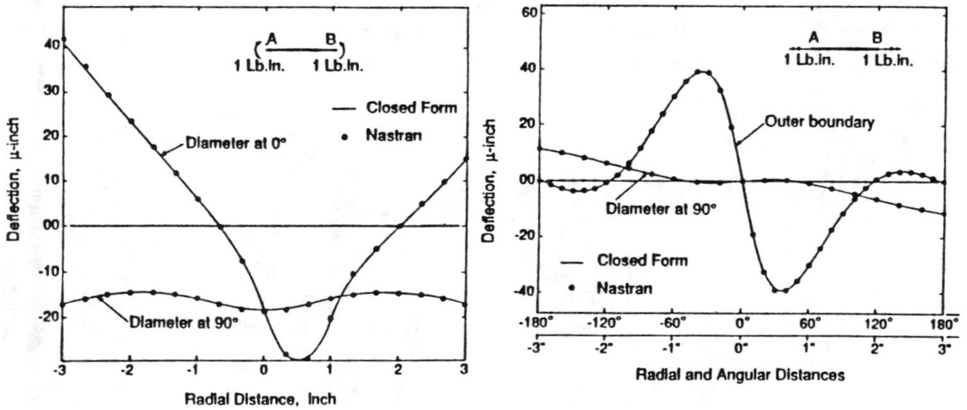

Case 1. Radial Moment Actuator

Case 2. Tangential Moment Actuator

Figure 10. Test run correlation between closed form and Nastran solutions for moment actuator influence functions

2.4 Deformable Mirror Analysis and Test Correlation

Using the closed form influence function capability for moment actuators, a trade-off analysis was performed for the 6-inch diameter deformable mirror concept development. This included the following.

- Number of actuators
- Placement of actuators
  - Actuator post locations
  - Actuator orientations
    - Radial moment only
    - Tangential moment only
    - Uncoupled radial and tangential moments
    - Coupled radial and tangential moments
- Clear aperture relative to the physical size of the mirror

The baseline configuration developed from this study corresponds to the schematic layout shown in Figure 8.

A prototype deformable mirror corresponding to the schematic of Figure 8 was detailed, developed and tested using a test set-up, which essentially consisted of a Zygo (Mark 3) interferometer, a desk top computer, and the deformable mirror assembly with necessary electronics and power supply. Figure 11 gives an illustrative comparison between the closed form and measured influence functions for actuator no. 2. Except for a small amount of rigid body piston and tilt in the measured data and different fringe densities, the two contour plots are nearly identical.

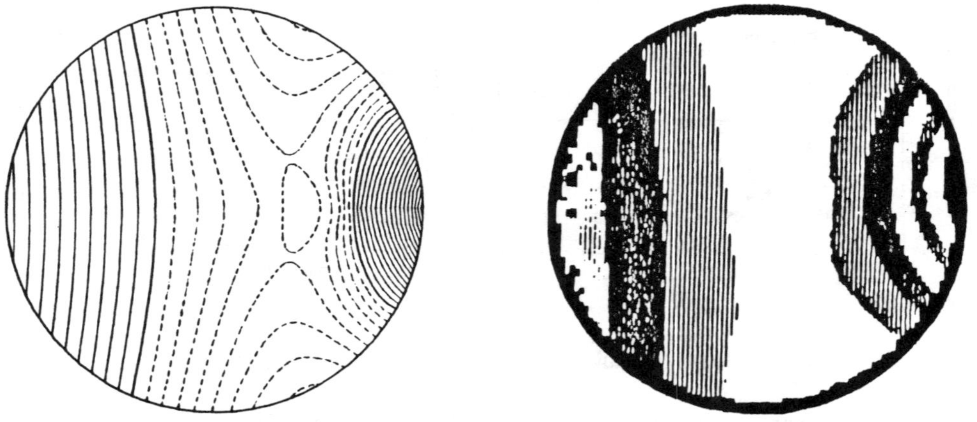

(a) Analytical (b) Measured

- The contour interval for analytical data is smaller than that for the measured data.
- Piston and tilt contents for the measured data are different from those of the analytical data.

Figure 11. A comparison of analytical and measure influence functions

Table 8 of reference 24 gives a comparison between the predicted and measured performances for Zernike shape generation in terms of the rms residual error associated with each shape. These shapes are define by formulas (23) given below.

$$\left.\begin{array}{ll} P_4 = A_4\,(2\rho^2 - 1) & P_8 = A_8\,(3\rho^3 - 2\rho)\sin\theta \\ P_5 = A_5\,(\rho^2)\cos 2\theta & P_9 = A_9\,(\rho^3)\cos 3\theta \\ P_6 = A_6\,(\rho^2)\sin 2\theta & P_{10} = A_{10}\,(\rho^3)\sin 3\theta \\ P_7 = A_7\,(3\rho^3 - 2\rho)\cos\theta & P_{11} = A_{11}\,(6\rho^4 - 6\rho^2 + 1) \end{array}\right\} \quad (23)$$

In that table, the smaller the residual error as a fraction of the input rms shape, the better is the generated shape. The predicted performance is the sum of the theoretical residual and the measurement noise. Since the noise dominated the two, the experimental set-up and associated errors limited the achievable performance levels. Nevertheless, the correlation between the predicted and measured residual errors was very good.

2.5 Actuator Meshing Characteristics

This section presents some results from an analytical study of actuator meshing characteristics by the present writer[27] for curved deformable mirrors, based on the characteristic length of shallow spherical shells. The study is applicable to triangular and square meshes of force actuators, typified by low spring rates, and includes the meshing error, actuator count, actuator stroke and force, and the effect of actuator pad size. It is shown in the paper[27] that the triangular mesh is somewhat better than the square mesh for equal areal density of spacing. Further, the pad has a beneficial effect on the meshing error, which decreases with increase in pad diameter. For approximately applying the results to light-weighted mirrors, a detailed study of the characteristic length of light-weighted mirrors is included in the paper. Also included is a comparison between the meshing error and the residual error for an actuator enforced change in spherical curvature based on finite element models[28]. All results are presented in the form of normalized parametric curves.

2.5.1 *Introduction*

One of the fundamental problems facing the designer of a deformable mirror is the determination of actuator spacing and the associated actuator count. Some of the factors influencing the selection of actuator spacing include spatial frequencies of the input surface and wave front errors, required correctibility, the region of effective influence of an actuator on a curved mirror and the matching of a surface generated by the superposition of influence functions of neighboring actuators with an ideally flat surface. The difference between these two surfaces has been described as meshing error in some active mirror programs (HIDAM, or **H**igh **D**ensity **A**ctuator **M**irror program, for example) with a design specification on its allowable rms value as a small percentage of the (average) flat surface. The meshing error is an inherent source of error, which greatly influences the deformable mirror correctibility. For the purpose of this study, force type of actuators are assumed to be placed on the back of curved mirrors on triangular and square meshes as illustrated in Figure 12.

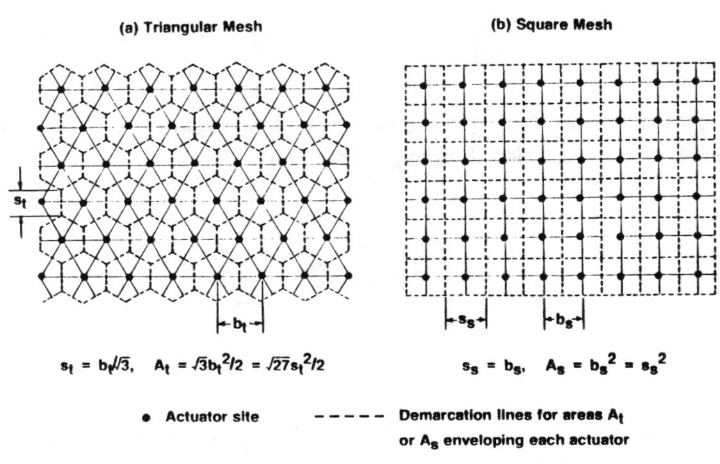

The actuator count depends on the enveloping area associated with each actuator. Interestingly, its shape is hexagonal for triangular array and square for square array.

Figure 12. Triangular and square actuator meshes.

The elastic behavior of most spherical and aspheric mirrors in response to applied loads is similar to that of a shallow spherical shell[8, 9, 10], which is schematically represented in Figure 13 with an apex load representing the actuator force P. When a shallow spherical shell of infinite extent is subjected to a transverse force, either concentrated or distributed over a small circular area, its deflection is mostly localized in the vicinity of the force as shown in Figure 14. The region of localization is dependent on a geometric parameter of the shell, dubbed as the characteristic length of the shell by E.Reissner[9]. The characteristic length for solid cross sections is given by

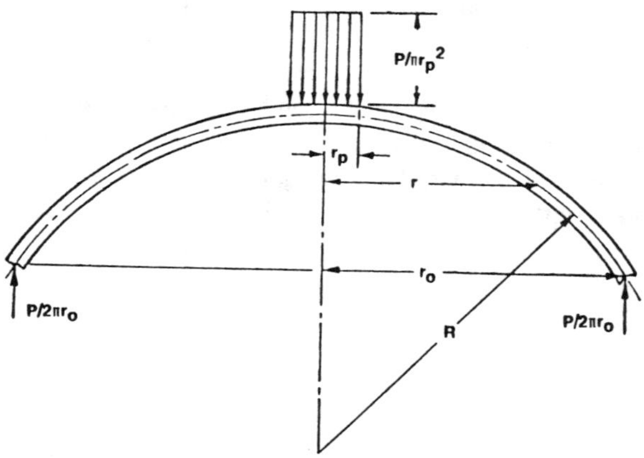

Figure 13. A thin shallow spherical shell with an apex force uniformly distributed over a circle of radius r_p

$$\ell = \sqrt{\frac{Rt_s}{\sqrt{12(1-\nu^2)}}} \tag{24}$$

where

ℓ = characteristic length,

R = vertex radius of curvature,

t_s = thickness of weight equivalent solid cross section,

ν = Poisson's ratio.

By considering the two partial differential equations (14) of the shallow spherical shell theory, it can be shown that the characteristic length is, in general,given by

$$\ell^4 = \frac{(Rt_s)^2}{12(1-\nu^2)} \frac{D/D_{sd}}{t_m/t_s} \tag{25}$$

where

D = flexural rigidity,

t_m = membrane thickness,

D_{sd} = flexural rigidity of weight equivalent solid section

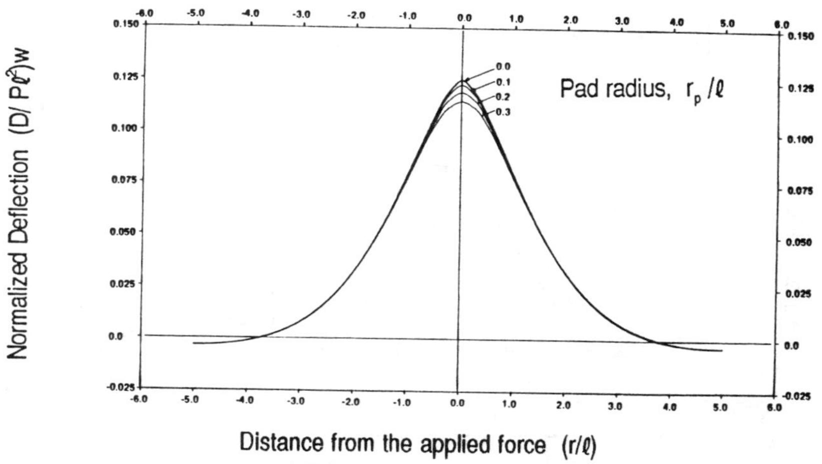

Figure 14. The deflection profile of a thin shallow spherical shell due to apex forces distributed over circular pads of different size. The profiles represent normalized influence functions for force actuators with pads of various sizes.

The deflection profile (in Figure 14) has the appearance of a localized bump in the shell and extends to a distance of about four characteristic lengths regardless of the pad diameter. The distributed pressure of the actuator pad over its diameter helps to somewhat flatten out this bump. An important consequence of the localization is that the shell yields essentially the same deflection bump regardless of the location of the force on the shell, except in the neighborhood of the outer boundary of a finite shell. Therefore, the deflection bump of Figure 14 may be regarded as an exact or near-exact representation of the normalized influence function of a force actuator in any location more than four characteristic lengths away from the boundary of a finite shell. Some analytical evidence exists to indicate that, with appropriate truncation, the same deflection bump may still be applicable at the boundary approximately. Therefore, we will assume the same normalized influence function to be approximately applicable also in the vicinity of the boundary.

When the influence functions of two adjacent actuators are superposed, the resulting deflection profile, in general, has a ripple as schematically shown in Figure 15. Its magnitude depends on the spacing between the actuators. For optimum spacing, the ripple should be a minimum. Extending the superposition concept to the influence functions of actuators in a two-dimensional array, the optimum spacing will yield the flattest resultant deflection surface in any given triangular or square cell of actuator meshes, schematically shown in Figure 12. The rms deviation from a flat deflection surface may then be regarded as a measure of the meshing error. This is an intrinsic error of an active mirror system and may be viewed as a design characteristic upon which depends the correctibility of the mirror. In practice, the correctibility is also affected by many other factors, such as the spatial and temporal frequencies of the input wave front error, non-uniform distortion due to thermal/structural loads, boundary and support conditions, etc. Nonetheless, the actuator meshing error is a very important consideration for nominal actuator spacing and the ensuing actuator count. From another perspective, if the actuator spacing results in a relatively large meshing error, the active mirror will most likely fail to achieve its correctibility goals.

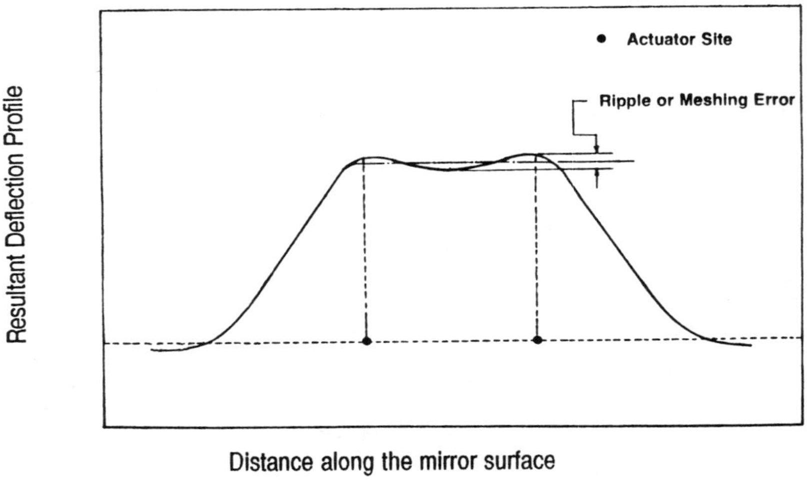

Figure 15. A schematic residual error ripple in the resultant deflection profile generated by the superposition of individual influence functions of two adjacent force actuators

2.5.2 *Influence Functions*

Based on E.Reissner's work[9], the influence functions, illustrated in Figure 14, are as follows for force actuators with and without pads.

Without a Pad ($r_p = 0$) :

$$w = \frac{P\ell^2}{D}\frac{\mathrm{kei}\,\xi}{2\pi} \tag{26}$$

With a Pad ($r_p > 0$) :

For $r \leq r_p$,

$$w = \frac{P\ell^2}{D}\frac{\mathrm{kei}'\xi_p\mathrm{bei}\xi - \mathrm{ker}'\xi_p\mathrm{ber}\xi - \dfrac{1}{\xi_p}}{\pi\xi_p} \tag{27}$$

For $r \geq r_p$,

$$w = \frac{P\ell^2}{D} \frac{\text{bei}'\xi_p \text{kei}\xi - \text{ber}'\xi_p \text{ker}\xi}{\pi \xi_p} \tag{28}$$

where ber, bei, ker, kei are Kelvin functions of zero order and ber', bei', ker', kei' are their respective derivatives with respect to the radial coordinate r. Further,

ξ	=	normalized radial coordinate,	P	=	applied force,
	=	r/ℓ.	w	=	deflection (influence function),
ℓ	=	characteristic length,	r_p	=	actuator pad radius,
D	=	flexural rigidity,	ξ_p	=	r_p/ℓ.

It may be noted that the origin of the coordinate system for these influence functions is local, that is, r is measured from the center of the actuator site under consideration.

2.5.3 *Meshing Characteristics for Padless Actuators*

The above influence functions may be used to determine the actuator meshing characteristics for both padless and padded actuators by a simple superposition algorithm that enables us to generate the resultant deflection surface in the smallest typical triangular or square area formed by actuators placed at its corners in Figure 12. From this, a great variety of useful information on the meshing characteristics can be derived. This includes :

- The residual rms error relative to an ideally flat superposition or the rms meshing error (ε_{rms}), which is expressed as a percentage of the resultant average (flat) deflection.
- Effect of actuator spacing ($\sqrt{A}/\ell$)
- Actuator count (N/A or number of actuators per unit mirror area)
- Effect of actuator pad size (r_p/ℓ)
- Resultant average (flat) deflection $[(D/P\ell^2)w]$ or force $[(\ell^2/wD)P]$
- Dependence on the mirror characteristic length ℓ

Using normalized parameters, the results are presented in this section for padless actuators and in the following section for padded actuators.

Meshing Error :

The variation of rms residual error or the meshing error with linear and areal (area-based) actuator spacings, shown in Figure 16 for zero pad radius, is linear on a log-log scale. Whereas the linear and areal spacings for the square mesh are equal, they are unequal for the triangular mesh. For a one-on-one comparison, the linear spacings are misleading, because the actuator count depends on the areal spacing. The comparison between the two types of meshes should, therefore, be based on equal areal spacings. On this basis, the following results are obtained from Figure 16.

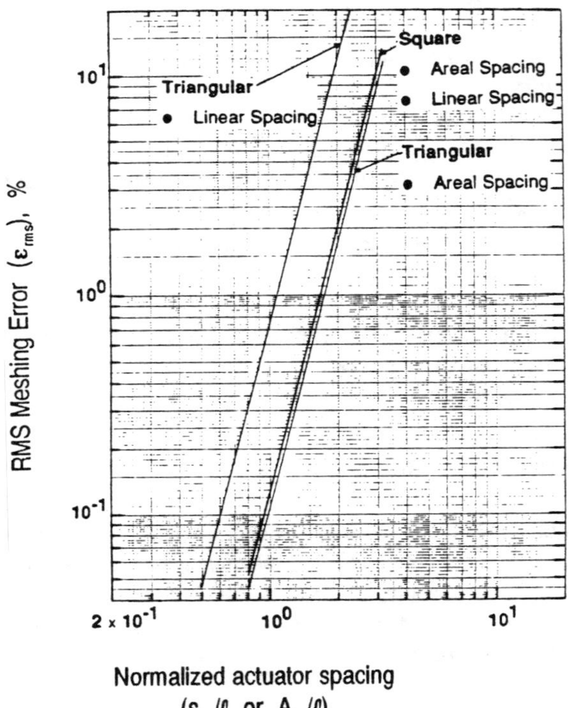

Figure 16. Variation of rms meshing error with actuator spacing for padless actuators

- o The triangular mesh gives a smaller residual error than the square mesh for a given areal spacing.
- o The difference between the two is small.
- o For a unit normalized spacing, which has been observed to be a common practice in some applications, the meshing error is quite small - of the order of 0.1 %. Changing the spacing to about 1.65 to 1.70 units increases the meshing error by one order - to 1 %. This may be acceptable in most applications and, therefore, may delineate the upper limit of acceptable actuator spacing for active mirror designs which are driven by the spatial frequency limit of the characteristic length. Beyond this spacing, the meshing error increases rapidly. An increase by one more order, to 10 %, results at approximately three units of normalized spacing.

Since the meshing error plots of Figure 16 are linear on a log-log scale, they may be functionally expressed as

$$\varepsilon_{rms} = c\ (\sqrt{A_a}\ /\ell\ )^{k} \tag{29}$$

where c and k are constants determined by a least square fit. Expressing ε_{rms} in percentage, we have c = 0.11506 and k = 3.98235 for square meshes. The least square fitting errors are 1.6 % and 1.9 %, respectively.

Actuator Count :

We are now positioned to develop some parametric data on the actuator count. Since the number of actuators for a mirror of area A may be taken as N = A/A_a, the actuator count per unit mirror area is given by

$$\frac{N}{A} = \frac{1}{A_a} = \frac{1}{\ell^2 (\sqrt{A_a}/\ell)^2} = \frac{1}{\ell^2 (\varepsilon_{rms}/C)^{2/k}} \tag{30}$$

Thus the number of actuators per unit mirror area depends on two variables - the characteristic length and the rms meshing error. Formula (30) may be used to generate data for the plots of

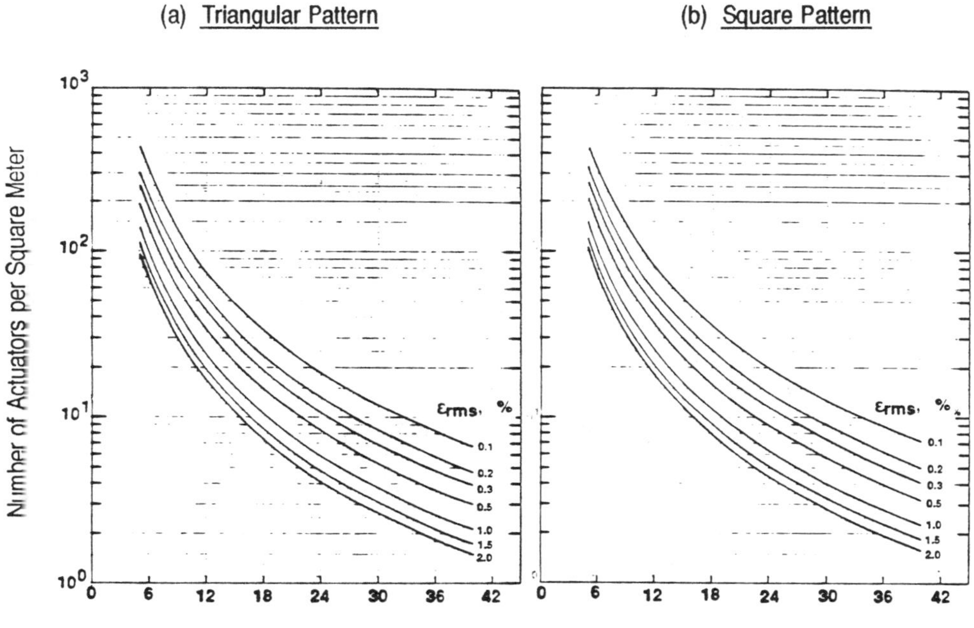

Figure 17. Variation of actuator count with characteristic length and a set of rms meshing errors for padless actuators

actuator count v. characteristic length for a parametric set of values of the meshing error. Alternately, such data may be generated directly by taking readings off Figure 16. The results are plotted in Figures 17 for triangular and square meshes. For convenience, the same data is replotted in Figure 18 as curves of actuator count v. rms meshing error for a parametric set of values of the characteristic length. It is seen that

- o for a given characteristic length and a given rms meshing error, the actuator count for the triangular mesh is slightly smaller than that for the square mesh, as expected.

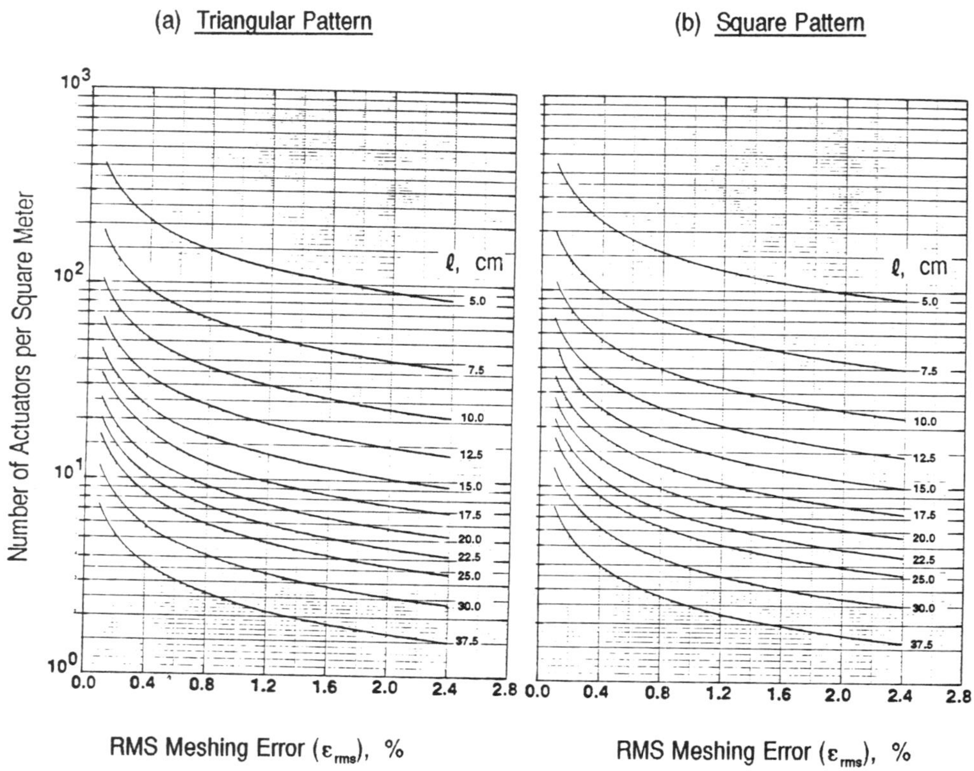

Figure 18. Variation of (padless) actuator count with rms meshing error for a set of characteristic lengths

- for a given rms meshing error, the characteristic length has a strong effect on the actuator count, which decreases rapidly with increase in the characteristic length.

- conversely, for a given characteristic length, the meshing error has a relatively moderate effect on the actuator count. It decreases sharply only for small meshing errors - up to about 0.5 %. Thereafter, the decrease is gradual.

A glance at formula (30) helps to explain the last two features. The actuator count varies inversely as the square of the characteristic length and inversely as the square root of the rms meshing error approximately, because $k \approx 4$.

To potential users of these charts, a word of caution here is in order. If the curves of Figures 17 and 18 yield a very small actuator count for some combination of the characteristic length and the rms meshing error, the governing criterion for actuator spacing may not be the characteristic length, but may be something else such as the spatial frequency of the input wave front error that is required to be corrected. In general, the actuator count and the associated actuator spacing should be selected to be compatible with the spatial frequency of the input error that the deformable mirror is required to compensate for.

Average Deflection :

Recalling that the meshing error is the ratio of the rms value of the residual ripple to an ideally flat deflection surface, it is useful to know how the magnitude of the flat surface varies with the actuator spacing. This is shown in Figure 19 as a plot of resultant normalized average deflection $w(D/P\ell^2)$ v. normalized actuator spacing ($\sqrt{Aa}/\ell$). It is interesting to note that the resultant average deflection

- is practically independent of the actuator mesh type.
- decreases with increase in the actuator spacing, because the number of actuators contributing to it also decreases at the same time.

One may also view the reciprocal quantity $P(\ell^2/wD)$ as a normalized actuator force.

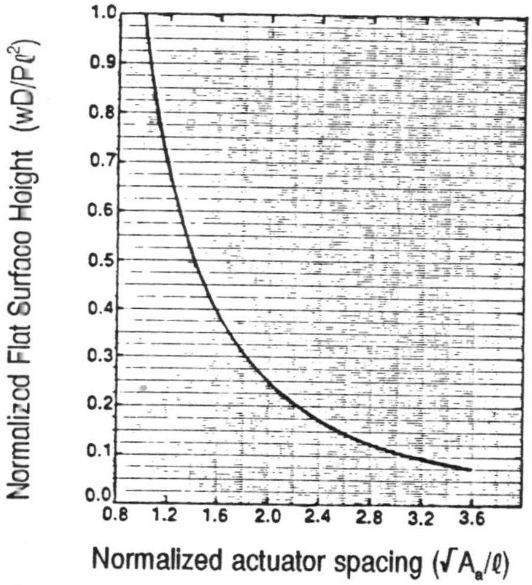

Figure 19. Variation of ideally flat surface height with actuator spacing for padless actuators located on triangular and square meshes

2.5.4 *Effect of Pad Size*

Similar meshing characteristic data can be developed by means of the influence functions represented by formulas (27) and (28) to include the effect of pad size. This has been reported in reference 28.

In view of its importance in the above meshing characteristics, a study of the characteristic length of curved light weighted mirrors of closed and open back constructions with ribbed structures is also presented in reference 28. This study closely parallels the flexural rigidity study[29], which will be briefly discussed later in section 4.0 .

3.0 Flexible Polishing Laps

Polishing convergence rates using the CCP (Computer Controlled Polishing) process are limited by the ability to maintain lap pressure variation over the work piece surface to a relatively small fraction of the applied pressure. A polishing lap initially fitted to any position on the aspheric surface does not maintain adequate fit when moved to another position unless the lap is flexible. It is impractical to depend on the compressibility and/or cold flow of the polishing pitch to provide adequate flexibility, because compressibility requires prohibitively large applied pressure and cold flow is too slow a phenomenon for significant instantaneous response.

The traditional CCP approach to correcting optical aspherics require the use of small tools, limited strokes and fixed orientation with respect to the work piece center. These conditions reduce the productivity of a polishing run, since less area per unit time is polished with small tools and limited strokes, while restrained lap orientation reduces the tool rotational velocity to nearly zero. The constraints also introduce spurious tool loads which can produce unpredictable results. Extended polishing runs and frequent stops to retest the surface are required. It is evident that an improved approach using flexible tools combined with larger strokes and tool rotation makes the CCP process more efficient and productive.

For proper working of a flexible polishing lap, it is necessary that the lap has appropriate flexural rigidity. If the polishing pitch is backed by a layer of highly compliant material only, the tool fits the optical surface by bending very easily, because the combined flexural rigidity of the pitch and the compliant layer is relatively small. However, very little smoothing of the surface occurs in this case, because the highly flexible tool does not ride the ridges, but merely maps the surface. Conversely, if the tool is very stiff, hydroplaning and edge effects due to misfit create high spatial frequency figure errors and may even ruin the basic figure. A proper balance in lap flexibility can be achieved by placing a thin layer of metal between the compliant material and the polishing pitch as shown in the schematic flexible lap concept of Figure 20. With appropriate selection of geometric and material parameters, such a design enables us to maintain the lap fit on the surface with acceptable levels of pressure variations as the tool moves from one position to another. Thus, surface smoothing can be achieved with enhanced convergence rates while maintaining surface figure.

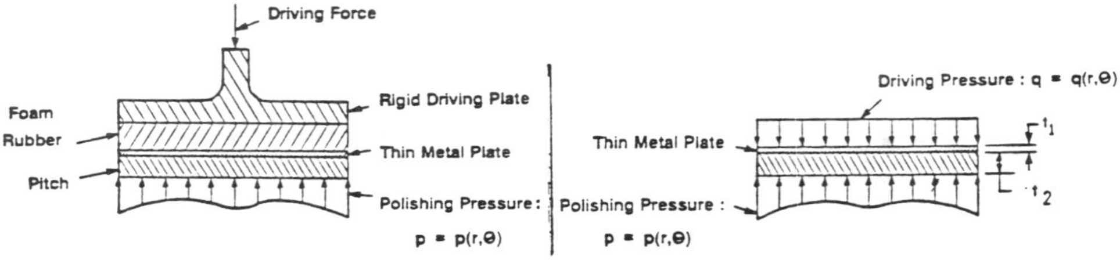

(a) Polishing lap schematic (b) Free-body diagram

Figure 20. The flexible polishing lap concept

3.1 Mathematical Model for Lap Pressure due to Misfit

A mathematical model has been developed[31] for the determination of change in lap pressure distribution as the tool is moved from one position to another, creating/changing the lap fit over the optical surface. This model is based on Kirchoff's thin plate theory, and permits a two-layered tool with equivalent bending and compressive stiffnesses for the thin metal plate and the polishing pitch. The compliant rubbery layer and the rigid driving plate above it can be ignored, because of the high flexibility of the compliant layer. This results in a "plate on a linearly elastic foundation" type of formulation as follows. Let

∇^2	=	Laplacian operator (eq. 12),	p	=	lap pressure at the optic interface,
w	=	lap deflection in bending,	k	=	lap compressive stiffness,
s	=	lap misfit function,	D	=	lap flexural rigidity.
q	=	applied pressure,			

Now, the net transverse compression of the pitch is directly proportional to the lap pressure. Therefore,

$$p = k(s - w) \tag{31}$$

Since the net pressure equals (p - q), Kirchoff's equation (11) becomes

$$\nabla^2 \nabla^2 w = (p - q)/D \tag{32}$$

Substituting for p from eq. (31) into eq. (32) and rearranging terms, we obtain the governing differential equation for the flexible lap problem.

$$\nabla^2 \nabla^2 w + \frac{kw}{D} = \frac{(ks - q)}{D} \tag{33}$$

This can be transformed into an equivalent eigenvalue problem for a free-free plate

$$\left.\begin{aligned} (\nabla^2 \nabla^2 - \lambda^4) w &= 0 \\ \lambda^4 D &= \frac{p - q}{s - p/k} \end{aligned}\right\} \tag{34}$$

The lap pressure p is obtained from the second of eqs. (34) as

$$p = \frac{s\lambda^4 D + q}{1 + \lambda^4 D/k} \tag{35}$$

The solution is valid for each of the infinite eigenvalues of the problem and any linear combination thereof. Further this solution is not limited to any specific shape. It is valid for laps of any shape, provided a corresponding free-free plate eigenvalue solution can be developed. Some such solution already exist in literature - for example, the solutions for circular[32] and rectangular[32] plates, which can be readily incorporated into the above lap pressure solution. This has been done by the present author for circular laps on conventional aspheric optics[31]. This paper also includes (a) a mathematical derivation of lap misfit equations, (b) graphical illustrations for some typical misfit shapes, and (c) test runs for comparison with Nastran analysis. Excellent correlation was obtained in all the test runs.

Figure 21 illustrates the contour plots of lap misfit shapes for the two types of analysis. In this test run, the lap was radially moved 2.5 inches outward from an initially fitted position at r = 45 inches on a parabolic surface with a 200-inch vertex radius of curvature and f-number of 1. The resulting misfit was represented by an 11-mode expansion plus rigid body piston and tilts. The misfit reconstructed with this expansion was within 1.04 % rms deviation of the original misfit, and is shown in the left part of Figure 21. The pressure distribution computed for the 11-mode representation from eq.(35) was applied to a Nastran model of the lap to produce a tool misfit shape. The Nastran generated misfit was within 3.3 % of the reconstructed misfit and is shown in the right part of Figure 21. It is evident that the correlation between the two analyses is very good.

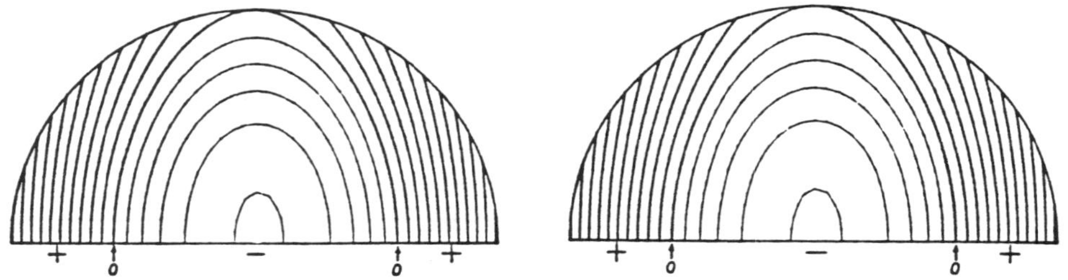

The rms error of this misfit relative to the original misfit is 1.04 %. Contour value = 7.5 x 10^{-6} inch.

(a) Reconstructed misfit, 11-mode expansion

The rms error of this misfit relative to the reconstructed misfit is 3.3 %. Contour value = 7.5 x 10^{-6} inch

(b) Misfit produced by Nastran analysis

Figure 21. Misfit contour plots for a test run, Circular lap on a conventional parabolic optic

The flexible polishing tools for cylindrical (conical aspheric) optics sometimes require different stiffnesses along two mutually perpendicular directions - relatively stiff parallel to the optical axis and compliant along the azimuthal direction. Figures 22 and 23 schematically illustrate the cross section

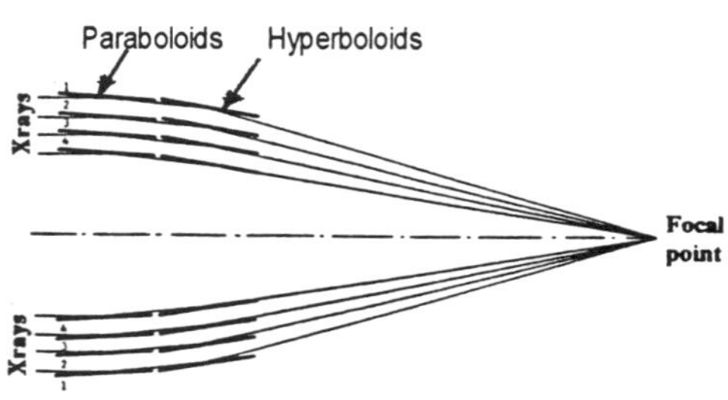

Figure 22. Schematic cross section of a cylindrical mirror system

- Stiff in the axial direction
- Compliant in the azimuthal direction
- Foam rubber layer not shown
- Rigid driving plate not shown

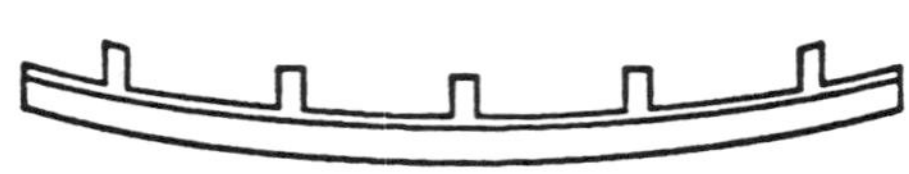

Figure 23. Schematic cross section of a two-layered tool with unequal stiffnesses in two directions

of a cylindrical mirror system and a two-layered polishing lap with unequal stiffnesses in two directions.

The above eigenvalue formulation for the analytical prediction of lap pressure distribution has been applied by the present author in another paper[32] to rectangular tools with unequal stiffnesses in two directions. The problem is considerably simplified by resolving the lap misfit into three components : $f_1(y)$, $f_2(z)$ and $f_3(y,z)$, where y and z denote the azimuthal and axial directions respectively. The misfit consists predominantly of the first component. A new one-dimensional eigenvalue algorithm was developed for laps with nonuniform cross sections. This permits analysis of lap designs with edges more compliant than the interior. The paper gives (a) the new eigenvalue algorithm, (b) misfit equations, (c) a number of graphical illustrations of misfit cases, and (d) several test runs for comparison with Nastran analysis. An illustrative lap pressure plot from one of the test runs for these two analyses is presented in Figure 24, which shows excellent agreement between them.

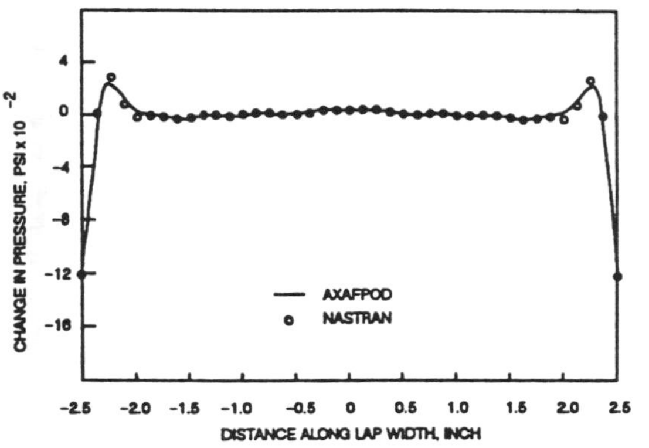

- o Paraboloid no. 1
  - \- AXAF mirror system
- o Lap width = 5 inches
- o Lap stroke = 1.5 inches

Figure 24. Correlation between the closed form and Nastran analysis for azimuthal lap pressure change

3.2 The Bridging Model for Smoothing

The mathematical model of a flexible polishing lap described in section 3.1 addresses the need for using large laps which are as stiff as possible without damaging the basic figure of the optic. That is, the lap is stiff but compliant enough to fit the surface by bending under the applied pressure. Once such a lap has been designed/analyzed by the model of section 3.1, it remains to determine how effective this lap is in smoothing the high frequency surface errors in the optic. The lap contacts the ridges of the surface under applied pressure and begins to wear them down under relative motion with the optic. The lap may be imagined to form a bridge over the ridges, wherefrom the process is sometimes known as "Bridging". This section describes a mathematical model for the Bridging process. The model is similar to that in section 3.1, except that it is based on flat plate eqs. (13) instead of eq. (11).

The spectral features of the optical surface error may be represented by a double Fourier series of the form

$$h = \sum_m \sum_n H_{mn} \tag{36}$$

$$\begin{aligned} H_{mn} = & \left(u_{mn} \cos\frac{\pi z}{a_m} + v_{mn} \sin\frac{\pi z}{a_m} \right) \cos\frac{\pi y}{b_n} \\ & + \left(u^{*}_{mn} \cos\frac{\pi z}{a_m} + v^{*}_{mn} \sin\frac{\pi z}{a_m} \right) \sin\frac{\pi y}{b_n} \end{aligned} \tag{37}$$

where u_{mn}, v_{mn}, u_{mn}^{*}, v_{mn}^{*} are amplitudes and a_m, b_n are half periods of the surface errors. It is further assumed that these periods are very small relative to the lap size (length and/or width) in the respective directions. With these assumptions, the differential equations of the Bridging problem have been solved by the present author to obtain the lap pressure distribution over the optical surface for the case of full lap contact with the surface. The case of partial contact was also solved earlier by this writer for a one-dimensional problem in which the surface error was dependent on only one variable (y or z). These works have not yet been published. However, some illustrative results will be presented here for lap pressure distributions due to high spatial frequency errors in optical

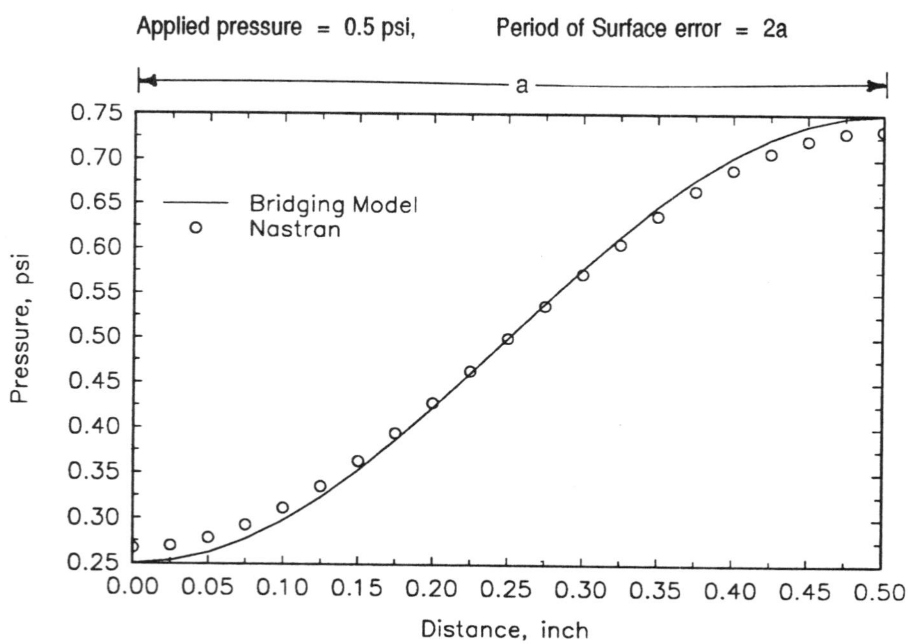

Figure 25. Correlation for one-dimensional lap pressure prediction between the Bridging Model and Nastran based pressure analysis

surfaces. Later we will consider the application of such a pressure distribution in material removal predictions via a solution of Preston's wear rate equation. Figures 25 and 26, respectively, show one- and two-dimensional lap pressure distributions obtained from the above mentioned solutions along with the corresponding data obtained from a Nastran based analysis. The agreement between the two types of analysis is excellent.

Applied pressure = 1 psi, Period (z axis) = 2a, Period (y axis) = 2b

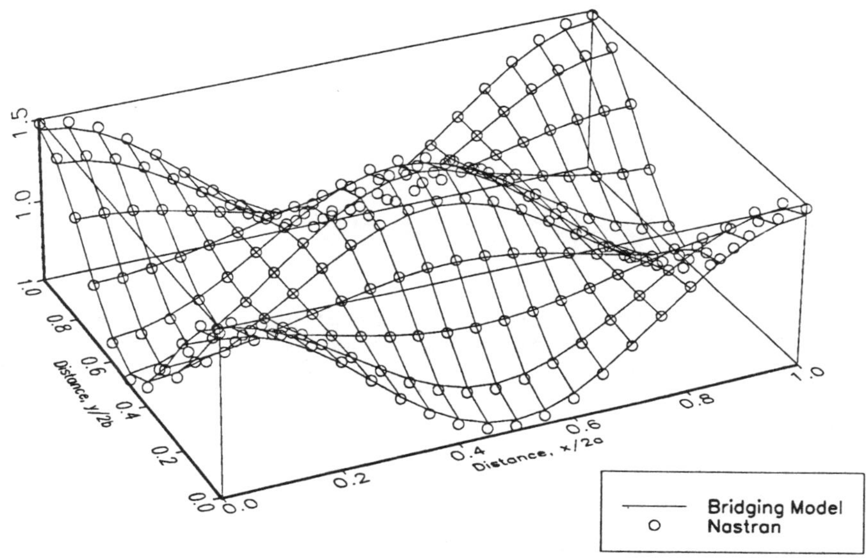

Figure 26. Correlation for two-dimensional lap pressure prediction between the Bridging Model and Nastran based pressure analysis

In order to solve for the material wear, the lap pressure distribution described above for high spatial frequency errors can now be applied to Preston's wear rate equation[33]

$$\frac{dh}{dt} = \pm \; C p \, v = \pm \; C p \, \frac{dS}{dt} \tag{38}$$

where

h	=	height of surface error,	t	=	time,
p	=	lap pressure,	v	=	relative velocity between the lap and the optical surface,
S	=	relative displacement between lap and the optical surface	C	=	a proportionality constant.

The negative sign in eq. (38) represents a reduction in surface height from the peak, while the positive sign represents an increase in the surface depth from the valley. The Preston coefficient C has the same dimensions as those of inverse pressure. This equation has been solved by the present writer for material wear using the lap pressure distribution of the Bridging model, but has not yet been published. In the meantime, we have obtained some experimental correlation for material wear prediction (a) in laboratory tests, and (b) with the measured data from the TMA (**T**echnology **M**irror **A**ssembly) program. This was a development and demonstration program for the fabrication and polishing technology of cylindrical X-ray optics for the AXAF (**A**dvanced **X**-ray **A**strophysics **F**acility) program currently underway. Excellent correlation has been obtained between analytical prediction of material wear and test measurements in both the cases.

4.0 Light Weighted Mirrors

The objective of light weighting a mirror is to improve its structural efficiency. In one method, light weighting is accomplished by shaping the mirror so that the cross section is solid but nonuniform in thickness[34]. A variety of shapes such as (a) concave flat back, (b) meniscus, (c) double concave, (d) single arch, (e) double arch, etc. have been proposed/used. In another method of light weighting, the mirror is made of a sandwich type of construction with two thin face sheets, which are held together by a core. The core is made of a two-dimensional maize like ribbed structure or a softer and lighter material like foam. This is also known as a closed back design. In a third type of light

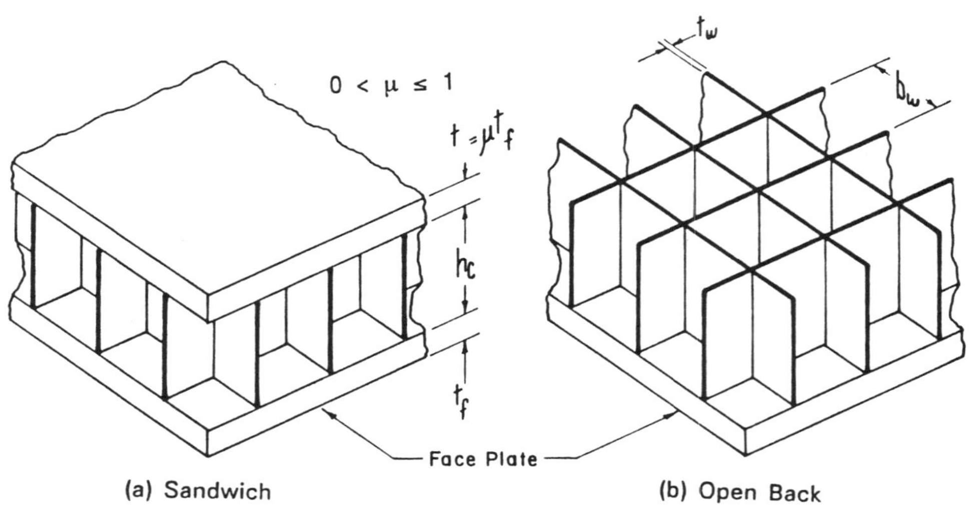

The square core/rib cell pattern shown here may be replaced by other types of cell patterns.

Figure 27. Light Weighted mirror concepts

weighting, the mirror is made of only one face sheet, which is stiffened by a similar two-dimensional ribbed structure. This is the open back design. In this paper we will limit our discussion to closed back (sandwich) and open back designs with core/rib structure as schematically illustrated in Figure 27. The core/rib cell patterns may be triangular, square, hexagonal, cylindrical or tubular as shown in Figure 28.

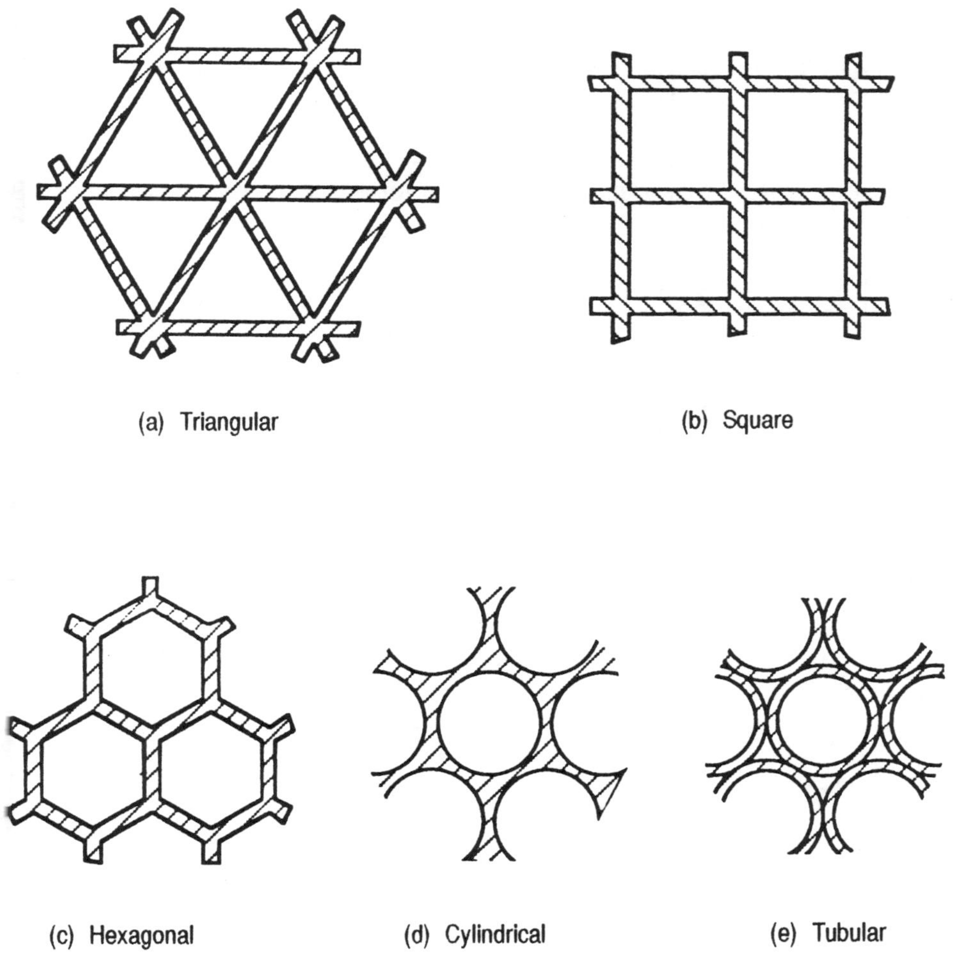

Figure 28. Core/rib cell patterns

In many applications, the improvement in structural efficiency means the maximization of its flexural rigidity for a specified mirror weight per unit area subject to the state-of-the-art fabrication constraints. The deformation of light weighted mirrors of the type shown in Figure 27 under mechanical (nonthermal) loads mostly depend on their flexural and transverse shear rigidities. The bending effect is usually predominant. The higher the flexural rigidity, the smaller the bending deformation. On the other hand, for minimum thermal bowing of the mirror, the parameter M_t/D, where M_t is the thermal moment should be minimized.

The details of light weighted core/rib cell patterns shown in Figure 28 are essentially not required, provided that the core/rib cell size is small in comparison with the mirror diameter. The core/rib cell structure may then be approximately treated as homogeneous, isotropic solids of reduced effective elastic moduli, proportional to $\kappa\eta$, where κ = core/rib effectiveness factor and η = core/rib solidity ratio.

A study of the flexural rigidity characteristics of light weighted mirrors of open back and sandwich types of constructions discussed above has been published by the present writer[29]. In part I of the paper, the relative merits of the two types of light weighted designs in the 10 to 30 kg/m^2 range (of mirror weight per unit area) are examined for ULE, fused silica, Zerodur and beryllium subject to the then representative state-of-the-art fabrication constraints. Part II of the paper addresses the same issue independently of any specific material. This is made possible by normalization of several parameters by the weight equivalent solid thickness of the mirror. Also included in the study are nonsymmetric sandwich cross sections with unequal front and rear face sheet thicknesses. The study shows that, **for given values of mirror weight per unit area and the core/rib solidity ratio**, there exists a transitional value of the normalized front face sheet thickness above which an open back configuration is superior to a closed back configuration. In understanding this seemingly counter intuitive result, it should be remembered that, under some **given** weight constraint as well as fabrication constraints, enough material may not be available to fashion two face sheets and a light weighted core. However, an open back design with rib stiffeners may still be possible. The study evaluates the mathematical limits in such situations and derives the transition point at which the two construction types have equal flexural rigidities. Above the transition point, the open back design has more flexural rigidity than the closed back design. Below the transition point, this trend is reversed. The paper also includes a lot of graphical data from which two normalized data charts are presented here in Figure 29.

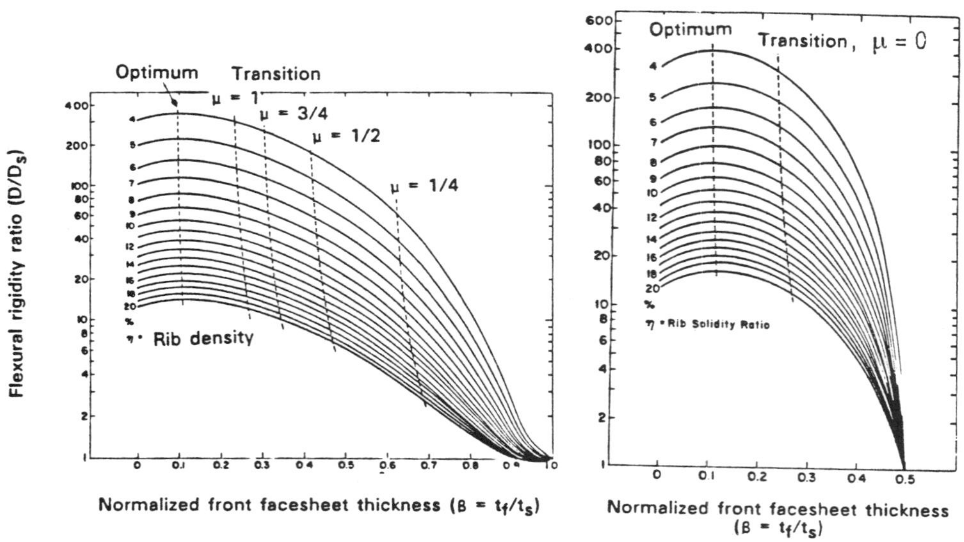

Open Back Design (μ = 0) **Closed Back Design (μ = 1)**

Figure 29. Normalized flexural rigidity charts

Figure 29 also includes loci of optimum and transitional values of the face sheet thickness ratio $\beta = t_f / t_s$, where t_f = front face sheet thickness and t_s = weight equivalent solid thickness. The transitional values are dependent on the rear face sheet thickness ratio $\mu = t_r / t_f$, where t_r = rear face sheet thickness. The transitional loci all lie to the right of the locus of optimum values, and also the smaller the rear face sheet thickness ratio μ, the further is the transitional locus from the optimum locus.

5.0 Closing Remarks

The presentation of this paper has been limited to works of this writer based on the plate and shell theories of elasto-mechanics. It has not been our intention here to present a comprehensive review of the field from much broader areas of works of so many people who have contributed to opto-structural mechanics over many decades. Rather, this is but a small attempt to illustrate that there are a variety of reoccurring problems in opto-structural mechanics, which can be tackled by the classical methods of elasto-mechanics effectively and advantageously without resorting to finite element modeling. Indeed, the problems of the plate and shell equations associated with most of the applied loading conditions discussed in Table 1 are mathematically tractable, provided that sufficient interest exists in the work place, with necessary support to accomplish it. Further, there are other types of mathematically tractable problems of interest (to the optomechanics community) which have not been discussed here. It is hoped that this paper serves to demonstrate the above viewpoint to some extent.

6.0 References

1. Pravin K. Mehta, "Nonsymmetric thermal bending of curved circular mirrors", Structural Mechanics of Optical Systems II, Editor : Alson E. Hatheway, Volume 748, pp.187-205, Proceedings of SPIE, 13-15 January 1987.

2. S. Timoshenko and S. Woinowsky-Krieger, THEORY OF PLATES AND SHELLS, McGraw-Hill Book Company, Inc., p. 291, 1952.

3. Bruno A. Boley and Jerome H. Weiner, THEORY OF THERMAL STRESSES, John Wiley & Sons, Inc., Chapter 12, 1967.

4. Karl Marguerre and Hans-Theo Woernle, ELASTIC PLATES, Blaisdell Publishing Company, Chapter 16, 1969.

5. Eric Reissner, "On the theory of bending of elastic plates", Journal of Mathematics and Physics, Volume 24, No. 4, pp.184-191, 1944.

6. Eric Reissner, "The effect of transverse shear deformation on the bending of elastic plates", Journal of Applied Mechanics, June 1945.

7. Eric Reissner, "On bending of elastic plates", Quarterly Journal of Applied Mathematics, Volume 5, No. 1, pp. 55-68, 1947.

8. Eric Reissner, "Stresses and small displacements of shallow spherical shells. I", Journal of Mathematics and Physics, Volume 25, pp. 80-85, 1946.

9. Eric Reissner, "Stresses and small displacements of shallow spherical shells. II", Journal of Mathematics and Physics, Volume 25, pp. 279-300, 1946. Also, Volume 27, p. 240, 1948.

10. Eric Reissner, "On the determination of stresses and displacements for unsymmetrical deformations of shallow spherical shells", Journal of Mathematics and Physics, Volume 35, pp. 16-35, 1959.

11. Harry Kraus, THIN ELASTIC SHELLS, John Wiley & Sons, Inc., pp. 229-240, 1967.

12. W. A. Bassali, "Thin circular plates supported at several points along the boundary", Proceedings of Cambridge Philosophical Society, Volume 53, pp. 525-535, 1957.

13. W. A. Bassali, "The transverse flexure of thin elastic plates supported at several points", Proceedings of Cambridge Philosophical Society, Volume 53, pp. 728-743, 1957.

14. R. Williams, "Deflection surface of a circular plate with multi-point support", Perkin-Elmer Engineering Report No. 9136, January 1968.

15. R. Williams and H.F. Brinson, "Circular plates on multi-point support", Journal of the Franklin Institute, Volume 297, No. 6, pp. 429-446, June 1974.

16. Jerry E. Nelson, Jacob Lubliner, Terry S. Mast, "Telescope mirror supports: plate deflections on point supports", Advanced Technology Optical Telescopes, Editors : Geoffrey Burbidge, Larry D. Barr, Volume 332, pp. 212-228, Proceedings of SPIE, 11-13 March 1982.

17. Pravin K. Mehta, "Flat circular optical elements on a 9-point hindle mount in a 1-g force field", Structural Mechanics of Optical Systems, Editor : Lestor M. Cohen, Volume 450, pp. 118-130, Proceedings of SPIE, 7-8 November 1983.

18. Roger A. Ross and Richard G. Mady, "Circular Plate Deflections", Unpublished report, 1967, (Private communication).

19. Pravin K. Mehta, "Nonsymmetric thermal bowing of flat circular mirrors", Optical Systems Engineering IV, Editor : Paul R. Yoder, Jr., Volume 518, pp. 155-171, Proceedings of the SPIE, 23-24 October 1984.

20. Pravin K. Mehta, "Nonsymmetric thermal bowing of curved circular mirrors", Structural Mechanics of Optical System II, Editor : Alson E. Hatheway, Volume 748, pp. 187-205, Proceedings of the SPIE, 13-15 January 1987.

21. Anthony Fuschetto, "Three-actuator deformable water-cooled mirror", Optical Engineering, Volume 20, No. 2, pp. 310-315, 1981.

22. R. H. Sawicki, "Continuously deformable mirrors", Electro-mechanical System Interaction with Optical Design, Editor : S. Gowrinathan, Volume 779, pp. 23-26, Proceedings of SPIE, Orlando

(Florida), 1987.

23. R. M. Scott, "New technique for controlling optical mirror shapes", Optical Engineering, Volume 19, No. 2, pp. 112-115, 1975.

24. P. K. Mehta, D. H. Kittel and M. E. Haller, "System wave front corrector", Final Report F30602-87-C-0195 (unclassified), Program manager : C. Roychoudhuri, prepared by Perkin-Elmer Corporation for RADC, Griffiss AFB, Rome, NY 13441, March 1989.

25. Pravin K. Mehta, "A new type of deformable mirror with diametral arrays of radial moment actuators, Technical Memorandum No. PKM-0111, Hughes Danbury Optical Systems, Inc., 31 May 1990.

26. Pravin K. Mehta, "Moment actuator influence function for flat circular deformable mirrors", Optical Engineering, Volume 29, No. 10, pp.1213-1222, 1990; Also, Advances in Optical Structure Systems, Editors : John Breakwell, Victor L. Genberg, Gary C. Krumwide, Volume 1303, pp. 2-19, Proceedings of SPIE, Orlando (Florida), 16-19 April 1990.

27. Pravin K. Mehta, "Actuator meshing characteristics for curved deformable mirrors", Technical Memorandum No. PKM-0109, The Perkin-Elmer Corporation, 11/16/1987.

28. Harold Levenstein and Joel Askinazi, "Design of light weight deformable mirrors, Report No. 13598, The Perkin-Elmer Corporation, March 1978.

29. Pravin K. Mehta, "Flexural rigidity characteristics of light-weighted mirrors", Structural Mechanics of Optical Systems II", Editor : Alson E. Hatheway, Proceedings of SPIE, Volume 748, pp. 158-171, Los Angeles (California), 13-15 January 1987.

30. Pravin K. Mehta and Robert E. Hufnagel, "Pressure distribution under flexible polishing tools - 1, Conventional aspheric optics", Advances in Optical Structure Systems, Editors : John Breakwell, Victor L. Genberg, Gary C. Krumwide, Volume 1303, pp. 178-188, Proceedings of SPIE, Orlando (Florida), 16-19 April 1990.

31. Arthur W. Leissa, VIBRATION OF PLATES, NASA SP-160, pp. 10-13 and 87-114, 1969.

32. Pravin K. Mehta "Pressure distribution under flexible polishing tools - 2, Cylindrical conical optics", Advances in Optical Structure Systems, Editors : John Breakwell, Victor L. Genberg, Gary C. Krumwide, Volume 1303, pp. 189-205, Proceedings of SPIE, Orlando (Florida), 16-19 April 1990.

33. Preston.....

34. Myung K. Cho, Ralph M. Richard, Daniel Vukobratovich, "Optimum mirror shapes and supports for light weight mirrors subjected to self-weight", Precision Engineering and Optomechanics, Editor : Daniel Vukobratovich, Volume 1167, pp. ___-___, Proceedings of SPIE, San Diego (California), 10-11 August 1989.